12th International Scientific Conference on Aeronautics, Automotive and Railway Engineering and Technologies (BulTrans-2020)

AA007011

Held online due to COVID-19

IOP Conference Series: Materials Science and Engineering
Volume 1002

Sozopol, Bulgaria
10 - 13 September 2020

ISBN: 978-1-7138-3315-4
ISSN: 1757-8981

Printed from e-media with permission by:

Curran Associates, Inc.
57 Morehouse Lane
Red Hook, NY 12571

Some format issues inherent in the e-media version may also appear in this print version.

Institute of Physics
Dirac House, Temple Back
Bristol BS1 6BE UK

Phone: 44 1 17 929 7481
Fax: 44 1 17 920 0979

techtracking@iop.org

Additional copies of this publication are available from:

Curran Associates, Inc.
57 Morehouse Lane
Red Hook, NY 12571 USA
Phone: 845-758-0400
Fax: 845-758-2633
Email: curran@proceedings.com
Web: www.proceedings.com

TABLE OF CONTENTS

AERONAUTICS

EXPERIMENTAL STUDY OF THE IMPACT OF THE NUMBER OF BLADES ON THE PROFILE DRAG OF UAV HELICOPTER ROTORS IN HOVER 1
F Panayotov, I Dobrev, F Massouh, M Todorov

COMPARATIVE STUDY OF THE ENERGY EFFICIENCY OF DIESEL AND HYBRID POWERED LIGHT AIRCRAFT 9
V S Serbezov, M I Nikolova

SUSTAINABLE DESIGN OF THE GUST SMALL WIND TURBINE 17
Karol Zawadzki, Anna Baszczyńska, Małgorzata Stępień, Angela Fliszewska, Szymon Molenda, Marcin Pisarski

OPTIMAL SECTOR MODELLING OF AIRSPACE BASED ON OPTIMIZATION ALGORITHMS 26
T D Todorov, S Simeonov

DYNAMICS, STRENGTH AND RELIABILITY OF VEHICLES

FUNCTIONALLY GRADED STRUCTURAL MEMBER WITH THREE LONGITUDINAL CIRCULAR CYLINDRICAL CRACKS LOADED IN TENSION: A FRACTURE ANALYSIS 35
V Rizov

THEORETICAL ANALYSIS OF FRACTURE IN NON-LINEAR ELASTIC FUNCTIONALLY GRADED BEAM OF LINEARLY CHANGING THICKNESS WITH THREE LONGITUDINAL CRACKS 42
V Rizov

EVALUATION OF THE LOAD-BEARING CAPACITY OF VARIOUSLY SHAPED STEEL-CONCRETE SLABS UNDER SHORT TERM LOADING 48
G L Vatulia, N V Smolyanyuk, A A Shevchenko, Ye F Orel, M O Kovalov

SIMULATION OF STEEL ROD ATMOSPHERIC CORROSION BY ALTERNATE IMMERSION IN SALT SOLUTION 57
Veselin Tsonev, Nikolay Nikolov, Kamen Penkov, Yordanka Marcheva

NUMERICAL INVESTIGATION ON THE EFFECTS OF THE AERODYNAMIC SHADING BETWEEN THE TOWER AND THE BLADE OF WIND TURBINE GENERATOR WITH HORIZONTAL AXIS 63
J A Genov, I A Angelov

RAILWAY ENGINEERING AND TECHNOLOGIES

CRACK RESISTANCE OF REINFORCED-CONCRETE SLEEPERS WITH ELASTIC RAIL FASTENING SYSTEMS WITHOUT BASE-PLATE 73
A A Plugin, S V Miroshnichenko, O V Lobiak, OA Kalinin, DA Plugin

INCREASING THE SAFETY OPERATION OF RAILWAY TANKS IN EMERGENCY SITUATIONS BY IMPROVING THE DESIGN OF BOILER BOTTOM PROTECTIVE ELEMENTS 85
Pavol St'astniak, Juraj Gerlici, Kateryna Kravchenko

LOADING ON THE CARRYING STRUCTURE OF A FLAT WAGON WHEN TRANSPORTING FIRING MILITARY EQUIPMENT 96
O Fomin, G Vatulia, A Lovska

RESEARCH INTO THE DYNAMIC LOADING ON THE CARRYING STRUCTURE OF A FLAT WAGON WITH LOWER CENTER OF GRAVITY WHEN TRANSPORTING FIRING EQUIPMENT 104
O Fomin, A Lovska, M Pavliuchenkov, S Sova

TRANSPORT INFRA-STRUCTURE

COMPARATIVE ANALYSIS OF THE ACOUSTIC EFFICIENCY OF CLASSICAL AND SONIC CRYSTAL NOISE BARRIERS 112
Elitsa Gieva, Ivelina Ruskova, Krasimir Nedelchev, Ivan Kralov

NUMERICAL INVESTIGATION OF THE MECHANICAL BEHAVIOUR OF A STRUCTURAL ELEMENT CONTAINING A SELF-HEALING AGENT 125
T Zhelyazov, R Ivanov

TRANSPORT MANAGEMENT AND LOGISTICS

DEVELOPMENT OF A METHOD FOR STUDYING THE IMPACT OF THE TIME RESERVE VALUE ON THE RELIABILITY OF THE TRAIN SCHEDULE BASED ON THE EPIDEMIOLOGICAL SIR MODEL 132
S Panchenko, A Prokhorchenko, O Dekarchuk, D Gurin, D Mkrtychian, V Matsiuk

STUDY OF THE OPERATION OF TAXI CARS IN AN EMERGENCY SITUATION 143
Dimitar Grozev, Ivan Georgiev, Dimitar Georgiev

GREEN TRANSPORT

DRIVING STRATEGY FOR MINIMAL ENERGY CONSUMPTION OF AN ULTRA-ENERGY-EFFICIENT VEHICLE IN SHELL ECO-MARATHON COMPETITION 155
Tsvetomir Gechev, Plamen Punov

ELECTROMAGNETIC ANALYSIS OF A CONTACTLESS CHARGING STATION FOR ELECTRIC VEHICLES 166
N L Hinov, N D Madzharov, D N Grozdanov

INFORMATION TECHNOLOGIES IN TRANSPORT

COMPARATIVE STUDY OF SELECTED CRYPTOCURRENCY 175
O Nakov, D Gotseva, Y Tomov

INTERNAL COMBUSTION ENGINES

STUDY OF HYDROGEN GASEOUS FUEL INFLUENCE ON THE THERMAL CONDITION OF A DIESEL ENGINE 181
E Dimitrov, M Peychev, Ph Michaylov, S Pantchev

OPERATION OF DIESEL ENGINE WITH FUELS TREATED WITH NANOPARTICLE ADDITIVES 187
Zdravko Ivanov, Veselin Mihaylov, Lech Sitnik, Zbigniew Sroka

EXPERIMENTAL STUDY OF THE FLAME PROPAGATION OF GASEOUS FUEL BASED MIXTURE USING OPTICAL VISUALISATION 193
R Hristov, S Stefanov, R Dimitrov

DEVELOPMENT OF A METHOD FOR AUTOMOTIVE ENGINE CONTROL UNIT MODIFICATION FOR ACHIEVING DIFFERENT OPERATING MODES USING SOFTWARE CORRECTION 200
N Nikolov, D Gotseva

AUTOMOTIVE ENGINEERING AND TECHNOLOGIES

FUEL CELL ELECTRIC VEHICLES. INVESTIGATION OF THE ENERGY BALANCE FOR OPTIMAL REFORMING PROCESS OF BIO-ETHANOL 207
Zlatina Dimitrova

ENERGY DEMAND ASSESSMENT FOR LONG TERM OPERATION OF HYBRID ELECTRIC VEHICLES 216
Lech J Sitnik, Zdravko D Ivanov, Zbigniew J Sroka

RANDOM VIBRATION ENDURANCE TEST OF AUTOMOTIVE COMPONENT USING VIRTUAL PROTOTYPING 227
Georgi Todorov, Konstantin Kamberov

DETERMINATION THE YAW MOMENT OF INERTIA OF AN SUV IN LABORATORY CONDITIONS 238
Dmytro M Yashchenko, Oleksii M Timkov, Ihor O Samburenko, Volodymyr S Hladchenko

SERVICE FOR OPTIMIZATION OF CHARGING STATIONS SELECTION FOR ELECTRIC VEHICLES USERS DURING LONG DISTANCES DRIVES: TIME-COST TRADEOFF 247
J Hassler, Z Dimitrova, M Petit, Ph Dessante

EXPERIMENTAL INVESTIGATION OF THE MOTORCYCLE BRAKING PROPERTIES WHEN RIDING ON DIFFERENT ROAD SURFACES 255
Dalibor Barta, Ján Dižo, Miroslav Blatnický, Jozef Vaňko

INVESTIGATION OF POWER SPLIT SCHEMES FOR MODERN HYBRID CARS TRANSMISSIONS 266
Boyko Gigov, Evgeni Dimitrov

EFFICIENCY OF AUTOMOTIVE ELECTRIC SUPERCHARGING COMPRESSORS 276
Pierre Podevin, Christelle Périlhon, Michael Deligant, Plamen Punov, Adrian Clenci, Guillaume De La Bourdonnaye

A STUDY ON CHARACTERISTICS OF VACUUM SOLENOID VALVES 288
Stoyan Stoyanov, Veselin Mihaylov

COMPARISON BETWEEN MODEL PREDICTIVE (MPC) AND MODEL REFERENCE ADAPTIVE CONTROLLERS (MRAC) FOR ELECTROHYDRAULIC STEERING SYSTEM IMPLEMENTED AS REAL-TIME SIMULINK® PROGRAM 295
Alexander Mitov, Jordan Kralev, Tsonyo Slavov, Ilcho Angelov

Author Index

IOP Conf. Series: Materials Science and Engineering 1002 (2020) 011001 doi:10.1088/1757-899X/1002/1/011001

Preface

The **BulTrans-2020** International Scientific Conference on Aeronautics, Automotive, Railway Engineering and Technologies is the twelfth international conference organized by the Transport Faculty at the Technical University of Sofia. The conference is part of the *Science Days* of the Technical University of Sofia and was held in the town of Sozopol, at TU Recreation Base Lazur on September 10-13, 2020. The conference was partially held virtually due to the COVID-19 pandemic.

The conference program included a plenary session and 7 thematic sessions. The duration of the plenary lectures was 30 minutes and of the thematic oral presentations - 15 minutes, followed by questions and answers sessions. The presentations of the participants present on-site and those of the online participants were screened in the conference hall. The conference was held successfully with *Zoom* videoconferencing software. The authors shared the screen and all attendees could discuss their work via the online platform. The chairpersons of all sections were present on-site in the conference hall, while some of the participants were online, including 13 presenting authors.

All accepted papers were subject to peer-reviewing by the scientific editorial board and invited reviewers. The papers were selected on the basis of their quality and their relevance to the conference topics. The review process, as well as all communication between participants and organizers was conducted through the web-based conference management system *process.bultrans.org*.

I would like to thank the reviewers and the scientific editorial board members for their time and advice, and the authors for taking into account the reviewers' recommendations and for improving the quality of the papers selected for presentation at the conference and for publication in this volume. I would also like to thank the conference committee, chairpersons, sponsors, and all other conference participants for their support of BulTrans-2020.

September, 2020

Prof. M. Todorov, PhD
Chairman of the Organizing Committee

IOP Conf. Series: Materials Science and Engineering 1002 (2020) 011001 doi:10.1088/1757-899X/1002/1/011001

Scientific editorial board

Reviewers

Peer review declaration

BulTrans-2020. International Scientific Conference on Aeronautics, Automotive and Railway Engineering and Technologies

All papers published in this volume of *IOP Conference Series: Materials Science and Engineering* have been peer reviewed through processes administered by the Editors. Reviews were conducted by expert referees to the professional and scientific standards expected of a proceedings journal published by IOP Publishing.

- **Type of peer review:** Double-blind
- **Conference submission management system:** process.bultrans.org
- **Number of submissions received:** 53
- **Number of submissions sent for review:** 47
- **Number of submissions accepted:** 34
- **Acceptance Rate (Number of Submissions Accepted / Number of Submissions Received X 100):** 64%
- **Average number of reviews per paper:** 1-2
- **Total number of reviewers involved:** 22
- **Any additional info on review process:** Verification and approval of the papers is carried out in four stages (Stage 1 – Title and abstract evaluation; Stage 2 – Full text evaluation by the Editor-in-Chief; Stage 3 – Review process; Stage 4 – Approval). Each paper goes plagiarism check and a double-blind peer review process. At least one review is being created by a specialist in the respective field, through a unified template. The authors are given access to the full text of the reviews. Upon completion of the review process, the papers and their reviews are submitted to the Editorial board for approval. At the discretion of the Editor-in-Chief, reports that have been refused publication in IOP Proceedings may be submitted for publication in the traditional BulTrans proceedings, published by Technical University of Sofia Academic Publishing House.
- **Contact person for queries:** Assoc. Prof. N. Pavlov, PhD, Technical University of Sofia, Bulgaria, E-mail: npavlov@tu-sofia.bg

Experimental study of the impact of the number of blades on the profile drag of UAV helicopter rotors in hover

F Panayotov[1,3], I Dobrev[2], F Massouh[2] and M Todorov[1]

[1]Technical University of Sofia, Department of Air Transport, blvd. Kliment Ohridski 8, 1000 Sofia, Bulgaria
[2]Arts et Métiers – ParisTech, LIFSE, blvd. L'Hôpital 151, 75013 Paris, France

[3] E-mail: fpanayotov@tu-sofia.bg

Abstract. Due to the motion of tip and root vortices of helicopter rotor blades and the inevitable blade-vortex interactions in multi-blade rotors, the profile drag of all individual blades is increased, when compared to the absence of any blade-vortex interaction. Different factors like the speed of rotation of the rotor and the number of blades greatly impact the magnitude of the profile drag increase. This effect is additionally worsened with the low-Reynolds operational conditions of small-sized UAV helicopter rotors. In addition, there is an initial upward motion of the tip vortex and thus there are three scenarios for the blade-vortex encounter, depending on whether the vortex emitted by the previous blade is still above, on or below the plane of rotation of the blades. All those uncertainties motivated the authors to carry-out this present experimental study. The results are shown in a convenient form. In hover, the profile drag increase is shown to be greater with increased number of blades and higher rotational speeds as there is less time for the preceding vortex to clear the path of the following blade.

1. Introduction

Due to their simplistic design, low-cost control systems and the great variety of on-board electronics and sensors, the rotary-wing unmanned aerial vehicle (UAV) systems have taken their place in our society. Their use has rapidly increased in all spheres of human activities, finding applications in the commercial, agricultural, leisure, security and military sectors. The rotary-wing systems have a unique hovering ability, as well as the capability to perform vertical take-offs and landings. Those properties make them quite suitable for various purposes, like on-spot observation, etc.

The rotors of small-sized rotary-wing UAVs operate in low-Reynolds flow conditions, as they must provide adequately high aerodynamic efficiency of the rotor. While gradually reaching the technological limitations in battery and fuel-cell technologies, in light-weight structural materials and in the field of miniaturization of electronic components, it is a must to fully explore the rotor aerodynamics in low-Reynolds flow conditions. The key is to understand the impact of all various geometrical parameters on the overall aerodynamics efficiency of small-size rotors. For that purpose the studies on the impact of various geometrical parameters on the aerodynamic performance of small-scale UAV rotors have intensified significantly in recent years.

The absence of precise mathematical models for the description of the complex nature of low-Reynolds flows, as well as for adequate predictions of the thickness of the boundary layers or for phenomena like laminar separation bubbles and blade-vortex interactions, makes all efforts for an analytical and computational evaluation of the aerodynamics performance of such rotors quite challenging. For those reasons the development of facilities and methodologies for experimental testing plays an important role in the further improvement of the aerodynamic efficiency of small-sized rotary wing propulsion systems.

Article [1] is a good example of an early pioneering work, in which the author estimates the variation of the lift and profile-drag for the tested airfoil for a wide range of Mach tip speeds and pitch angle settings of the blades. The study also reveals the hover performance of a large-scale helicopter rotor with blades of NACA0012 airfoil. It was found out that the studied airfoil generates larger lift and lower profile-drag than expected, when compared with the previously obtained experimental two-dimensional airfoil data.

In [2], the authors study the aerodynamic efficiency of a variety of airfoils with different thickness and camber settings. Their lift and drag coefficients as well as their surface pressure and flow-field characteristic are evaluated. The authors conclude that for Reynolds below 10^6, the lift and drag characteristics cannot be assumed as constant. It is shown that when the Reynolds number of the flow falls below 10^5, the laminar separation bubble grows in size reaching a diameter of between 15 and 40 percent of the chord length, while gradually moving toward the trailing edge, which causes early stall at lower angles of attack. The turbulent boundary-layer is shown to increase its thickness, which results in the increase of the airfoil profile drag.

In [3], the authors show the impact of various geometrical parameters like thickness, camber, chord length, blade twist, taper ratio and rotor solidity, as well as the effect of winglets for different sets of airfoils both symmetrical and non-symmetrical. Shown is an optimum set of parameters for UAV rotor blades for their efficient operation in low-Reynolds flow conditions.

In [4], the authors study the impact of taper and twist, as well as the effects of sharpening of the leading and trailing edges of the blades. In total, 8 rotor configurations are tested for Reynolds of approximately 35000. It is shown that sharpening of the leading and trailing edges of the blades reduces the thickness of the trailing wake and thus lowers the profile drag, which results in a better aerodynamic performance. The negative linear blade-twist is shown to improve the rotor aerodynamic performance [4, 5].

In [5], the authors compare experimental and numerical data for small-sized hovering rotors with highly-twisted and untwisted blades. The experimental data for that study comes from both thrust measurements in steady operational regimes and PIV flow-field measurements. This work reveals the inadequacy of the BEMT, when used for the prediction of the aerodynamic performance of rotors, operating at low-values of the coefficient of thrust and in low-Reynolds flow conditions. It also reveals that in cases of very low thrust settings the highly-twisted blades can become inferior to the untwisted blades in terms of aerodynamic efficiency. This is due to the fact that for the same trust setting, the highly-twisted blades will generate higher inflow and pressure drag, which contributes to the overall profile drag of the rotor.

In [6-8], the authors reveal the importance of geometry optimization as a way to improve the aerodynamic performance at low-Reynolds flow conditions. Those works are an excellent example of complex aerodynamic analysis, which incorporates both numerical and experimental data.

In [9, 10], the authors study the static aerodynamic performance of model helicopter rotors for different sets of blades, angular velocities and pitch settings. In both works the respective authors present in detail the helicopter rotor test stands and the applied methodologies, which permit the evaluation of the aerodynamic performance of the tested hovering rotors. In [10] is shown a methodology for the experimental evaluation of the profile drag of the airfoil of the rotor blades.

From all reviewed works, so far only in [3] the authors show the impact of the variation of the number of blades on the aerodynamic efficiency of the rotor. Apart from [3], there are no other experimental studies, known to the authors of this paper, which evaluate purely the impact of the number of blades on the aerodynamic characteristics of the hovering rotor. The present experimental study aims to reveal that impact by comparing 3 rotor configurations, which have the same Goettingen 417 airfoil, twist setting, taper ratio, chord length, planform and radius. In such a way it is assured that only the rotor solidity is being altered. A fourth configuration of two-bladed rotor with NACA0012 airfoil is also being tested in order to compare the aerodynamic performance between thin nonsymmetrical and thick symmetrical airfoils at those low-Reynolds flow conditions.

What is even more important in this present experimental work is that by increasing the number of blades in the rotor setup, a more complex variable is being introduced, which is the Blade-Vortex Interaction (BVI). This phenomenon has a complex mathematical predictability, as there are three scenarios for the blade-vortex encounter, depending on whether the vortex emitted by the previous blade is still above, on or below the plane of rotation by the time the following blade of the rotor passes through that same angular position. The BVI is not constant and varies quantitatively with the variation of the angular velocity of the rotor and the pitch angle of the blades. All those uncertainties have motivated the authors to undertake the present experimental study.

2. Experimental setup

Different series of measurements are performed on the specially built helicopter rotor test stand. Figure 1 shows all four rotor configurations that are tested. All four rotor configurations have non-tapered (rectangular) untwisted blades. One of those two-bladed rotor configurations has the NACA0012 airfoil, while the other three have the Goettingen-417 (Goe417) airfoil. All performed experimental tests are static, thus both the speed of rotation of the rotor and the pitch angle of the blades are kept constant. Those parameters are changed incrementally between each test sessions, as described in section 3. Experimental Methodology.

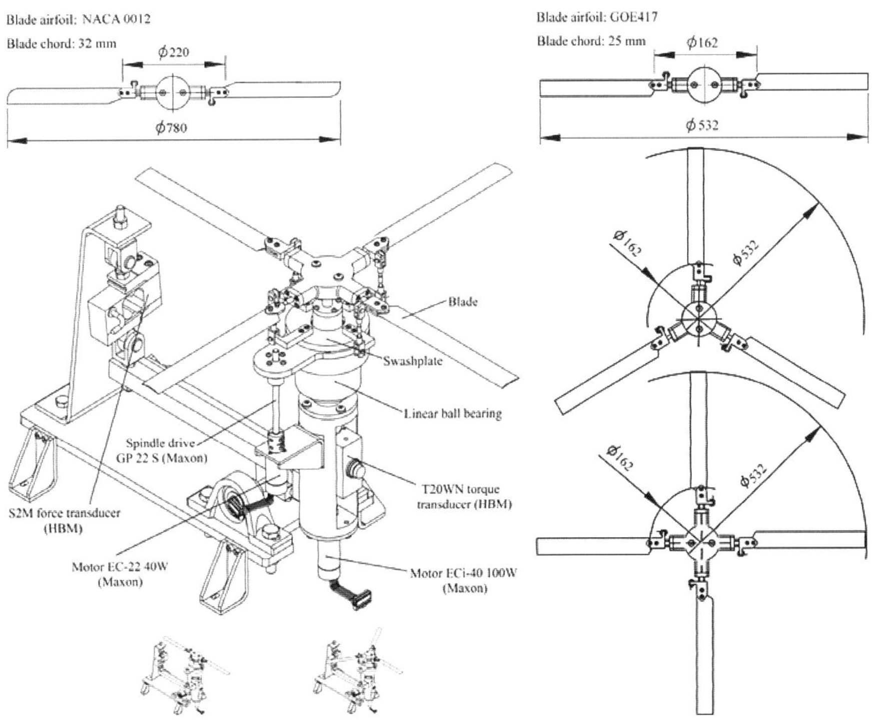

Figure 1. Helicopter rotor test stand with all 4 rotor configurations.

2.1. Test stand description

Figure 1 shows the helicopter rotor test stand, which allows for the possibility to measure the rotor thrust T, N, the torque Q, Nm and the angular speed of rotation Ω, rad/s. The stand is equipped with the following measuring devices:

- Torque transducer HBM T20WN with nominal (rated) torque of 1 Nm, accuracy class 0.2 and output voltage ±10 V,
- Force transducer HBM S2M with nominal (rated) force of 10 N, accuracy class: 0.02 and output voltage of 2 mV/V.

The tested rotor is driven with a brushless electric motor Maxon ECi-40 with nominal power of 100 W. It allows for rotational speeds of up to 4190 RPM. The electric motor has an in-built Hall sensor, which allows for measurement of the rotational speed. The motor is driven with Maxon ESCON 50/5 servo controller, which permits precise regulation of the rotational speed.

The pitch angle of the blades is varied with a spindle-driven mechanism Maxon GP22S with a gear ratio of 1:370. The spindle is driven by the EC-22 electric motor with nominal power of 40 W.

The electric motors are controlled automatically according to a programmed test schedule. This allows for multiple operational regimes to be evaluated in one test session without the need to constantly stop, re-adjust and re-start of the test stand.

The data acquisition from the force and torque transducers is ensured by the HBM PMX measuring amplifier system, shown in figure 2. The data acquisition system allows for the automatic gathering of all measured experimental data, which is stored in a well-structured database.

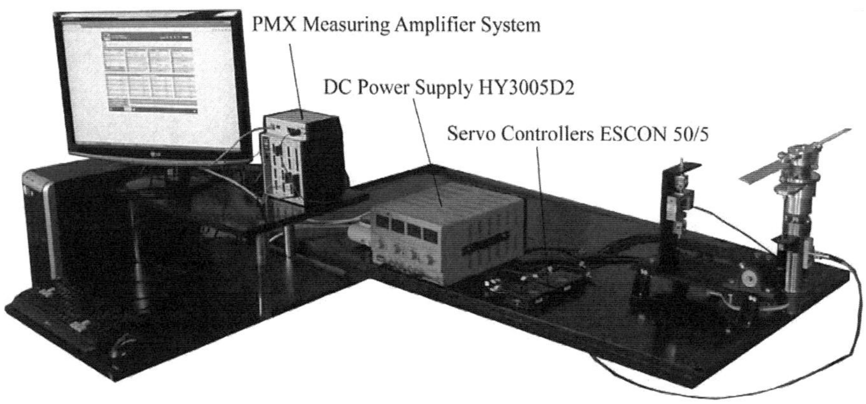

Figure 2. Helicopter rotor test stand setup.

3. Experimental Methodology

The experimental study is performed as follows: for each test session for all four rotor configurations the pitch angle θ, deg is set from 2 deg to 16 deg with an increment of 2 deg, while the speed of rotation of the rotor is varied from 500 RPM up to 1700 RPM with an increment of 100 RPM. Each test session is repeated 10 times and the results are averaged. Additionally, for all four rotor configurations, a study of the profile drag at zero thrust is performed. For that purpose, for each speed of rotation the pitch angle of the blades is varied until the thrust of the rotor T=0 N is equal to zero. Thus, the torque at zero thrust Q_0, Nm is obtained. After that, all blades are removed, in order to measure the no blades torque Q_{nb}, Nm. Then, the torque due to profile drag Q_{pd}, Nm can be obtained by subtracting Q_{nb} from Q_0. If Q_{pd} is multiplied by the angular speed of rotation Ω that gives the power consumption due to profile drag P_{pd}, W.

The coefficient of profile drag C_{pd} can be computed with (1):

$$C_{pd} = \frac{(Q_0 - Q_{nb})\Omega}{\rho A \Omega^3 R^3} = \frac{Q_{pd}\Omega}{\rho A \Omega^3 R^3} = \frac{P_{pd}}{\rho A V_{tip}^3}, \tag{1}$$

where ρ is the air density, kg/m^3; $A=\pi R^2$, m^2 is the area of the rotor; R, m is the radius of the rotor; $V_{tip}=\Omega R$, m/s is the tip speed of the blades.

The rotor solidity σ [11] is obtained with (2):

$$\sigma = \frac{N_b c}{\pi R}, \tag{2}$$

where N_b is the number of blades of the rotor configuration and c, m is the blade chord.

The coefficient of profile drag at zero thrust C_{D0} [12] is derived with (3):

$$C_{D0} = \frac{8C_{pd}}{\sigma}. \tag{3}$$

The coefficient of thrust of the rotor C_T [11] is computed with (4):

$$C_T = \frac{T}{\rho A \Omega^2 R^2}. \tag{4}$$

The coefficient of thrust of the rotor C_P [11] is obtained with (5):

$$C_P = \frac{Q}{\rho A \Omega^2 R^3}. \tag{5}$$

The coefficient of thrust of the rotor FM [11] is derived with (6):

$$FM = \frac{C_T^{3/2}}{\sqrt{2}C_p}. \tag{6}$$

The Figure of Merit computed with (6) is shown to have been introduced as a nondimensional measure of hovering efficiency for helicopter rotors, which provides a common base for comparison between rotors with different geometries.

In order to ensure comparability between the results for helicopter rotors with different number of blades and thus with different rotor solidity, the following dimensionless parameters are used:

- C_T/σ – blade loading coefficient,
- C_P/σ – power coefficient.

The coefficient of lift of the rotor C_{Lr} is computed with (7):

$$C_{Lr} = \frac{2T}{\rho(0.7\Omega R)^2 S}, \tag{7}$$

where $S = N_b c$, m^2 is the total area of all blades of the rotor.

The coefficient of drag of the rotor C_{Dr} is obtained with (8):

$$C_{Dr} = \frac{2\left[Q/(0.7R)\right]}{\rho(0.7\Omega R)^2 S}, \tag{8}$$

4. Results and discussions

Figure 3 presents the variation of the coefficient of profile drag at zero thrust C_{D0} from the Reynolds number of the flow for all 4 rotor configurations. It can be observed that Goe417 airfoil has a lower profile drag compared to NACA0012, which is expected as the latter airfoil has a two times higher relative thickness. It can be seen that as the number of blades increases the profile drag increases too. This can be contributed to the more frequent blade-vortex encounters, as with larger number of blades there are more BVIs, which alter the inflow distribution along the radius of the blades and especially at the tip part. That results in a redistribution of the effective angle of attack along the blades, which alters the zones with positive and negative thrust contribution, all within the total sum of zero rotor thrust. PIV experimental studies will be performed with the same rotors, in order to confirm that observation.

Figure 4 shows the hover polar for all tested rotor configurations for Reynolds number of approximately 42000. As expected, lower disk loading of the rotor achieves better hovering efficiency. Thus, the four-bladed rotor configuration requires the least power for each produced unit of thrust. It can be seen that for very low values of the blade loading coefficients the two-bladed rotor with NACA0012 airfoil has superior hovering efficiency. However, at those low power setting there are no practical helicopters or rotary-wing UAVs, which will attempt to perform hover.

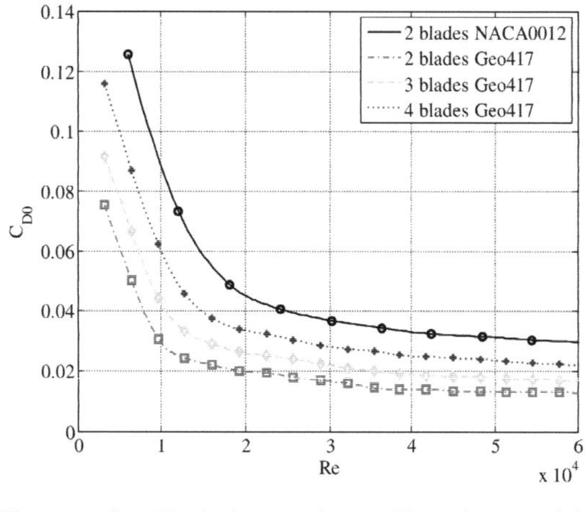

Figure 3. Variation of profile drag with Reynolds.

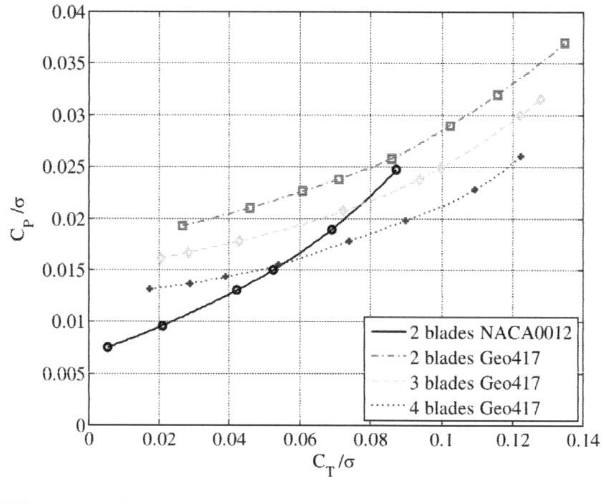

Figure 4. Hover polar for all tested rotor configurations for Re≈42000.

Figure 5 presents the dependency of the Figure of Merit on the blade loading coefficient for all 4 rotor configurations. The larger rotor solidity allows for higher power loading settings of the rotor before reaching stall conditions. That is why a larger number of blades allow for a higher value of the Figure of Merit. The two-bladed rotor with Goe417 airfoil is superior to the one with NACA0012 airfoil at those low-Reynolds flow conditions.

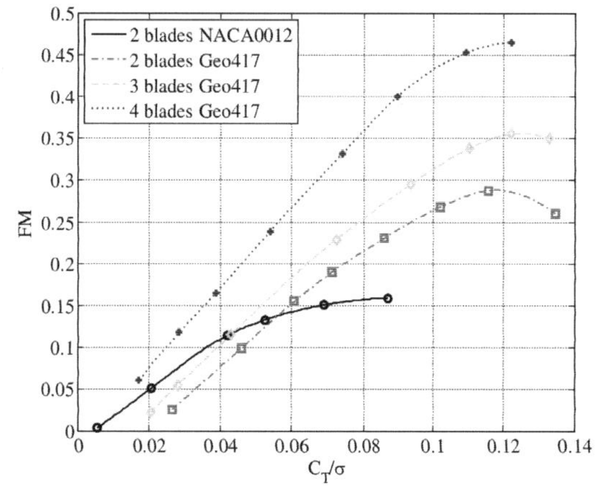

Figure 5. Figure of Merit against the blade loading coefficient for all 4 rotors for Re≈42000.

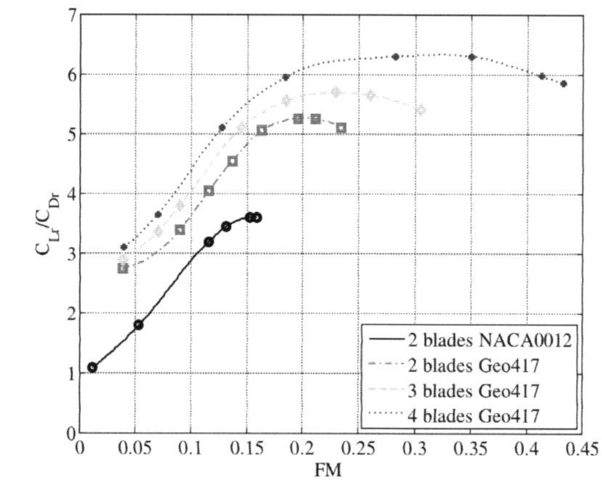

Figure 6. Rotor aerodynamic efficiency in hover for all 4 configurations for Re≈42000.

On figure 6 is shown the aerodynamic quality of the blades of the rotor against the Figure of Merit for all 4 tested configurations. Those results clearly show that rotors having blades with higher aerodynamic quality can achieve better hover efficiency. It can be noticed that a larger number of blades and hence rotor solidity, allows for a broader operational range at maximum aerodynamic efficiency in hover. For similar hover performance the two-bladed rotor configuration with Geo417 airfoil has a superior aerodynamic quality when compared to the rotor with the thicker symmetrical NACA0012 airfoil.

Figure 7 is presents the aerodynamic polar of the airfoil of the rotor blades for each one of the 4 tested configurations. It seems that as the number of blades is increased, the more frequent BVIs contribute for more favorable pressure redistribution along the radial span of the blades, which results in a higher overall coefficient of lift of the airfoil. PIV experimental studies have to be performed with the same rotor configurations, in order to obtain a greater insight of this effect.

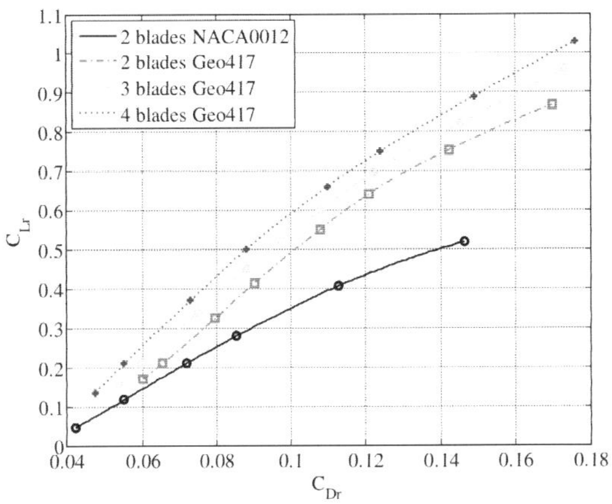

Figure 7. Airfoil aerodynamic polar for all tested configurations for Re≈42000.

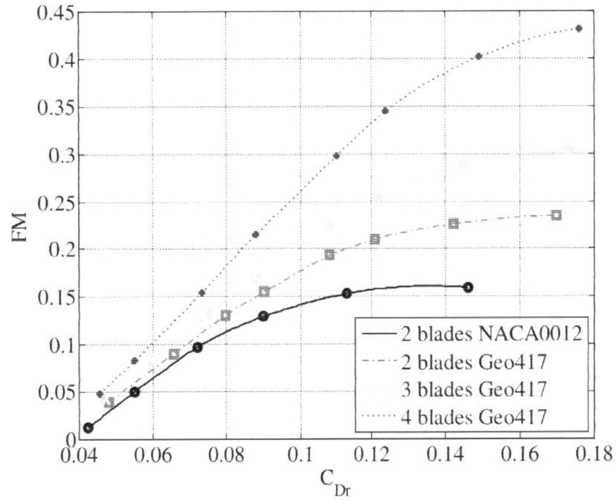

Figure 8. Figure of Merit against the coefficient of drag for all 4 configurations for Re≈42000.

Figure 8 presents the hover efficiency against the coefficient of drag for all tested rotor configurations. This data representation allows for the evaluation of the amount of generated drag for each rotor configuration in their entire hover operational range. As previously demonstrated, with the increase of the number of blades higher aerodynamic efficiency of the rotor can be achieved. Hence, what can be observed in the case of the three- and four-bladed rotors is that less units of drag are generated for each unit of hover efficiency of the rotor.

5. Conclusion

Although the profile drag increases with a larger number of blades, the Goe417 airfoil achieves higher coefficient of lift and hence better aerodynamic quality in both the three- and four-bladed configurations, when compared with the two-bladed rotor.

By increasing the rotor solidity with a larger number of blades, the rotor achieves better aerodynamic and hover efficiency, generating more lift and less overall drag for identical operational regimes.

The three- and four-bladed rotor configurations require less power for each produced unit of thrust, when compared to the two-bladed rotor, as a lower disk loading of the rotor achieves higher hovering efficiency.

A larger number of blades and hence rotor solidity, allows for a broader operational range at maximum aerodynamic efficiency in hover.

Rotors, having blades with better aerodynamic quality, can achieve better hover efficiency.

PIV flow-field measurements have to be performed in order to fully study the nature of the BVIs for those low-Reynolds ranges.

Acknowledgements

This work was supported by the European Regional Development Fund within the Operational Program "Science and Education for Smart Growth 2014 - 2020" under the Project CoE "National center of mechatronics and clean technologies" BG05M2OP001-1.001-0008.

References

[1] Carpenter P J 1958 Lift and Profile-drag Characteristics of an NACA 0012 Airfoil Section as Derived from Measured Helicopter-rotor Hovering Performance NACA TN 4357

[2] Winslow J, Otsuka H, Govindarajan B and Chopra I 2018 Basic understanding of airfoil characteristics at low Reynolds numbers *Journal of Aircraft* **55**(3) pp 1050–61
[3] Benedict M, Winslow J, Hasnain Z and Chopra I 2015 Experimental investigation of micro air vehicle scale helicopter rotor in hover *Int. J. of Micro Air Vehicles* pp 231–55
[4] Ramasamy M, Johnson B and Leishman J G 2008 Understanding the aerodynamic efficiency of a hovering micro-rotor *Journal of the American Helicopter Society* **53**(4) pp 412–28
[5] Ramasamy M, Gold N P and Bhagwat M J 2010 Rotor hover performance and flowfield measurements with untwisted and highly-twisted blades *36th European Rotorcraft Forum* **1** pp 770–84
[6] Hnidka J, Rozehnal D and Maňas K 2020 Optimization of SUAV's propeller in a hover *MATEC Web Conf.* **313** 00045
[7] Hnidka J and Rozehnal D 2019 Calculation of the maximum endurance of a small unmanned aerial vehicle in a hover *IOP Conf. Ser.: Mater. Sci. Eng.* **664** 012002
[8] Kądrowski D, Kulak M, Lipian M, Stępień M, Baszczyński P, Zawadzki K and Karczewski M 2018 Challenging low Reynolds - SWT blade aerodynamics *MATEC Web Conf.* **234** 01004
[9] Lee B, Byun Y, Kim J and Kang B 2011 Experimental hover performance evaluation on a small-scale rotor using a rotor test stand *Journal of Mech. Science and Technology* **25**(6) pp 1449–56
[10] Panayotov F, Dobrev I, Massouh F, and Todorov M 2018 Experimental study of a helicopter rotor model in hover *MATEC Web of conf.* **234** 01002
[11] Leishman J G 2006 *Principles of Helicopter Aerodynamics* (Cambridge University Press) chapter 2 p 69

Comparative study of the energy efficiency of diesel and hybrid powered light aircraft

V S Serbezov[1] and M I Nikolova

Department of Aeronautics, Technical University of Sofia, 8 Kl.Ohridski Blvd., 1000 Sofia, Bulgaria

[1] E-mail: vserbezov@ tu-sofia.bg

Abstract. In recent times many research efforts are aimed at the development of aircraft hybrid propulsion technology as a means to minimize the aircraft environmental impact. The objective is to provide an additional assessment of the energy efficiency of the hybrid powered light aircraft in comparison with current state of the art Diesel powered aircraft. Typical 4-seat propeller airplane is chosen as a prototype and its flight performance model is created using open source data. The comparison is performed for a realistic flight profile including the power demands for go-around and flight to alternate destination airport. Parallel hybrid architecture is considered. A parametric study with variation of the mass, efficiency and energy capacity parameters of the hybrid power train is performed in order to determine the parameter regions where the hybrid power train has advantage over the Diesel one. The present study is based on the previous experience of the authors in assessing the energy efficiency of alternative power sources for the aviation.

1. Introduction

Presently the electric hybridization of the aircraft power plants is envisioned as a necessary step towards fully electric aircraft propulsion. Profound study of the subject is performed by Brelje and Martins [1]. The main conclusion of their work is that successful implementation of electric propulsion, including hybrid electric, remains challenging and should be an object of multidisciplinary design analysis and optimization. More practical efforts in this direction are undertaken in the horizon 2020 program MAHEPA, targeted at designing and testing real hybrid aircraft [2].

A massive effort in the field of aircraft hybrid-electric propulsion was undertaken by a team of researchers from FH Aachen University of Applied Sciences and RMIT University [3-7]. Booth serial hybrid [3, 6] and parallel hybrid [4, 5] architectures were investigated in regard with the sizing of the hybrid propulsion system (HPS). Some important conclusions, concerning the optimum wing loading and power loading of the hybrid aircraft were drawn. As baseline aircraft for the studies were chosen Cessna-172S and Cirrus SR-22 light aircraft. In [7] the research was outspread also on larger aircraft and unmanned aircraft.

Researchers working on MAHEPA also published the results of their studies on sizing serial HPS for light commuter and for regional passenger aircraft [8]. Additional degree of freedom in this study was the inclusion different energy management strategies during the flight.

The objectives of the present work are somehow different from those of the publications listed above. The main objective is not to size and optimize the HPS, but to compare its energy efficiency with that of the state of the art present days technology. Another objective is to assess the viability of

HPS as a drop-in replacement of the conventional aircraft engines for light aircraft. For this reason the approach chosen in this work is to create a flight performance model of an existing light airplane with an internal combustion engine (ICE), modify the model for a hypothetic HPS, and compare the results of the booth modes over typical flight scenario.

2. Baseline aircraft and hybrid system description

2.1. Baseline aircraft

As a baseline aircraft Diamond DA-40 NG light aircraft is chosen. It is diesel-powered, single engine, composite structure aircraft. The combination of the low weight and smooth aerodynamic shape of the composite structure and the diesel engine with very low specific fuel consumption puts the DA-40 NG among the most energy efficient airplanes of its class [9]. The Austroengine AE-300 turbo diesel engine of the aircraft is rated at 123.5 kW of takeoff power, with dry mass of 186 kg. The turbo diesel possesses also good high-altitude performance, which contributes to the good cruise performance of the aircraft. The airplane is capable of taking 84 kg of diesel fuel in its main tank.

Among the disadvantages of the diesel engine of the baseline aircraft are the relatively low specific power of 0.66 kW/kg, together with relatively high nitrogen oxides and particulate matter (PM) emissions and higher prize, compared to the aviation gasoline engines.

2.2. HPS structure.

A general assumption of this study is that the HPS replaces the diesel engine without any 'external' changes of the aircraft. This includes:

- The propeller remains the same;
- Aerodynamic performance is unchanged (outer shape of the airplane is generally unchanged);
- Maximum take-off mass of the aircraft is unchanged;
- Useful load is unchanged;
- The power ratings of the power plant (take-off, maximum continuous, idle ratings) remain as of the diesel engine, in accordance with DA-40 NG Aircraft flight manual (AFM).

Another assumption is that a serially produced aviation ICE should be used as part of the HPS to reduce the development costs and time. A logical choice is the Rotax 914 turbocharged spark ignition engine, rated at 84.5 kW at takeoff and installation mass of roughly 75 kg. As alternative engine Rotax 915, rated at 104 kW was considered, but it was rejected, mainly because of its higher specific fuel consumption. It should be stated that the ICE optimization for the HPS is a separate, complex task, which is not an object of the current research.

Given these assumptions the practically feasible HPS may look as the one shown on figure 1. It is arguably the simplest form of parallel HPS. The freewheel unit is added as a safety feature in case of ICE failure. The permanent magnet electric motor is mounted directly on the propeller shaft.

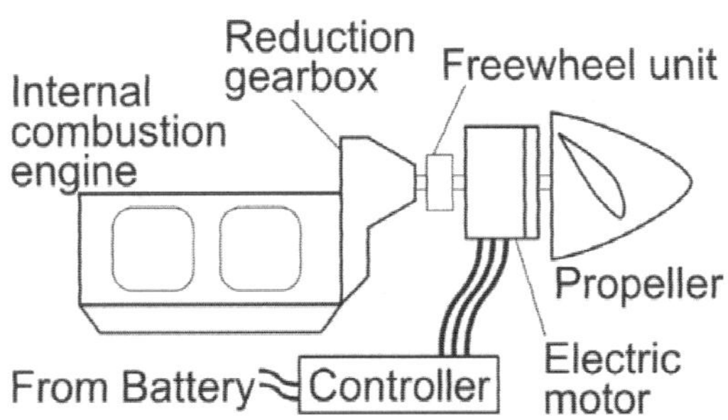

Figure 1. Simplified scheme of the parallel HPS investigated in the present work.

2.3. Sizing and energy management strategy of the HPS.

The main specific parameters, which are needed for the sizing of the HPS are taken from the reviewed research works [2-5] and are presented in table 1.

Table 1. Specific properties of the electric components of the HPS.

Electric motor specific power	$P_{sp.EM}$	kW/kg	6
Motor power electronics specific power	$P_{sp.PE}$	kW/kg	15
Electric efficiency of the system	η_{el}	–	0.85 – 0.95
Battery specific energy	$E_{sp.Bat}$	Wh/kg	200 – 500

With regard to the relatively low energy density of the batteries the following energy strategy is adopted:

- At take-off and climb the ICE (Rotax 914) provides its take-off or maximum continuous power respectively. The electric motor provides the rest of the power needed to reach the power of the original diesel engine;
- At cruise the ICE provides the required power. If the required power is higher than the maximum continuous power of the ICE at the current ambient conditions, the ICE works at maximum continuous power and the electric motor provides the excess power needed.
- At descent the electric motor is not working.

With regard to this strategy the electric motor should provide around 40 kW at take-off and climb. Also an electric motor power reserve should be considered as a safety feature in case of ICE failure. The electric power reserve needs special study, but in this work a guess of 20% is accepted, resulting in electric motor maximum power P_{EMmax} of 50 kW. With regard to the specific power of the motor and its electronics (table 1) the weight of the electric part of the HPS is roughly 12 kg.

The sizing of the battery is made with the presumption that the electric energy will be used for take-off and climb, but also for eventual go-around and climb to safe altitude. An initial guess, providing enough safety margin is to have enough stored electric energy for 30 minutes of the electric motor full power ($t_{FullPWR}$). In this case the mass of the batteries is determined as follows:

$$m_{Bat} = \frac{P_{EMmax} t_{FullPWR}}{\eta_{el} E_{sp.Bat}} \quad (1)$$

For the values of $E_{sp.Bat}$ and η_{el} in table 1 the mass of the batteries will be between 147 kg ($E_{sp.Bat}$=200 Wh/kg, η_{el}=0.85) and 52.6 kg ($E_{sp.Bat}$=500 Wh/kg, η_{el}=0.95).

With regard to the assumptions stated in p.2.2. the power plant mass equilibrium between the baseline power plant (diesel) and the HPS is expressed in equation 21:

$$m_D + m_{Dfuel} = m_{ICE} + P_{EMmax}/P_{sp.EM} + P_{EMmax}/P_{sp.PE} + m_{Bat} + m_{HPSfuel} \quad (2)$$

where m_D=186 kg is the mass of the baseline diesel engine, m_{Dfuel}=84 kg is the baseline aircraft fuel tank capacity, m_{ICE}=75 kg is the mas of the ICE of the HPS and $m_{HPSfuel}$ is the fuel capacity of the HPS retrofitted aircraft. For the stated values of the m_{Bat} the HPS retrofitted aircraft fuel will be between 36 kg and 130 kg (nearly as much as in the avgas powered version of Diamond DA 40).

3. Aircraft flight performance model

The aircraft flight performance model is realized in Matlab and is based on the model described in [10]. The model is adapted for propeller aircraft, by dividing the power plant sub-model into two parts for the ICE performance and the propeller performance. The generic features of the flight model are described below:

The system of ordinary differential equations (ODE) that describes all of the stages of the flight are:

$$m\frac{dV}{dt} = F_x - c_d\frac{\rho V^2}{2}S - mg\sin(\gamma) \quad (3)$$

$$\frac{dh}{dt} = V\sin(\gamma) \tag{4}$$

$$\frac{dL}{dt} = V\sqrt{1 - \sin^2(\gamma)} \tag{5}$$

$$\frac{dm}{dt} = -psfcP_{ICE} \tag{6}$$

where V is the air speed, t is the time, F_x is the engine thrust force, ρ is the density of the air, S is the wing area, m is the mass of the aircraft, γ is the trajectory angle, h is the altitude above the mean sea level, L is the distance and *psfc* and P_{ICE} are respectively the power-specific fuel consumption and the shaft power of the ICE.

To solve the ODE system, at every step of the numerical solution an algebraic system consisting of the following equations is to be solved:

$$c_l = \frac{2mg\sqrt{1-\sin^2(\gamma)}}{\rho V^2 S} \tag{7}$$

$$c_d = c_{d0} + \frac{c_l^2}{\pi A e} \tag{8}$$

Equation 7 expresses the aerodynamic lift coefficient c_l in the case of rectilinear flight (true for cruise and acceptable for climb and descend flight phases). Equation 8 is the classical expression of the airplane drag polar, where c_d is the aerodynamic drag coefficient, c_{d0} is the drag coefficient at zero lift, A is the wing aspect ratio and e is the Oswald factor.

The trajectory angle in the phases of climb and descend with constant instrumental air speed (IAS) is calculated by solving together equation 1 and the following equation:

$$\sin(\gamma) = \frac{dh}{dt}\frac{1}{V} = \frac{dh}{dV}\frac{dV}{dt}\frac{1}{V} \tag{9}$$

The derivative dh/dV can be calculated, bearing in mind that constant IAS means $\rho V^2/2$ = const and ρ is a function of the altitude h.

For all of the sub-parts of the model the ambient conditions are presented in terms of altitude above the sea level and deviation from the international standard atmosphere (ISA) temperature at that altitude.

3.1. Propeller model

The input parameters for the propeller performance model are the shaft power and the rotational speed provided by the engine, the true airspeed (TAS) of the airplane and the ambient conditions.

The main output parameter is the effective thrust of the propeller, taking into account the interference with the parts of the airplane. The methodology of estimating the effective propeller parameters based on the isolated propeller performance chart and the airplane body characteristic areas is taken from Torenbeek [11, pp.190-194].

The propeller performance charts are represented in the model as look-up tables. As far as there are no freely available charts for the original DA-40 NG propeller –MTV-6-R-/190-69, in this study generic characteristics (Hamilton Standard PDB 6101 [11, p.193]) are used.

3.2. ICE models

The models are based on look-up tables and polynomial approximation. The engine performance data is taken from the respective Operation manuals (Austro Engine Operation Manual Doc.No.: E4.01.01, Rev.20 and OPERATORS MANUAL FOR ROTAX® ENGINE TYPE 914 SERIES Rev.0).

3.2.1. Baseline engine. The input parameters for the baseline engine model are the required power as a fraction of the maximum available power and the ambient conditions. The influence of the airspeed ram effect on the engine power is neglected.

The output parameters are the shaft power, the shaft rotational speed, and the fuel flow to the engine.

3.2.2. ICE of the HPS. The ICE sub-model is constructed as a supplement to the baseline engine model. In accordance with the energy management strategy of the HPS it modifies the output of the baseline model in the following way:

- At take-off mode it recalculates the fuel flow and calculates the required shaft power of the electric motor P_{EM}, based on the difference between available power of the baseline engine and the HPS engine;
- At modes between the maximum continuous (max cont.) of the baseline engine and max cont. of the HPS engine, in addition to the fuel flow and P_{EM} the supplement model modifies the shaft speed to be equal to the HPS engine speed at max cont.. These modes of operation generally correspond to high speed cruise flight;
- If the required power is less than the available HPS engine max cont. power at the current ambient conditions, the sub-model modifies only the fuel flow to match that of the Rotax 914 at given power and shaft speed.

It should be noted that the shaft speed modification at high-speed cruise modes affect the performance of the propeller so the overall performance of the aircraft is not exactly the same, but this difference is negligible.

3.3. Airplane aerodynamic characteristics

The coefficients c_{d0} and e, required in equation 8, are initially estimated, using the methodologies given in [11, section 5.3.2.] for 'clean' configuration of the airplane and [11, appendix G] for take-off and landing configurations (flaps extracted at 20° and 42° respectively).

The fine tuning of the coefficients for airplane in cruise configuration is achieved by comparing the DA-40 mathematical model results for the true airspeed (TAS) with the data given in chapter 5 of the Flight Manual (AFM).The tuning is made by varying the values of the coefficients until the difference between the calculated and AFM value of TAS at given altitude, engine power setting and ambient conditions is minimized to a sufficient level (in this case 3%). The results of the tuning and verification are shown in table 2.

Table 2. Verification of the drag coefficient tuning.

	AFM, ISA conditions	Initial estimate c_{d0}=0.033, e=0.8		Final c_{d0}=0.031, e=0.825	
Power setting, %	TAS, m/s	Calculated TAS, m/s	Difference, %	Calculated TAS, m/s	Difference, %
		Flight altitude 610 m			
92	69.96	68.95	-1.45	70.60	0.91
45	49.39	46.40	-6.05	48.10	-2.60
		Flight altitude 1829 m			
92	72.54	72.10	-0.60	73.80	1.74
45	50.42	48.80	-3.20	50.50	0.17
		Flight altitude 4267 m			
92	77.68	74.00	-4.74	76.00	-2.16
45	51.96	49.50	-4.73	52.90	1.81

The drag coefficients for the airplane in take-off configuration are tuned and verified by comparing the model results for take-off run and take-off distance to 15 m above the runway with the data in section 5.3.7 of the DA-40 NG AFM for two altitudes, at ISA. The values c_{d0}=0.040, e=0.56 are adopted.

For the airplane in landing configuration there is no appropriate data in the AFM to check the drag coefficients. On the other hand the final approach and landing phase has minor influence on the overall energy efficiency of the flight, so c_{d0}=0.050, e=0.5 are adopted without further verification.

It should be noted that the obtained drag coefficients produce good model results only together with the propeller and engine sub-models presented above. They are not the 'real' aerodynamic drag coefficients of DA-40 NG.

3.4. Flight scenario

For the purpose of comparing energy efficiency of aircraft with different power plants it is assumed that flight at maximum range with maximum fuel and take-off mass is most characterizing. The flight scenario includes also energy reserve (fuel and electric) for go-around and 30 min navigation reserve.

The flight model is constructed to follow close as practical the AFM procedures in terms of IAS, calibrated for sensor errors (CAS), engine power settings (PL), wing high-lift devices configuration, etc. Airspeeds for take-off rotation (V_R), initial climb to 50ft/15m (V_{50}) and landing approach (V_{ref}) are calculated as functions of the airplane mass, in accordance with the AFM. The phases of the flight scenario are presented in table 3.

Table 3. Modelled phases of the flight.

	Phase description	PL,%	Airspeed	Flaps pos., deg
1	Take-off run	100	0 -> V_R	20
2	Climb to 15m above runway	92	V_{50}	20
3	Climb to safe altitude 300m	92	CAS=72 kts	20
4	Acceleration (in horizontal flight)	92		0
5	Climb to cruise	92	CAS=88 kts	0
6	Cruise at selected altitude	45..75	Depending on PL	0
7	Top of descend deceleration	0 (idle)		0
8	Descend	0 (idle)	CAS=88 kts	
9	Deceleration	0 (idle)		20
10	Final aproach		V_{ref}	42

The phases of the reserve flight are the same except that the take-off run and climb to 15m are replaced by acceleration from V_{ref} to CAS=72 kts, the cruise altitude is always set to 1000 m above the sea level and reserve flight duration is set to 30 minutes.

To calculate the flight at maximum range the following calculation procedure is used:

- Reserve flight is reversely calculated, starting from the landing with empty fuel tanks (m_{fuel} = 0 kg). The aircraft mass at beginning of reserve flight is determined;
- Descend from cruise altitude and final approach are reversely calculated, starting with the aircraft mass at beginning of reserve flight and determining the aircraft mass at the top of descend and the descend distance and time;
- Take-off and climb are calculated. Aircraft mass flight distance and time at end of climb are determined;
- Cruise is calculated, starting with aircraft mass equal to that at end of climb and ending when precalculated mass at top of descend is reached;
- Finally the data for all flight phases is brought together.

In the case of HPS the electric energy taken from the battery E_{el}, for every flight phase is also calculated.

3.5. Energy efficiency calculations

As in [9] the energy intensity is used as a metric to characterize the transport efficiency of the airplane:

$$Ei = \frac{m_{fuel} H_u}{1000 m_{pld} L_{flt}} \tag{10}$$

where m_{fuel} is the mass of the fuel spent for the flight in kg, H_u – the net heat of combustion of the fuel that is used in MJ/kg, m_{pld} – the payload in kg and L_{flt} – the distance of the flight in km. The dimension of Ei could be either MJ·(ton·km)$^{-1}$ or kJ·(kg·km)$^{-1}$. In the case of HPS the energy intensity equation becomes:

$$Ei = \frac{E_{el} + m_{fuel} H_u}{1000 m_{pld} L_{flt}} \tag{11}$$

4. Simulation results and discussion

By simulating flights at different altitudes and different engine power setting it is found that minimum energy intensity of the baseline airplane Ei_{min}=7.09 kJ/(kg.km) is achieved at values of PL=52%, at cruise altitude of 4267m (14000 ft). For this reason the simulation of the baseline airplane with these parameters is selected as a reference and the simulations of the HPS airplane are made for the same conditions.

4.1. Take-off and climb

The described HPS at PL=52% uses only the ICE. This means that the electric energy is used only at take-off and climb phases of the flight. For this reason the take-off and climb are analyzed first. It should be noted that the size of the battery do not affect the flight performance at this stage of the flight and only the electrical efficiency plays role. The simulation results are given in table 4:

Table 4. Simulation results for take-off and climb to 4267 m.

	Baseline airplane	HPS, η_{el}=0.85	HPS, η_{el}=0.95
Flight distance, km	67.243	67.396	67.423
Duration, h	0.378	0.379	0.379
Fuel consumed, kg	9.735	6.876	6.881
Electric energy, kWh	-	18.65	16.71
Energy intensity, kJ/(kg·km)	19.19	16.55	16.24
Relative fuel economy, %		29.37	29.31
Relative energy economy, %		13.72	15.36

The results show significant reduction of the fuel consumption and to less extent of the energy consumption during the take-off and climb phases of the flight. The electric efficiency of the HPS do not affect strongly these results. It can be concluded that even HPS with current state of technology will have positive local ecological effect in the airport areas.

4.2. Flight at maximum distance

The results for the flight at maximum range are presented in table 5:

Table 5. Simulation results for maximal distance flight at 4267m and PL=52%.

	Baseline airplane	HPS, η_{el}=0.85, $E_{sp.Bat}$=200Wh/kg	HPS, η_{el}=0.95, $E_{sp.Bat}$=200Wh/kg	HPS, η_{el}=0.95, $E_{sp.Bat}$=500Wh/kg
Flight distance, km	1395.52	446.90	645.67	1316.50
Duration, h	6.349	1.894	2.815	5.867
Fuel consumed, kg	74.816	25.240	40.766	119.921
Energy intensity, kJ/(kg.km)	7.089	7.468	8.635	12.185
Relative increase of the fuel consumption per hour, %	-	13.08	22.89	73.47

Because of the lower fuel consumption of the baseline diesel engine at long distances it is superior to the HPS with gasoline ICE. In case of electrically efficient HPS with high energy density battery, it may allow higher fuel capacity and reach the maximum range of the diesel airplane, but at the cost of much higher fuel consumption. At shorter distances the relative weight of the climb and descend phases of the flight where the HPS surpasses the efficiency of the diesel at cruise, the energy intensity of the HPS airplane is much closer to the diesel one. A general conclusion can be drawn, that the optimization of the ICE will continue to play very important role in the HPS.

5. Conclusions

In this study the energy efficiencies of a light airplane with diesel engine and with HPS are compared. The diesel is superior to HPS at long distances. The HPS has distinct advantage over diesel at take-off and climb, which may contribute to the reduction of the emissions in the airport areas. In terms of energy intensity the simulations show that the perfection of the electric part of the HPS and the energy density of the battery do not play as important role as the efficiency of the ICE of the HPS. It can be concluded that the improvement and optimization of ICS will continue to play important role in the light aviation even with the coming of the airplanes with HPS.

Acknowledgements

This work was supported by the European Regional Development Fund within the Operational Program "Science and Education for Smart Growth 2014 - 2020" under the Project CoE "National center of mechatronics and clean technologies" BG05M2OP001-1.001-0008.

References

[1] Brelje B J and Martins J R 2019 Electric, hybrid, and turboelectric fixed-wing aircraft: A review of concepts, models, and design approaches *Progress in Aerospace Sciences* **104** pp 1–19

[2] Trainelli L and Perkon I 2019 MAHEPA–A Milestone-Setting Project in Hybrid-Electric Aircraft Technology Development *Proc. More Electric Aircraft Conf. (Toulouse, France)*

[3] Ludowicy J, Rings R, Finger D F and Braun C 2018 Sizing studies of light aircraft with serial hybrid propulsion systems *67 Deutscher Luft-und Raumfahrtkongress DLRK 2018*

[4] Rings R, Ludowicy J, Finger D F and Braun C 2018 Sizing studies of light aircraft with parallel hybrid propulsion systems *67 Deutscher Luft-und Raumfahrtkongress DLRK 2018*

[5] Rings R, Ludowicy J, Finger D, Braun C and Bil C 2019 Sensitivity analysis of general aviation aircraft with parallel hybrid-electric propulsion systems *APISAT 2019: Asia Pacific International Symposium on Aerospace Technology (Gold Coast, Australia)* pp 1687–700

[6] Ludowicy J, Rings R, Finger D F, Braun C and Bil C 2019 Impact of Propulsion Technology Levels on the Sizing and Energy Consumption for Serial Hybrid-Electric General Aviation Aircraft *APISAT 2019: Asia-Pacific International Symposium on Aerospace Technology (Gold Coast, Australia)* pp 1111–24

[7] Finger D F, Braun C and Bil C 2020 Comparative assessment of parallel-hybrid-electric propulsion systems for four different aircraft *Journal of Aircraft* pp 1–11

[8] Trainelli L, Salucci F, Rossi N, Riboldi C E D and Rolando A 2019 Preliminary sizing and energy management of serial hybrid-electric airplanes *Proc. XXV AIDAA Congress (Roma, Italy)*

[9] Serbezov V S 2019 Assessment of the fuel efficiency of unmanned cargo aircraft, based on general aviation aircraft *IOP Conf. Ser.: Mater. Sci. Eng.* **664** 012006

[10] Serbezov V 2019 Numerical modelling of turbofan engine deterioration as a factor in the airlines fuel conservation *Aviation* **23**(3) pp 91–6

[11] Torenbeek E 1982 *Synthesis of Subsonic Airplane Design* (Springer, Dordrecht)

Sustainable design of the GUST small wind turbine

Karol Zawadzki[1], Anna Baszczyńska, Małgorzata Stępień, Angela Fliszewska, Szymon Molenda and Marcin Pisarski

Institute of Turbomachinery, Lodz University of Technology, 90 - 924 Lodz, 219/223 Wolczanska Street, Poland

[1] E-mail: karol.zawadzki@edu.p.lodz.pl

Abstract. One of the increasingly important criteria, when constructing a Small Wind Turbine (SWT), is its sustainable design. It means constructing the SWT in such a way that it can provide a cheap, easy and safe access to clean energy at a specific location for predicted number of consumers. This paper highlights three issues. First, benefits from the application of the rapid prototyping technology are discussed. 3D printers can be used as a durable and relatively cheap means of SWT components production. Also, the possibility of using recycled plastic for 3D-printed components is of big importance. Secondly, the modular construction of the small wind turbine is outlined. This facilitates the product versatility, proneness to adjustments and ease of maintenance. Last but not least, the production costs for individual units as well as multi-unit production are reviewed to show the financial benefits of SWTs and the payback of such investments over the years. The article concludes with a set of recommendations for wind turbine design, conformal with the suggestions of the UN Sustainability Report and Sustainable Development Goal 7.1. The utilization of described concepts is demonstrated basing on the GUST 1.6 m diameter horizontal axis SWT.

1. Introduction

Securing access to energy is one of the contemporary challenges standing in the way of stable growth of developing countries. Beginning with fulfilment of basic human needs, through any type of technological advancements improving the quality of life – the access to energy is undeniably a vital aspect. Still, the most commonly used resources for energy generation base on fossil fuels – depletable, environment-degrading and increasingly expensive. It is estimated that 2.8 billion people use, as cooking fuels, traditional solid energy sources like wood or coal, while 1.2 billion still have no access to electricity at all [1]. However, basing on the "Renewables 2019" report [2], it is indicated that at the end of 2018 more than 26% of global electricity production was generated from renewable energy, mostly solar and wind power. A report created by the Global Tracking Framework, a multi-agency study led by the World Bank and the International Energy Agency, indicates the course to achieve universal energy access, double the use of renewable energy and improve energy efficiency [1]. These targets strongly resonate with the Sustainable Development Goals (SDGs) introduced by the United Nations [3], particularly SDG 7.1: to ensure (by 2030) universal access to affordable, reliable, and modern energy services. Rural areas of Sub-Saharan Africa are particularly affected by the lack of easily available sources of clean energy. Therefore, to meet SDG 7.1, special efforts should be made to provide access to the renewable energy sources (RES) in the mentioned regions [3]. In this context, one of the most promising types of energy sources are small wind turbines (SWTs) [4]. However, in order to construct the SWT that meets the Sustainable Development Goals, numerous specific conditions (like local weather, profitability in the long-term use, financial resources and knowledge) should be considered at the design stage.

1.1. Sustainability aspects of the SWT design

To provide sustainability of power generating systems it is necessary to consider a variety of environmental and economic conditions. A wide range of SWT types enables to match an appropriate device for diverse, specific needs, regardless of climatic zone. Developing countries and remote village areas, which are still often deprived of electricity sources, require simple, easily maintainable and relatively cheap solutions. There are various methods that can be utilized to increase sustainability and significantly reduce costs of SWT installations, especially in Sub-Saharan Africa regions.

1.1.1. 3D-print. Nowadays, the 3D-printing technology is eagerly applied in multiple branches of science and industry. This type of additive manufacturing is suitable not only for rapid prototyping process [5], but also becomes increasingly important in the small-batch manufacturing [6]. Objects produced with this method can be characterized by small part size and complex shape. The specification of 3D-printing technology allows to produce highly customized elements in low volumes and with high repeatability. The variety of available manufacturing materials includes polymers, resins or even super alloys, such as nickel-based chromium alloys [6]. What is more, 3D-printing is considered as a manufacturing process with a high potential for production resource and energy reduction, with a possibility of decreasing the production-related CO_2 emission [7, 8]. Regarding the available methods, one of the most popular 3D-printing techniques is the Fused Deposition Modelling (FDM) [9]. It uses a thermoplastic material in form of a filament. It is melted and applied layer by layer to build a complete model. Thermoplastics are materials that can be recycled, which is a valuable feature from the point of view of economics and ecology [10]. This makes unwanted or wasted products possible to be reused, according to the circular economy idea aiming "to close the loop of product lifecycles through greater recycling and reuse, and bring benefits for both the environment and the economy" [11].

1.1.2. Modular assembly. Efficient arrangement of the components and ease of assembly of complex mechanisms, such as SWT, is best achieved with the modular design of their subsystems. This term refers to the collection of distinct blocks that combined, build a complete system. Every separate block or sub-assembly has a different function and can be freely modified without the influence on the whole structure [12]. Introducing the modularity into the design process provides numerous advantages in terms of sustainability of the final product. The device can be easily customized, the production process is economic and the lead-time is reduced. Moreover, the product can be easily maintained and repaired [13]. The modular construction allows to customize the device for multiple sustainable purposes along its lifecycle. The concept of modularity in engineering pioneered in the 1960s [14]. Modular product design (MDP) and its benefits for product sustainability have been widely studied for several years and many design methods have significantly evolved. Traditional methods of modular product design include matrix-based and decoupling algorithms [15], cluster-identification algorithms for product redesign [16] or fishbone diagrams which identify relationships between particular modules [17]. Modularization is considered to be an effective method to compromise between environmental consciousness and providing a wide assortment of products [13].

1.1.3. Profitability estimation. Determining the costs (and, consequently, the payback time) of an SWT is an important aspect from the point of view of a single consumer. It allows to assess the overall profitability of the investment. For instance, the approximate payback time of a 50-kW turbine operating at an average wind speed of 6.6 m/s is less than 10 years (taking into account the price of energy 0.25 EUR / kWh) [18]. Considering the SWTs, the time required to fully recover costs of their construction can take from 6, to even 30 years [19]. For instance, according to the American consumers' claims, the SWT payback time can be estimated to 15 years, as for example in the case of the 400-Watt rooftop HAWT located in Rockford (northern Illinois, USA) [20]. In addition to that, SWTs also present numerous non-financial benefits. They have a high potential for the sustainable development, mainly considering distributed generation and low-cost technologies [4]. According to [4], the investment costs in case of SWTs are remarkably smaller than for big wind turbines. What is more, SWTs also contribute to decreasing the dependence of countries on foreign energy supply and simultaneously develop many benefits for the domestic economy [21].

2. Overview of the GUST SWT

The Team GUST (Generative Urban Small Turbine) designed and constructed a small wind turbine for the purpose of energy generation and use by individual consumers (prosumers) living at locations in Sub-Saharan Africa: Potiskum (Nigeria), Calvinia (South Africa), Atsbi (Ethiopia), and the Kesses Region (Kenya). Four versions of the SWT were investigated in terms of their efficiency:

- "3 blades 1.00" (3-bladed version with the average chord 0.08 m, shown in figure 1(left).
- "4 blades 1.00" (4-bladed version with the average chord equal to 0.08 m, figure 1(right).
- "3 blades 1.33" (3-bladed version with the average chord increased by 33% and equal to 0.1067 m).
- "4 blades 1.33" (4-bladed version with the average chord increased by 33% and equal to 0.1067 m).

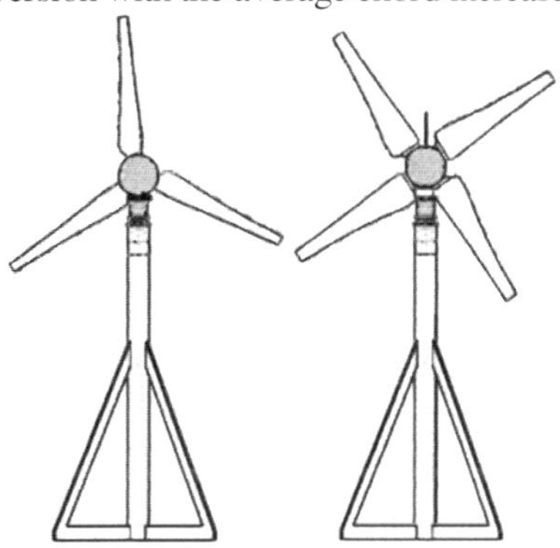

Figure 1. 3- and 4-bladed variants of the GUST SWT.

All aforementioned versions of the rotor have fixed-pitch blades, whose geometry is based on the NREL S826 aerofoil. The swept area of the rotor is equal to 2 m^2 and the rated rotational speed is equal to 335 RPM. The GUST SWT can operate in the wind speed range from 3 m/s to 15 m/s, with the rated wind speed equal to 5 m/s. A synchronous, permanent-magnet, off-the-shelf generator is used Considering the outcomes from the wind tunnel tests at the International Small Wind Turbine Contest, the nominal power of the turbine is equal to around 750 W [22, 23].

In order to make the prototype affordable and cost-effective for the application in the Sub-Saharan region, special emphasis on sustainability criteria was placed. The modular assembly of mechanical parts and use of simple manufacturing techniques play an important role in the design process. The Team took the profitability of the constructed SWT into consideration as well. The cost estimation and payback time for four variants of the wind turbine was calculated. Basing on the obtained results, the version most attractive for consumers from the financial point of view was chosen (see Section 3.3).

3. Sustainable design of the GUST SWT

This chapter will discuss the most important aspects of the sustainable design of the GUST small wind turbine.

3.1. 3D-print

Considering all benefits of using the 3D-printing technology, the importance of biodegradability and recyclability of numerous filament types is essential from the point of view of sustainability. This section focuses on the importance of 3D-printed components regarding the sustainable design of the GUST wind turbine.

Acrylonitrile Butadiene Styrene (ABS) is a type of thermoplastic material that was used in manufacturing components for GUST SWT. The granules of virgin ABS material are used to make the filament. With its use, the 3D-printed models can be manufactured. To recycle the waste material, the mechanical shredder brakes the model to obtain again the small granules of ABS. The re-use of the filament, however, changes its mechanical and thermal properties. Although the recycled material shows decreased tensile strength after the first cycle, the further deterioration of mechanical properties does not appear with consecutive recycle series. What is more, combining the recycled granules with the virgin material gives a possibility to create a blend, which has good mechanical properties and meets the criteria of sustainability [24].

Due to the benefits of additive manufacturing, all GUST wind turbine components indicated in figure 2 are created by means of 3D-printing technology separately and are recyclable due to the use of ABS filament.

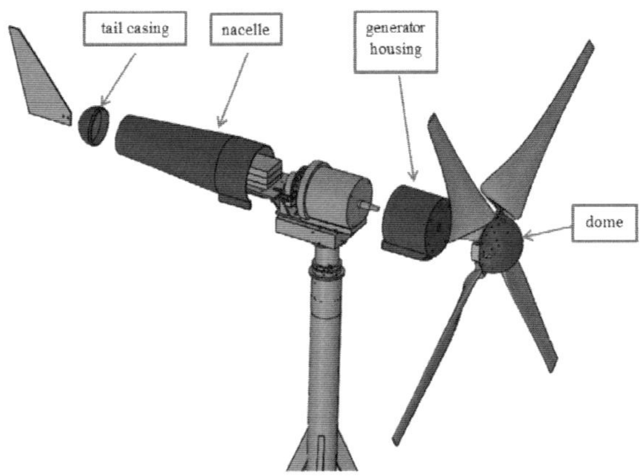

Figure 2. 3D-printed elements (marked blue) of the GUST SWT.

The recyclability and biodegradability are, however, not the only necessary criteria to be met. All 3D-printed parts should be resistant to stresses and weather conditions (especially temperature and rainfall) that may appear during the outdoors operation at a particular location. The statistics regarding the temperature and humidity in two locations typical for the Sub-Saharan Africa (Potiskum and Atsbi) are presented in table 1.

Table 1. Selected weather conditions in two locations in Sub-Saharan Africa.

Parameter	Potiskum [25]	Atsbi [26]
Average daily temperature, °C	35	17
Annual rainfall, mm	713	778

ABS is a material, that is impact-resistant and can withstand high temperatures (up to around 100°C). The "glass transition", which is the moment of liquefaction for this plastic is achieved at the temperature around 100°C [27]. However, the ABS material can soak with water if exposed to moisture or heavy rain conditions for a long time. To prevent this and decrease the negative impact of humidity, the appropriate postprocessing – like the application of the epoxy glue and wax – can be applied. Also such measures as the infill level not less than 40% and wall thickness not less than 3 shell layers increase noticeably the water resistance of the 3D-printed elements [28].

Still, the 3D-printed parts are relatively easy to manufacture, comparing to another manufacturing methods (CNC machining or injection moulding). In all, the use of filament-made parts in the GUST small wind turbine design meets the following sustainability criteria:

- low costs and simplicity in design,
- eco-friendliness (recyclability and biodegradability),
- resistance to weather conditions (for selected filaments),
- possibility to manufacture almost everywhere (3D printers are usually easy to transport due to reduced dimensions).

One should consider also drawbacks of the additive manufacturing. The process is relatively sensitive to printing conditions and settings (temperature, extrusion velocity, etc.). If it ends up with poor accuracy of the 3D-printed parts, it may e.g. provoke problems with the assembly of SWT components. Also, the long printing time can be a noticeable issue [29].

3.2. Modular assembly

The GUST wind turbine is designed in a way, that permits to construct and improve every block separately. This approach provides the possibility of cooperation of numerous design teams, responsible for different modules. The construction of SWT can be divided into the following principle modules (see figure 3 left):

- yawing system,
- rotor,
- generator,
- electronic system.

The cooperation of all wind turbine modules is presented in the flow chart (figure 3 right).

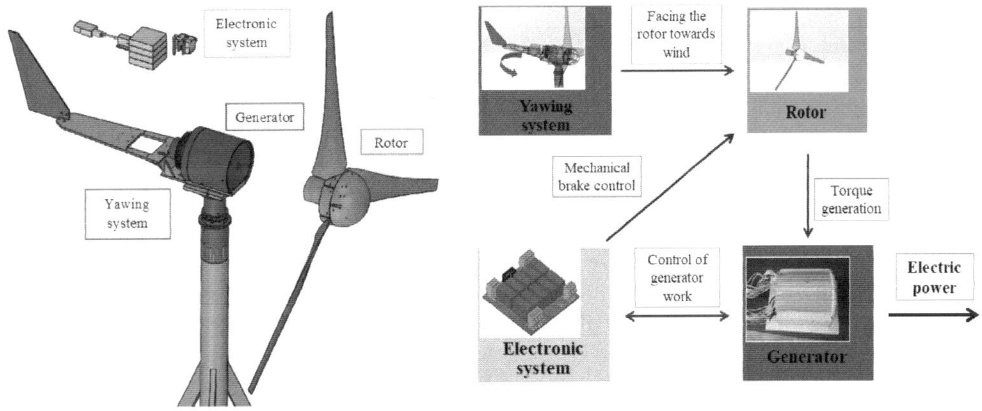

Figure 3. Sub-systems in the GUST SWT (left) and the interdependence of its modules (right).

3.2.1. Yawing system. The yawing system is composed of a turntable and a tail. The purpose of implementation of this module is to face the wind turbine towards the oncoming wind. It is especially significant in geographical regions, where the wind direction is unstable and often varies. Considering for instance the location of Potiskum, main directions of the wind are north and south, however east and west air currents can also transport an important amount of wind kinetic energy [23].

3.2.2. Rotor. The rotor module is responsible for the conversion of the wind energy into torque, transported further to the generator. The module consists of a previously mentioned set of blades, a dome and a spinner. These parts are directly connected to the shaft, which transports the torque to the generator. The number of blades can be easily changed (e.g. three-bladed rotor can be traded for a four-bladed one) without significant interference with other sub-systems. Also some extra features such as blades with winglets [30] or a diffuser [31] could be employed, if necessary (e.g. for particular environmental conditions).

3.2.3. Generator and electronic system. The electronic system is based on the worldwide available components which are easily accessible. This decreases price of the modules and their complexity, so that only the defective part is replaced in case of a failure. Another advantage is the compact dimensions of the modular electronic system. Printed circuit boards (PCBs) are designed to be easily made in electronic-circuit manufacturing companies. The system can operate for 5 years with full load, according to the estimations based on thermal optimisation of the PCBs and their individual components. The SWT is designed to operate fully autonomously, with low maintenance requirements. Moreover, the turbine is equipped with the RGB diode in the tail in order to warn flying objects/planes and inform about the state of the turbine.

3.3. Cost estimation

The overall cost of GUST SWT construction is estimated at 2052.06€ and 2121.48€ (for a 3-bladed and 4-bladed variant, respectively). The most expensive parts are the electronic components (650.00€) and the generator (564.00€). Among the mechanical parts, the manufacturing of one turbine blade and the nacelle generates the highest expenditure (70.00€ and 87.00€, respectively). Other costs include the elements of the braking system, as well as the hub + rotor system. The breakdown of the manufacturing costs into the main subsystems of a wind turbine is shown in figure 4.

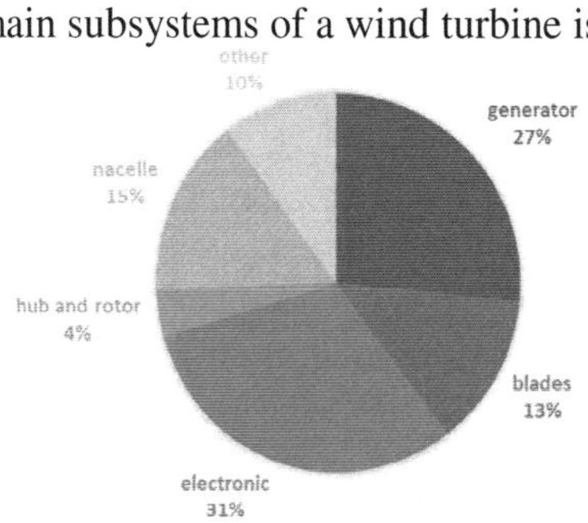

Figure 4. Sub-systems in the GUST SWT.

To assess the profitability of the GUST wind turbine in the long-term use, the expected Annual Energy Production was calculated (see table 2). Having information about the annual energy gain, the income associated with the use of the SWT was estimated, considering the following assumptions:

- the lifecycle of the GUST wind turbine is estimated to be 20 years [19],
- the cost of energy is for the Sub-Saharan Africa, being equal to 0.18€/kWh (2018) [32].

The energy production, as well as the estimation of the SWT payback period, were calculated for all variants (see Chapter 2). Results are presented in table 2.

Table 2. Cost estimation of the GUST SWT.

	3 blades 1.00	3 blades 1.33	4 blades 1.00	4 blades 1.33
Turbine lifetime (years)	20			
Investment cost (€)	2052.06		2121.48	
Cost of 1 kWh (€/kWh)	0.18			
Energy production (1 year, kWh)	742.90	806.25	796.15	835.17
Energy production (20 years, kWh)	14858.03	16125.07	15923.06	16703.44
Income (1 year, €)	133.72	145.13	143.31	150.33
Income (20 years, €)	2467.42	2902.51	2866.15	3006.62
Simple payback (years)	15.35	14.14	14.80	14.11
NPV (Net Present Value, €)	134.49	320.95	221.81	336.65
IRR (Internal Rate of Return)	2.67%	3.56%	3.05%	3.58%

The Annual Energy Production is of the highest magnitude for the 4 blades 1.33 version of the GUST SWT and is equal to approximately 835.17 kWh. This yields almost 16125 kWh during turbine lifecycle. Other variants produce similar outcomes. The income after 1 year is of the lowest value for the 3 blades 1.00 version. Basing on the aforementioned outcomes, the average payback time is estimated to approximately 14-15 years for every GUST SWT model. This is a satisfactory result, comparing with existing and commercially available small wind turbines. Considering for instance the Zephyr Airdolphin SWT (3-bladed model with the rotor swept area equal to 2.54 m^2), the payback time is equal to 63 years. For another commercial small wind turbine, Whisper 200 (3-bladed rotor with the swept area 5.73 m^2), the financial return period is 25 years [19].

Net Present Value (NPV) and Internal Rate of Return (IRR) are the economic indicators presenting profitability of the entire investment in wind energy harvesting system. NPV is the return on investment cost [33], assuming discount rate equal 2% and the operation time 20 years. For all four configurations of the GUST wind turbine, the IRR (which may be understood as an interest rate from the investment [33]) gives values approximating at 3%. This gives positive results and guarantees the profitability of the investment. The NPV value is the lowest for the "3 blades 1.00" variant (134.49€) and the highest for the "4 blades 1.33" version (336.65€). Considering the aforementioned results, the most profitable business option for the consumer is the variant "4 blades 1.33". It is the most cost-effective and most attractive for the potential customers.

Taking into account the overall price of the GUST SWT, the manufacturing costs may be decreased substantially by the production of a greater number of units. Manufacturing turbine components repeatedly without major modification brings the noticeable decrease of the production process costs. It is possible due to the modular assembly of the GUST wind turbine, which consists of many components manufactured on a large or mass scale. Costs of the manufacturing process of a single model of a wind turbine in series production are estimated using the following formula [18]:

$$P_n = B_n n^{lnT_f/ln2} \quad (1)$$

where: P_n – costs of the n-th unit; B_n – costs of the first unit; n – number of units; T_f – technology factor.

Technology factor of industrially manufactured products usually varies between 0.80 and 0.96. For example, the factor for the aircrafts industry varies between 0.80 and 0.90. On the other hand, for the wind turbine the best technology factor range varies between 0.90 and 0.95 [18]. The greater the number of mass-produced wind turbines, the lower their unitary cost is. It is showed in figure 5 depending on different values of the technology factor.

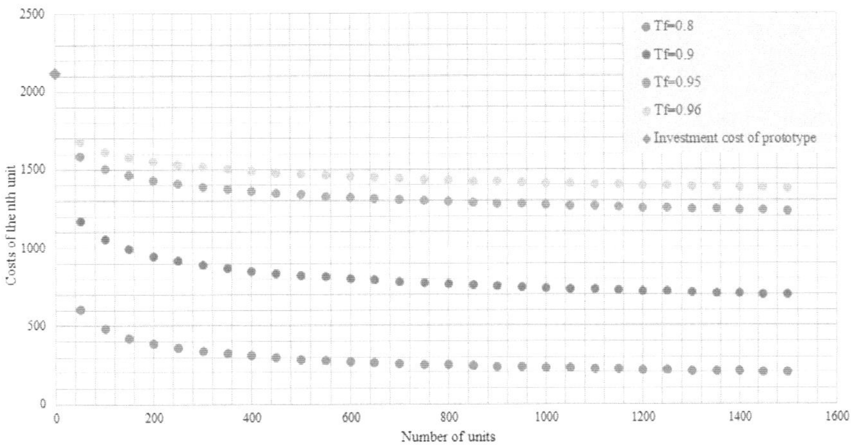

Figure 5. The cost of a single unit of SWT in mass production.

One can observe that, for example, for $T_f = 0.95$ the price of the construction of one SWT produced in 1500 units reaches only the 58% of the total cost of manufacturing of a single wind turbine prototype.

4. Conclusions and comments

The presented article highlights the economical aspects of SWT installations and proposes solutions that visibly increase sustainability of the product. The utilized methods and cautious consideration of specific environmental conditions result in a proposal of a 1.6 m diameter SWT, suitable for multiple locations in Sub-Saharan Africa.

Among all the variants analysed under the described assumptions, the four-bladed turbine with extended chord length (marked as "4 blades 1.33") appeared to be the most profitable. With its ability to produce over 850 kWh of electrical energy annually, the device would provide a simple investment payback after 14.11 years. The income from a power generation of an individual SWT would be equal to 150.33 € after one year and 3006.62 € after 20 years.

Implemented design methods, such as 3D printing and modular construction, ensure that the installation will be easily maintainable, reducing necessity for highly qualified staff during basic repairs. Separate modules can be effortlessly replaced without a risk of causing whole system failure. The GUST SWT design is, therefore, an example of practical implementation of the SDG 7.1, targeted at the developing countries of Africa.

References

[1] Angelou N et al. 2013 *Global tracking framework (English) Sustainable energy for all* (Washington, D.C. World Bank Group)
[2] REN21 2019 *Renewables 2019 Global Status Report* (Paris: REN21 Secretariat) Available: https://www.ren21.net/reports/global-status-report/
[3] *The Sustainable Development Goals Report 2019* Available: https://unstats.un.org/sdgs/report/2019/The-Sustainable-Development-Goals-Report-2019.pdf
[4] Wen T W, Palanichamy C and Ramasamy G 2018 Small wind turbines as partial solution for energy sustainability of Malaysia *Int. J. of Energy Economics and Policy* **9**(2) pp 257–66
[5] Lipian M, Kulak M and Stepien M 2019 Fast track integration of computational methods with experiments in small wind turbine development *Energies* **12**(9) 1625
[6] Berman B 2012 3-D printing: The new industrial revolution *Business Horizons* **55**(2) pp 155–62
[7] Gebler M, Schoot Uiterkamp A J M and Visser C 2014 A global sustainability perspective on 3D printing technologies *Energy Policy* **74** pp 158–67
[8] Kreiger M and Pearce J M 2013 Environmental life cycle analysis of distributed three-dimensional printing and conventional manufacturing of polymer products *CS Sustainable Chem. Eng.* **1**(12) pp 1511–19
[9] Masood S H 2014 Advances in fused deposition modeling *Reference Module in Materials Science and Materials Engineering Comprehensive Materials Processing* **10** pp 69–91
[10] Pakkanen J, Manfredi D, Minetola P and Iuliano L 2017 About the use of recycled or biodegradable filaments for sustainability of 3D printing. In: Campana G, Howlett R, Setchi R, Cimatti B (eds) *Sustainable Design and Manufacturing 2017* pp 776–85 *Smart Innovation, Systems and Technologies* **68** (Springer, Cham)
[11] European Commission 2019 *Closing the loop: Commission delivers on Circular Economy Action Plan* Brussels
[12] Tseng M M, Wang Y and Jiao R J 2018 Modular Design. In: Chatti S, Laperrière L, Reinhart G, Tolio T *The International Academy for Production* (eds) *CIRP Encyclopedia of Production Engineering* (Springer, Berlin, Heidelberg)
[13] Bryant C R, Sivaramakrishnan K L, Van Wie M, Stone R B and McAdams D A 2004 A modular design approach to support sustainable design *Proc. of the ASME 2004 Int. Design*

Engineering Technical Conf. and Computers and Information in Engineering Conf. **3d** *8th Design for Manufacturing Conf. (Salt Lake City, Utah, USA)* pp 909–18

[14] Koga T and Aoyama K 2009 Modular design method for sustainable life-cycle of product family considering future market changes *Proc. of the ASME 2008 Int. Design Engineering Technical Conf. and Computers and Information in Engineering Conf.* **5** *(Brooklyn, New York, US)* pp 3–10

[15] Kusiak A and Wang J 1993 Efficient organizing of design activities *Int. J. of Production Research* **31**(4) pp 753–69

[16] Newcomb P J, Bras B, Rosen D W 1998 Implications of modularity on product design for the life cycle *ASME. J. Mech. Des. September* **120**(3) pp 483–90

[17] Ishii K J, Juengel C and Eubanks C F 1995 Design for product variety: key to product line structuring

[18] Hau E *Wind Turbines Fundamentals, Technologies, Application, Economic*

[19] Ugur E, Elma O, Selamogullari U S, Tanrioven M and Uzunoglu M 2013 Financial payback analysis of small wind turbines for a smart home application in Istanbul/Turkey *Proc. of Int. Conf. on Renewable Energy Research and Applications (ICRERA) (Madrid)* pp 686–9

[20] Lombardo T, Rooftop Wind Turbines: Are They Worthwhile? engineering.com, Available: https://www.engineering.com/3DPrinting/3DPrintingArticles/ArticleID/9556/Rooftop-Wind-Turbines-Are-They Worthwhile.aspx

[21] Chagas C C M, Pereira M G, Rosa L P, da Silva N F, Freitas M A V and Hunt J D 2020 From Megawatts to Kilowatts: A review of small wind turbine applications lessons from the US to Brazil *Sustainability, MDPI* **12**(7) pp 1–25

[22] Kadrowski D, Kulak M, Lipian M, Stepien M, Baszczynski P, Zawadzki K and Karczewski M 2018 Challenging low Reynolds - SWT blade aerodynamics *MATEC Web. Conf.* **234** 01004

[23] Lipian M, Czapski P and Obidowski D 2020 Fluid–structure interaction numerical analysis of a small, urban wind turbine blade *Energies* **13**(7) 1832

[24] Mohammed M I, Wilson D, Gomez-Kervin E, Tang B and Wang J 2019 Investigation of closed-loop manufacturing with acrylonitrile butadiene styrene over multiple generations using additive manufacturing *ACS Sustainable Chem. Eng.* **7**(16) 13955–69

[25] https://weatherspark.com/y/68821/Average-Weather-in-Potiskum-Nigeria-Year-Round

[26] Jacob M, Lanckriet S, Van Vooren S and Nyssen J *Dogu'a Tembien's Tropical Mountain Climate. Geo-trekking in Ethiopia's Tropical Mountains* (Springer, Cham) pp 45–61

[27] Yazdi M H, Lee-Sullivan P 2009 Determination of dual glass transition temperatures of a PC/ABS blend using two TMA modes *J Therm Anal Calorim* **96** pp 7–14

[28] *3D printing of water proof objects for underwater applications*, retrieved from: https://docs.beobachtung3d.com/waterproof-3d-printed-objects-for-underwater-applications.html Accessed on: 20.06.2020

[29] Xinyan D 2017 Application of three-dimensional printing technology in the manufacture of wind turbine generator equipment *Chemical Engineering Transactions* **62** pp 1153–8

[30] Kulak M, Lipian M and Zawadzki K 2020 Investigation of performance of small wind turbine blades with winglets *Int. J. of Numerical Methods for Heat & Fluid Flow*, in press

[31] Lipian M, Dobrev I, Karczewski M, Massouh F and Jozwik K 2019 Small wind turbine augmentation: Experimental investigations of shrouded- and twin-rotor wind turbine systems *Energy* **186** 115855

[32] *South Africa's petrol and electricity prices vs the world* Retrieved from: https://businesstech.co.za/news/energy/306592/south-africas-petrol-and-electricity-prices-vs-the-world/ Accessed on: 21.06.2020

[33] Asma Arshad 2012 Net Present Value is better than Internal Rate of Return *Interdisciplinary J. of Contemporary Research in Business* **4**(8) pp 211–19

IOP Conf. Series: Materials Science and Engineering 1002 (2020) 012004 doi:10.1088/1757-899X/1002/1/012004

Optimal sector modelling of airspace based on optimization algorithms

T D Todorov[1] and S Simeonov

Department of Aeronautics, Faculty of Transport, Technical University of Sofia, 8 Kl. Ohridski Blvd, 1000 Sofia, Bulgaria

[1] E-mail: teodosiy.dtodorov@gmail.com

Abstract. Reasonable assessment and effective management of air space resources is the base for ensuring the safety and efficiency of air traffic management systems. Our research is focussed on optimal airspace organization that is an index to measure the ability of the airspace system to deliver services to meet air traffic demand. Optimal airspace organization is a key factor in reducing the workload for air traffic controllers. The purpose of the study is to present a method for optimal sector modeling with direct routes applicable to free route airspace with the help of mathematical modelling of airspace. A computational model is implemented in a numerical computing environment. As a result, the following was achieved: defining the horizontal boundaries of sectors, subjected to the following constraints; minimum time spent in a sector and balancing workload distribution of air traffic controllers in the sectors. After final optimization a minimum value of the optimality criterion is obtained which is 8.5 times lower than the initial value.

1. Introduction

Diseases such as COVID-19 have enormous impact on the European air traffic network. In the context of the COVID-19 pandemic, EUROCONTROL publishes a regular comprehensive assessment of the latest traffic situation in Europe and provides a comparison with the same period in 2019. The scenarios are for a gradual increase in traffic by -15% by February 2021 from current levels. A prerequisite for this is that air transport is one of the most popular means of transport and plays a major role in the economy of the country. This allows for the liberalization of air transport and the emergence of low-cost operators, combined with overall economic growth in European countries. The significant increase in the number of flights has resulted in an increased air traffic congestion and manifest contradiction between limited airspace resources and traffic demands. Forecasts indicate that global air traffic over Europe will increase by 2030. For this purpose, the effective management of air traffic resources is a key aspect of optimization Air traffic resources include two parts, namely, air traffic controllers and system resources, such as airspace [1]. Reasonable assessment and effective management of air traffic controllers and airspace resources are the base for ensuring the safety and efficiency of air traffic management systems.

Airspace capacity estimation is a major technique in the implementation of effective management and rational allocation of airspace resources. This assessment is directly linked to the delimitation of sectors and the planning of airspace resources. Airspace capacity assessment is carried out at an acceptable level of air traffic controller workload [2]. Each air traffic management system assesses the level of airspace workload it can absorb and from here comes a variety of developed capacity assessment models to meet the needs of any air navigation service provider.

Studies on airspace capacity estimation date back to the theoretical research on runway capacity estimation conducted by Bowen and Pearcey in the late 1940s [3]. Since then, scientists and air traffic management experts have been proposing various theories and models for airspace modelling and capacity assessment. The proposed 2D algorithms for optimal airspace organization utilize real aircraft route data [4] and achieve an even distribution of air traffic controllers' workload. Some authors further develop these algorithms using metaheuristic and genetic algorithms [5] and others statistical data analysis [1]. International Civil Aviation Organization (ICAO) recommends using the Directorate of Operation Research and Analysis Task (DORATASK) method for aerospace capacity evaluation through normative documents for different countries. The DORATASK method is an analytical method used for evaluation and workload analysis [6]. This method is based on Fast-Time Simulation (FTS). The ATC Capacity Analyzer (CAPAN) method is Fast Time Simulation (FTS) developed by Eurocontrol [7]. For the purposes of this study, a macroscopic workload model for capacity assessment is used [8]. This model quantifies the workload depending on the geometric characteristics of the sector, the direction of the air traffic flows and the frequency of conflicts between aircrafts [9].

Nowadays, air transport is built around strictly structured airspace and the use of an automated air traffic management system with centralized control (human - operator). Therefore, as traffic increases, airspace is divided into small sectors in order to avoid saturation. Boundary determination requires those constraints to be taken into account which limits the minimum size of the sectors. Necessary communication when aircraft enters and exits too many sectors also leads to an increase in the overall workload of the air traffic controller.

Mathematical simulation consists of evaluating the sectorization using simulation software and simulated traffic. This approach is essentially local and produces global sectorization, which is a juxtaposition of individually optimized regions of control [5].

The purpose of this study is to analyze and determine the optimal division of sectors of the free route airspace through uniform workload distribution of air traffic controllers based on mathematical modelling. The analysis identifies problems arising from the design of the route network and the determination of the boundaries of the sectors where the criterion of workload is minimized. Finding an optimal solution to these problems can be achieved through a combination of mathematical optimization methods.

2. Optimization process

Optimization in an air traffic management system is a purposeful activity in the sense of obtaining the best result of the state of the system under certain conditions. The object of optimization in the study is the process of airspace organization. The optimization process consists of creating mathematical modelling of a transportation network with optimal routes. A selection of mathematical indicators for quantitative assessment of the workload of air traffic controllers has been made and a criterion for optimality for balanced workload has been determined.

2.1. Transportation network model

Initially, mathematical modelling of a transportation network with direct routes is performed, which is applicable to free route airspace. The design of the transport network allows for flight planning on the shortest route between entry and exit waypoints. The announced waypoints and possible direct routes of the aircraft are used for input data. The waypoints can be defined as entry and exit, as well as to determine the parity of the flight level for the flight of these waypoints – even and odd. This allows one input even waypoint to be connected to another output even waypoint as well as an odd input waypoint to an odd output waypoint. It is also assumed that these direct routes are one-way ones. On one direct route an aircraft will be able to fly either only on even or odd flight, depending on the route directions. In this way, the transport network can be optimized and its horizontal projection determined.

The horizontal projection of the transportation network is considered as a graph, whose nodes correspond to the intersection points of the possible routes and whose links represent these routes. Since each waypoint and each intersection point can be represented by two coordinates (x, y), the model is called a 2D model [5]. Aircraft flows passing through intersection points create conflicts between these aircraft, which must be detected and resolved by air traffic controllers.

Let $V = \{v_1, v_2, \dots, v_n\}$ be a finite set, the elements of which are the intersection points and $E = \{e_1, e_2, \dots e_m\}$ be a finite set, the elements of which are the routes. The connecting function $f_G: E \rightarrow V \times V$ compares on each couple ordered of intersection points. Then $G(V, E, f_G)$ is a finite oriented multigraph, in which each waypoint $v_i \in V$ is represented by a point in the plane, and each route $e \in E$ is such that $f_G(e) = (v_i, v_j)$ with a track starting at point v_i and ending at v_j. For $i \neq j$ it follows that $f(e_i) \neq f(e_j)$ and $G(V, E, f_G)$ is a finite oriented graph. Since function f_G is unique, it follows that each couple ordered pair of $V \times V$ could be found at most only once in the graph and therefore the set E is defined as a subset of $V \times V$, i. e. the final oriented graph is a relation over the Cartesian product $V \times V$. In this way, each direct route is unambiguously determined by a single pair of two waypoints. Mathematical modelling of the transportation network allows us to determine and plan the optimal flight route.

2.2. Determining the number of sectors

The optimal organization of airspace requires an assessment of aircraft performance. When dividing the airspace into sectors, the minimum time spent in a sector and the maximum speed of aircraft flying through the sector shall be assessed.

Let airspace *S* be divided into *n* number of sectors $s_1, s_2, \dots s_n$. Therefore,

$$S = s_1 + s_2 + \cdots + s_n = \sum_{j=1}^{n} s_j, \tag{1}$$

where *S* is the area of airspace, s_j is the area of sector *j*, and *n* is the number of sectors.

Initially, it is assumed that the sectors have the shape of an elementary rectangle and $s_1 = s_2 = \cdots = s_n$, then $S = n.s_j$. Thus s_j can be represented as:

$$s_j = (V_m t)^2, \tag{2}$$

where V_m is the maximum speed of the aircraft and *t* is the minimum time spent in a sector *j*.

If $V_m = const$ then $s_j = f(t)$, and if $S = const$ then the number of sectors depends on s_j. Therefore $n = f(t)$.

The flight time spent in a sector is considered in the interval $[t_{min}, t_{max}]$ at a given function $n = f(t)$. Points $t_0, t_1, \dots, t_n$ are used to divide the interval $[t_{min}, t_{max}]$ into subintervals $[t_{j-1}, t_j]$, $j = 1,2, \dots, n$. In each subinterval a random point $\xi_j$is chosen, as $t_{j-1} \leq \xi_j \leq t_j$ and the product s_j is found from the values of the function at this point $f(\xi_j)$ and the length of each sub-interval $\Delta t_j = t_{j+1} - t_j$, such as:

$$s_j = (V_m t)^2 = f(\xi_j) \times \Delta t_j, j = 1,2, \dots n. \tag{3}$$

Therefore, the sum of these products can be represented as:

$$S = \sum_{j=1}^{n} f(\xi_j) \times \Delta t_j. \tag{4}$$

In the expression (3), $s_j$are the areas of the elementary rectangles at $j = 1,2, \dots, n$. If $n \to \infty$, then it follows that $\Delta t_j \to 0$ and $n = f(t)$ is a linear function.

For each sector, a point called a sector centre is chosen at random. The number of randomly generated sector centres *n* is equal to the total number of calculated sectors *K*.

2.3. *Optimality criterion*

Air traffic controllers provide air traffic control, ensure that aircraft are on the right routes and resolve conflicts that arise in the transportation network. When aircraft moves from one sector to another, they coordinate with adjacent air traffic services unit. It is assumed that the air traffic controller workload is formed by the following main activities: monitoring, conflict detection and resolution and coordination [9]. In the case of a model, based on the transportation network, the monitoring of the workload is proportional to the flow, the air traffic controller workload in resolving conflicts arising from the passage of flows through vertexes and coordination depends on the flows crossing the boundaries of the sectors. The aim is to determine the optimal division of sectors of the free route airspace by uniform workload distribution of air traffic controllers for each sector and minimize air traffic flows crossing the boundaries of these sectors. This aim is expressed by the optimality criterion on uniform workload of air traffic controllers *J*, with the following restrictions:

- convexity – the resulting sectors must be convex in relation to the routes of the aircraft;
- minimum determined time spent in a sector.

The following mathematical indicators were used to quantify the workload. The monitoring workload indicator $w_1(k)$ in sector *k* is as follows:

$$w_1(k) = \sum_{l_{ij} \in S_k} \frac{l_{ij}^k}{V_{ij}} f_{ij}, \tag{5}$$

where l_{ij} is the length of the route from the entrance waypoint *i* to the exit waypoint *j*, l_{ij}^k is the length of the route from *i* to *j* in sector *k*, V_{ij} is the maximum speed of the aircraft on the route from *i* to *j* and f_{ij} is the air traffic flow intensity from *i* to *j*.

The conflict workload indicator $w_2(k)$ in sector *k* is as follows:

$$w_2(k) = \sum_{r \in S_k} p_r, \tag{6}$$

where *r* is an intersection point, S_k is an area of sector *k* and p_r is traffic density,

$$p_r = \sum_{f_{ij} \in r} f_{ij}. \tag{7}$$

The coordination workload indicator $w_3(k)$ in sector *k* is as follows:

$$w_3 = \sum_{i \oplus j \in S_k} f_{ij}, \tag{8}$$

where f_{ij} is the air traffic flow intensity from *i* to *j*.

The total workload indicator $w(k)$ in sector *k* is defined as the sum of these three indicators:

$$w(k) = w_1(k) + w_2(k) + w_3(k). \tag{9}$$

As an objective function, determining the optimal division of sectors of the free route airspace by uniform total workload distribution of air traffic controllers for each sector:

$$J = \sum_{k=1}^{K} |w(k) - W_{av}| \rightarrow min, \tag{10}$$

where W_{av} is the average workload.

The total workload W in K sectors is defined as:

$$W = \sum_{k=1}^{K} w(k), \tag{11}$$

where W is the total workload and K is the total number of sectors.

An initial model of the airspace is determined by randomly generating the location of n number of points, centers of Voronoi cells, for which the value of the optimality criterion is calculated. The optimization process consists of n number of iterations by randomly generating a new location of the centers of the Voronoi diagram at the same traffic intensity and with minimum time spent in a sector. For each iteration, the value of the optimality criterion is calculated. The minimum value of the optimality criterion determines the optimal solution of the divided airspace with even distribution of the workload.

A computer program is developed to incorporate the optimization algorithm in order to minimize calculation time by orders of magnitude and also to eliminate the possibility of human error. For that purpose, a proprietary programming language in a multi-paradigm numerical computing environment was used. MATLAB 2019 was chosen mainly because of the friendly user interface and the fact that it is common programming language used by engineers. Although full vectorization of the code was not achieved, loops were used only when necessary. This further reduces the required processing time. The use of computer systems also eases the data management and the visualization of the final results.

3. Results

A geometric algorithm was used to calculate the intersections between simulated orthodromes. The algorithm is applied to the transportation network in figure1, for which 598 intersection points are determined and the distances of the possible routes along the network are calculated.

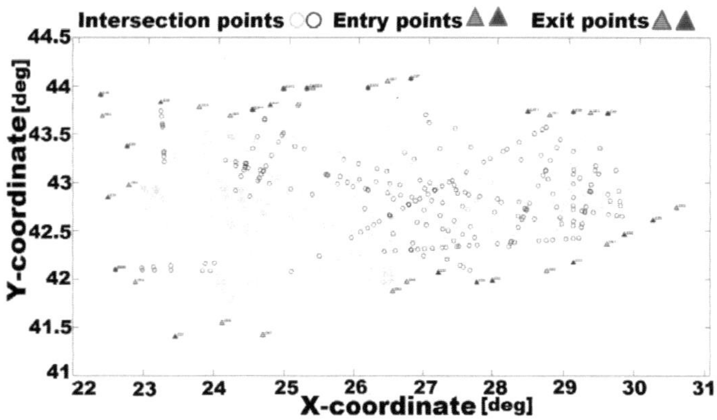

Figure 1. Intersections of direct routes.

The determination of the sum of products (4) is obtained using the value of the function at the midpoints ξ_j of the sub-intervals (3). The calculated dependence of the number of sectors on the total area of the airspace $S = 97804\ \mathrm{NM}^2$ and the maximum speed of aircraft flying through the sector $V_{max} = 460$ kt is presented graphically in figure 2.

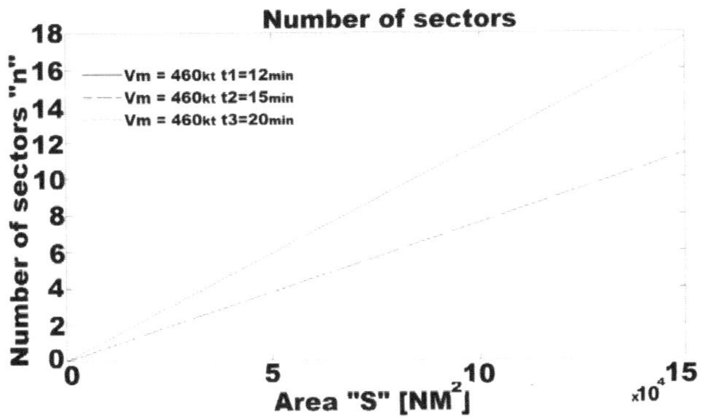

Figure 2. Determination of the total number of sectors for $t_{min} = 12$ min, $t_{min} = 15$ min, $t_{min} = 20$ min and $V_{max} = 460$ kt.

Interval $[t_{min} = 12$ min, $t_{max} = 20$ min$]$ is divided into subintervals, as $t_0 = 1$ and $t_n = 60$. Then 12 sectors at $t_{min} = 12$ min, 8 sectors at $t_{min} = 15$ min and 4 sectors at $t_{min} = 20$ min were calculated.

In the considered airspace, 8 points were chosen at random, which are the centers of 8 Voronoi cells. Since the formation of the Voronoi diagram is preceded by Delaunay's triangulation, the initial distribution is relatively good and covers the whole space in a balanced way. The Voronoi diagram shown in figure 3 is a 2D projection applied to the 598 intersection points taken from figure1 in 8-sector airspace.

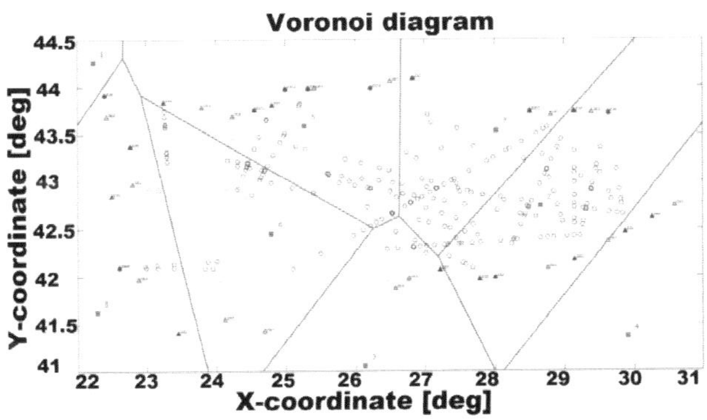

Figure 3. Initial airspace model with 8 sectors, 598 intersections, 114 aircraft per hour and $t_{min} = 15$ min.

For the initial airspace model (before optimization), the mathematical expectation of the distribution of intersection points 74.75, the standard deviation 29.80 and the optimality criterion *J*=5430.38 are calculated.

The optimization process consists of 100 iterations with air traffic intensity of 114 aircraft per hour and $t_{min} = 15$ min. In figure4, optimal solution for aerospace divided into 8 sectors, containing 598 intersection points, 114 aircraft per hour and interval of $t_{min} = 15$ min is graphically presented.

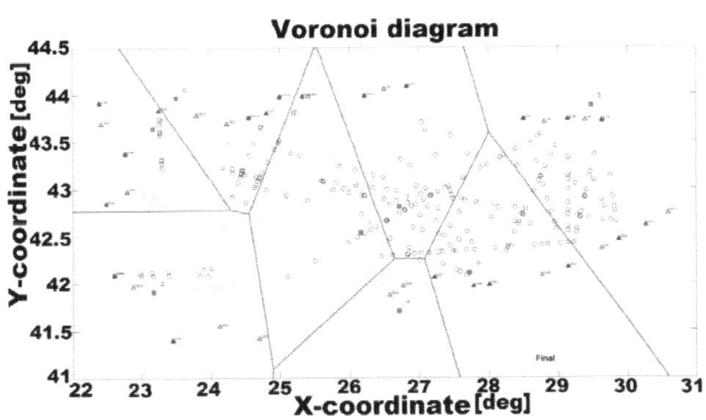

Figure 4. Optimal solution of airspace division into 8 sectors, with 598 intersections and 114 aircraft per hour and $t_{min} = 15$ min.

The value of the optimality criterion for this iteration is $J = 636.58$, which is reduced 8.5 times compared to the initial airspace model. The value of the standard deviation from the mean distribution of the intersection points is 23.56 compared to the initial iteration of 29.80. The optimal solution in figure 4 has changed the boundaries of the sectors, improving the distribution of the intersections by 7.9 %.

The estimate of the distribution of workload indicators for sector 1 and sector 8 is presented in figure 5 and figure 6. For every iteration after the optimization process is done, the algorithm sums the values for $\omega_1(k)$, $\omega_2(k)$ and $\omega_3(k)$. The lowest summation value must be found. This is achieved by first taking the sum of all values for $\omega_1(k)$, $\omega_2(k)$ and $\omega_3(k)$ for the first iteration, than the algorithm continues to probe values as follows; if the current summation value is a number, smaller than the lowest summation value of all previous iterations this new value is set as the new "check for lower" value in the algorithm. This allows the algorithm to find the optimality criterion on uniform workload of air traffic controllers for the summation values of $\omega_1(k)$, $\omega_2(k)$ and $\omega_3(k)$3 for the whole iteration cycles.

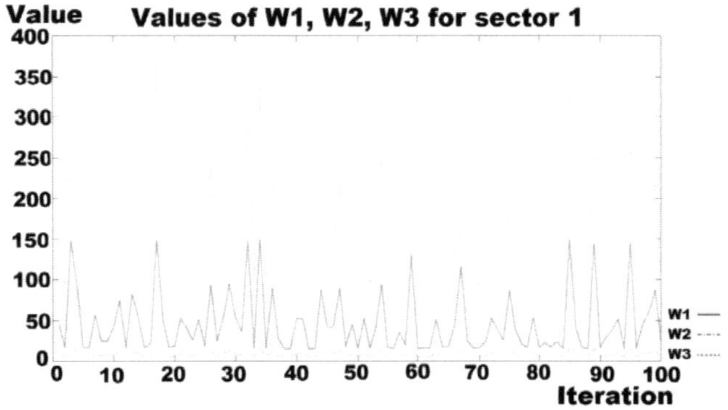

Figure 5. Distribution of workload indicators for sector 1.

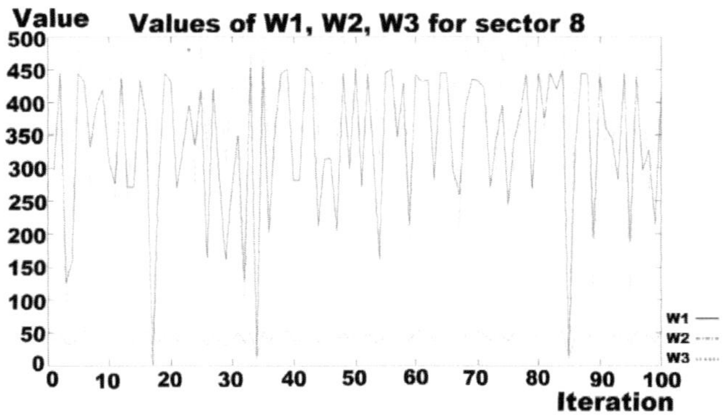

Figure 6. Distribution of workload indicators for sector 8.

The value of the standard deviation of the monitoring workload is 714.81 and decreases 2.20 times compared to the initial iteration of 1571.02. The value of the standard deviation of the conflict workload indicator is 66.12, but increases by 2.36% compared to the initial iteration of 64.57. This increase is due to the redistribution of air traffic by sector as a result of the redistribution of intersections. The value of the standard deviation of the coordination workload is 129.74 and decreases 1.64 times compared to the initial iteration of 212.79. When applying this method, a decrease in the total workload indicator $w(k)$ is observed. The value of the standard deviation is 826.80 and decreases 2.20 times compared to the initial iteration of 1820.92.

The sectors modelled with this method are convex with respect to the routes of the aircraft, which is obtained by means of the Voronoi diagram. The better value of the standard deviation from the average workload shows a more uniform distribution of intersections and air traffic by sectors.

Therefore, the obtained minimum value of the optimality criterion shows that the airspace model of figure 4 is optimal in terms of balanced workload distribution.

4. Analysis of the obtained results

The results of the allocation of airspace to sectors show that as the flight time spent in the sector decreases, the number of sectors increases and the size of sectors s_j decreases. The application of this algorithm has its limitations as it must be taken into account that air traffic controllers need sufficient volume of airspace and sufficient time to service its traffic. Therefore, the size of the sectors cannot be reduced indefinitely. The sectors obtained using the herein presented method are convex geometric figures. The mathematical method using Voronoi diagram satisfies the practical requirements for the sectors.

The results of solving the presented optimization problem show that the set goal for determining sectors with uniform workload has been solved successfully. The presented indicators for the types of workload of the air traffic controllers adequately reflect the real practical problems in the definition of the sectors. The repeated reduction of the criterion for optimality and the better value of the standard deviation from the mean workload shows an even distribution of the workload by sectors. Therefore, the obtained minimum value of the optimality criterion shows that the airspace model of figure 4 is optimal in terms of balanced workload distribution.

5. Conclusion

This study proposes a method for determination of sector boundaries through mathematical modelling of a transport network with direct routes. The method is relevant and applicable to free route airspace. The analysis of the results of the conducted experiments shows that with the used mathematical optimization algorithms an optimal sector modelling is obtained according to the balanced distribution of the workload of the air traffic controllers. The applied optimization methods allow for even distribution of the total workload by reducing the monitoring and coordination workload in each sector. The minimization of the value of the optimality criterion 8.5 times compared to the initial value is obtained by optimal distribution of the intersections and aircraft flows. This is achieved by using the random search method to change the location of the generated points on the centres of the Voronoi diagram.

The presented method allows for repeated resolution of the problem of sector modeling in real time depending on the changes in traffic.

The presented algorithm for sector modelling can be improved by adding weights to the indicators of the types of workload in the general criterion of optimality. The results obtained are encouraging and other numerical optimization methods can be tested, such as correlation and regression data analysis, genetic methods, and others.

References

[1] Zhang M, Shan L, Zhang M, Liu K, Yu H and Yu J 2016 Terminal airspace sector capacity estimation method based on the ATC dynamical model *Kybernetes* **45**(6) pp 884–99

[2] Majumdar A, Ochieng W, McAuley G, Lenzi J and Lepadatu C 2005 The factors affecting airspace capacity in Europe: A framework methodology based on cross sectional time-series analysis using simulated controller workload data *Proc. of the 6th USA/Europe Air Traffic Management R&D Seminar*

[3] Bowen E and Pearcey T 1948 Delays in the flow of air traffic *J. of Royal Aeronautical Society* **52**(448) pp 251–8

[4] Basu A, Mitchell J and Sabhnani G 2010 Geometric algorithms for optimal airspace designandairtraffic controller workload balancing *J. of Experimental Algorithmics* **14** pp 75–89

[5] Delahaye D, Puechmo S 2013 *Modeling and Optimization of Air Traffic* (Wiley) p 352

[6] Jaurena R 2009 *Guide for the application of a common methodology to estimate airport and ATC sector capacity for the SAM region* Lima, Peru (Electronic Materials)

[7] Dowling F 1998 *EEC Note No. 4/98 - Sector Capacity Assessment For Dublin ACC* (Electronic Materials) EUROCONTROL
[8] Welch J, Andrews J, Martin D and Sridhar B 2007 Macroscopic workload model for estimating en route sector capacity *7th Seminar Barselona, Spain,* paper 37 pp 1–10
[9] Todorov T and Petrov P 2017A study of sector configurations capacity for air traffic service MATEC Web Conf. **133** 01003
[10] Dimitrov D and Slavov S 2015 Methods for dynamic sectorization of the Bulgarian airspace *Voenen Zhurnal* **122** pp 66–74 (In Bulgarian)

Functionally graded structural member with three longitudinal circular cylindrical cracks loaded in tension: a fracture analysis

V Rizov

Department of Technical Mechanics, University of Architecture, Civil Engineering and Geodesy, 1046 Sofia, Bulgaria

E-mail: v_rizov_fhe@uacg.bg

Abstract. The present paper is concerned with fracture analysis of a structural member of circular cross-section with three longitudinal circular cylindrical cracks. The cracks are located arbitrary in radial direction of the structural member. Beside, the three cracks have different lengths. A circular notch is cut-out in the mid-span. The structural member is loaded in tension by two axial forces applied at the ends. The material is functionally graded in radial direction. The mechanical behaviour of the material is non-linear elastic. The fracture is studied in terms of the strain energy release rate. The solutions to the strain energy release rate derived for three cracks are verified by considering the balance of the energy. A parametric investigation of the fracture is carried-out. The effects of locations of cracks and material gradient on the fracture are evaluated.

1. Introduction

The properties of functionally graded structural materials change continuously along one or more directions in the volume of structural member [1-3]. The functionally graded materials are widely used in aeronautics and automotive engineering since they combine the best qualities of their constituent materials [4]. Fracture strongly influences the functioning of structural members and components made of functionally graded materials [5]. Thus, analysis of fracture is an important part of the design process of functionally graded components.

The present paper is devoted to the problem of longitudinal fracture of a functionally graded non-linear elastic structural member of circular cross-section with three circular cylindrical longitudinal cracks loaded in tension. It should be noted that the previous papers deal with fracture analyses of functionally graded members with one longitudinal crack [6, 7]. However, the change of material properties in radial direction is a premise for appearance of more longitudinal cracks. Hence, the present paper contributes towards this topic by developing a fracture analysis of a structural member with three concentric longitudinal cracks.

2. Deriving of the strain energy release rate for the longitudinal cracks

A functionally graded structural member with three longitudinal cracks is shown in figure 1. The cross-section of the structural member is a circle of radius, R_4. The structural member is loaded in centric tension by axial forces, F, applied at the ends. The length of the structural member is $2l$. The member is functionally graded in radial direction. The appearance of more longitudinal cracks is related to the fact that the material properties change in radial direction. The material has non-linear elastic behaviour. The cracks are concentric circular cylindrical surfaces. The lengths of cracks 1, 2 and 3 are $2a_1$, $2a_2$ and $2a_3$, respectively. The radiuses of the fronts of cracks 1, 2 and 3 are R_1, R_2 and R_3, respectively. The three cracks are located symmetrically with respect to the middle of the beam (figure 1). The cracked portion of the structural member is divided by the three cracks into four concentric crack arms (a crack arm is part of the beam located between two neighboring concentric circular cracks). The cross-section of crack arm 1 is a circle of radius, R_1. The length of crack arm 1 is $2a_1$. The crack arm 2 has a ring-shaped cross-section of internal and external radiuses, R_1 and R_2, respectively (this crack arm is located between cracks 1 and 2). The length of crack arm 2 is $2a_1$. The crack arm 3 has a ring-shaped cross-section of internal and external radiuses, R_2 and R_3, respectively (this crack arm is between cracks 2 and 3). The length of crack arm 3 is $2a_2$.

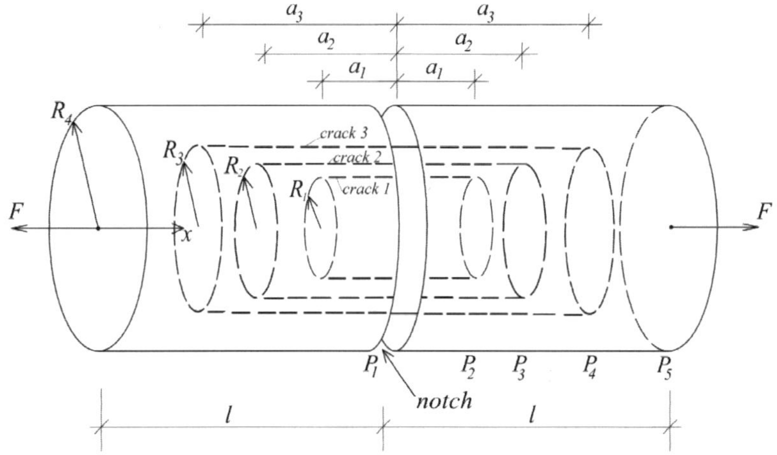

Figure 1. Geometry and loading of a functionally graded structural member of circular cross-section with three longitudinal cracks.

The crack arm 4 has a ring-shaped cross-section of internal and external radiuses, R_3 and R_4, respectively (this crack arm is between crack 3 and the surface of the structural member). The length of crack arm 4 is $2a_3$. A circular notch of depth, $R_4 - R_3$, is cut-out in the crack arm 4 in the middle of the structural member. The notch divides the crack arm 4 into two symmetric parts each of length, a_3. It is obvious that the two parts of crack arm 4 are free of stresses.

The longitudinal fracture behaviour is studied in terms of the strain energy release rate, G. Only half of the structural member, $l \leq x \leq 2l$, is analyzed due to the symmetry (figure 1). First, the strain energy release rate is obtained assuming an elementary increase, da_1, of the length of crack 1. For this purpose, the strain energy release rate is written as [8]:

$$G = 2\frac{dU^*}{l_{cf1}da_1}, \tag{1}$$

where U^* is the complementary strain energy, l_{cf1} is the length of the front of crack 1. It should be noted that the right-hand side of (1) is doubled in view of the symmetry (figure 1). The length of the front of crack 1 is expressed as

$$l_{cf1} = 2\pi R_1. \tag{2}$$

By combining of (1) and (2), one obtains

$$G = \frac{dU^*}{\pi R_1 da_1}. \tag{3}$$

The complementary strain energy is written as

$$U^* = U_1^* + U_2^* + U_3^* + U_{P_2P_3}^* + U_{P_3P_4}^* + U_{P_4P_5}^*, \tag{4}$$

where U_1^*, U_2^*, U_3^*, $U_{P_2P_3}^*$, $U_{P_3P_4}^*$ and $U_{P_4P_5}^*$ are the complementary strain energies cumulated in crack arms 1, 2, 3, in the internal parts of structural portions, P_2P_3 and P_3P_4, and in the un-cracked structural portion, P_4P_5, respectively (figure 1). It should be specified that the internal part of structural portion, P_2P_3, has a circular cross-section of radius, R_2. The length of structural portion, P_2P_3, is $a_2 - a_1$. The cross-section of the internal part of structural portion, P_3P_4, is a circle of radius, R_3. The length of structural portion, P_3P_4, is $a_3 - a_2$. The length of the un-cracked structural portion is $l - a_3$. The complementary strain energy in crack arm 1 is expressed as

$$U_1^* = a_1 \int_0^{R_1} u_{01}^* 2\pi R dR, \tag{5}$$

where u_{01}^* is the complementary strain energy density in this crack arm. The following formula is used to calculate u_{01}^* [6]:

$$u_{01}^* = \sigma\varepsilon - u_{01}, \tag{6}$$

where σ is the stress, ε is the strain, u_{01} is the strain energy density.

In the present paper, the mechanical behaviour of the material is treated by using the following non-linear stress-strain relation [9]:

$$\sigma = \frac{\varepsilon}{b + f\varepsilon}, \tag{7}$$

where b and f are material properties. The distribution of b in radial direction is described as

$$b = b_0 + \frac{b_1 - b_0}{R_4^g} R^g, \tag{8}$$

where b_0 and b_1 are the values of b in the centre of the member cross-section and the surface of the member, g controls the material inhomogeneity in radial direction.

In principle, the strain energy density is equal to the area enclosed by the stress-strain curve. Thus, the strain energy density in the crack arm 1 is obtained by integrating of (7)

$$u_{01} = \frac{1}{f}\left[\varepsilon - \frac{b}{f}\ln\left(\varepsilon + \frac{b}{f}\right) + \frac{b}{f}\ln\frac{b}{f}\right]. \tag{9}$$

By substituting of (7) and (9) in (6), one derives

$$u_{01}^* = \frac{\varepsilon^2}{b + f\varepsilon} - \frac{1}{f}\left[\varepsilon - \frac{b}{f}\ln\left(\varepsilon + \frac{b}{f}\right) + \frac{b}{f}\ln\frac{b}{f}\right]. \tag{10}$$

The magnitude of ε is determined by using the following equation for equilibrium of the axial forces in the crack arms:

$$N_1 + N_2 + N_3 = F, \tag{11}$$

where N_1, N_2 and N_3 are axial forces in crack arms 1, 2 and 3, respectively. The axial forces are expressed as

$$N_1 = \int_0^{R_1} \sigma_{ca1} 2\pi R dR, \quad N_2 = \int_{R_1}^{R_2} \sigma_{ca2} 2\pi R dR, \quad N_3 = \int_{R_2}^{R_3} \sigma_{ca3} 2\pi R dR, \tag{12}$$

where σ_{ca1}, σ_{ca2} and σ_{ca3} are the stresses in crack arms 1, 2 and 3, respectively. After substituting of (7), (8) and (12) in (11), the equation for equilibrium is solved with respect to ε by using the MatLab computer program.

The complementary strain energies in crack arms 2 and 3, in the internal parts of member portions, P_2P_3 and P_3P_4, and in the un-cracked member portion, P_4P_5, are obtained as

$$U_2^* = a_1 \int_{R_1}^{R_2} u_{02}^* 2\pi R dR, \tag{13}$$

$$U_3^* = a_2 \int_{R_2}^{R_3} u_{03}^* 2\pi R dR, \tag{14}$$

$$U_{P_2P_3}^* = (a_2 - a_1) \int_0^{R_2} u_{0P_2P_3}^* 2\pi R dR, \tag{15}$$

$$U_{P_3P_4}^* = (a_3 - a_2) \int_0^{R_3} u_{0P_3P_4}^* 2\pi R dR, \tag{16}$$

$$U_{P_4P_5}^* = (l - a_3) \int_0^{R_4} u_{0P_4P_5}^* 2\pi R dR, \tag{17}$$

where the complementary strain energy densities are found by (10). For this purpose, ε is replaced with the strain in the corresponding part of the member.

By substituting of (4), (5), (13), (14), (15), (16) and (17) in (3), one derives the following solution to the strain energy release rate:

$$G = \frac{2}{R_1}\left(\int_0^{R_1} u_{01}^* R dR + \int_{R_1}^{R_2} u_{02}^* R dR - \int_0^{R_2} u_{0P_2P_3}^* R dR \right). \tag{18}$$

The integration in (18) is carried-out by the MatLab computer program.

Formula (3) is applied also to drive the strain energy release rate for crack 2. For this purpose, da_1 and R_1 are replaced with da_2 and R_2, respectively. Analogically, the strain energy release rate for crack 3 is found by replacing of da_1 and R_1 with da_3 and R_3.

The strain energy release rate is derived also by considering the balance of the energy for verification. The longitudinal displacement of the application point of force, F, that is needed when analyzing the balance of the energy is determined by the integrals of Maxwell-Mohr. It should be noted that the strain energy release rates obtained by considering the balance of the energy are exact matches of these found by using (1). This fact is a verification of the fracture analysis developed in the present paper.

3. Numerical results

The solutions to the strain energy release rate obtained in previous section of the paper are applied here in order to evaluate the effect of material gradient on the longitudinal fracture of the structural member. The strain energy release rate is presented in normalized form by using the formula $G_N = Gb_0 / R_4$. The material gradient is characterized by b_1 / b_0 ratio (this is the ratio of the values of b at the surface and the centre of the member cross-section). The locations of cracks 1, 2 and 3 in radial direction are characterized by R_1 / R_4, R_2 / R_4 and R_3 / R_4 ratios, respectively. It is assumed that $R_4 = 0.010$ m, $f = 0.3b_0$ and $g = 0.6$.

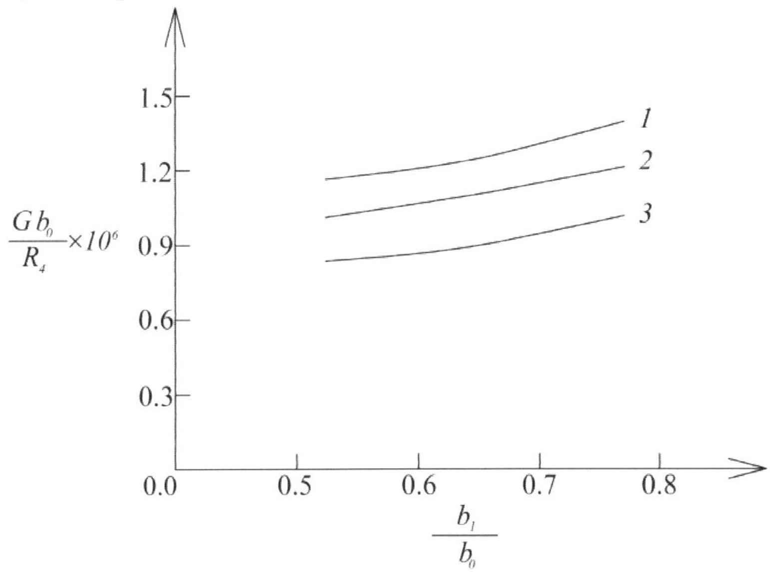

Figure 2. The strain energy release rate in normalized form plotted against b_1 / b_0 ratio (curve 1 – at $R_1 / R_4 = 0.1$, curve 2 – at $R_1 / R_4 = 0.2$ and curve 3 – at $R_1 / R_4 = 0.3$).

The strain energy release rate in normalized form is plotted against b_1 / b_0 ratio in figure 2 at three R_1 / R_4 ratios by using the solution to the strain energy release rate for crack 1.

One can observe in figure 2 that the strain energy release rate increases with increasing of b_1 / b_0 ratio. It can also be observed in figure 2 that the strain energy release rate decreases with increasing of R_1 / R_4 ratio.

The effect of location of crack 2 on the fracture behaviour is evaluated too. For this purpose, the strain energy release rate in normalized form is plotted against R_2 / R_4 ratio in figure 3 at three values of F by applying the solution to the strain energy release rate for crack 2. The curves in figure 3 indicate that the strain energy release rate decreases with increasing R_2 / R_4 ratio.

The effect of the location of crack 3 on the fracture behaviour of the beam is illustrated in figure 4 where the strain energy release rate in normalized form is plotted against R_3 / R_4 ratio by using the solution to the strain energy release rate for crack 3. The curves in figure 4 show that the strain energy release rate decreases with increasing R_3 / R_4 ratio.

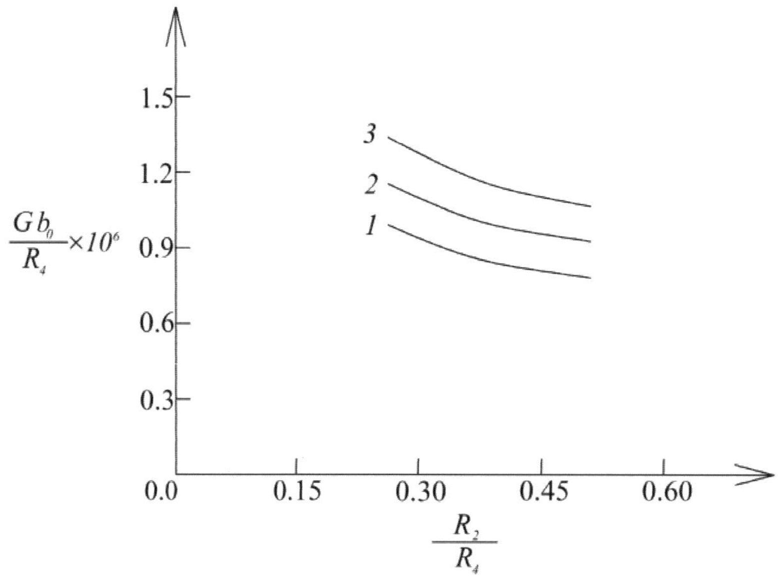

Figure 3. The strain energy release rate in normalized form plotted against R_2/R_4 ratio (curve 1 – at $F = 10$ N, curve 2 – at $F = 12$ N and curve 3 – at $F = 14$ N).

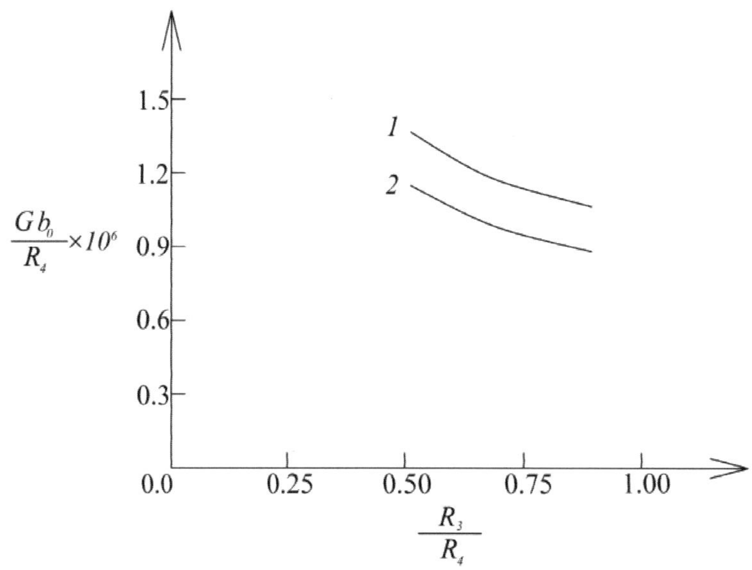

Figure 4. The strain energy release rate in normalized form plotted against R_3/R_4 ratio (curve 1 – at non-linear behaviour of the material and curve 2 – at linear-elastic behaviour).

The strain energy release rate obtained assuming linear-elastic behaviour of the functionally graded material is also plotted in figure 4. The linear-elastic solution is derived by substituting $f = 0$ in the non-linear solution since at $f = 0$ the non-linear stress-strain relation (7) transforms in the Hook's law assuming that $1/b$ is the modulus of elasticity (therefore, the non-linear solution can be applied also for linear-elastic behaviour as a particular case). It can be observed in figure 4 that the material non-linearity leads to increase of the strain energy release rate.

4. Conclusions

Fracture behaviour of a structural member of circular cross-section with tree longitudinal circular cylindrical cracks loaded in centric tension is analyzed in terms of the strain energy release rate (the appearance of more longitudinal cracks is related to the fact that the material properties change in radial direction). The structural member is functionally graded in radial direction. The material has non-linear elastic behaviour. The solutions to the strain energy release rate derived in the present paper are used to evaluate the effect of the locations of the cracks on the fracture behaviour. It is found that the strain energy release rate decreases with increasing of R_1/R_4, R_2/R_4 and R_3/R_4 ratios. The effect of material gradient is evaluated too. The analysis reveals that the strain energy release rate decreases with increasing of b_1/b_0 ratio. These findings can be applied in preliminary design of functionally graded non-linear elastic structural members with considering of the longitudinal fracture. The analysis of the strain energy release rate developed in the present paper can be used also to check for growth of existing longitudinal cracks in structural members. For this purpose, the calculated strain energy release rate has to be compared with the fracture toughness. A crack growth will occur if the calculated strain energy release rate exceeds the fracture toughness.

References

[1] Bohidar S, Sharma R and Mishra P 2014 Functionally graded materials: A critical review *Int. J. of Research* **1**(7) pp 289–301
[2] Zhang Y, Sun M and Zhang D 2011 Designing functionally graded materials with superior load-bearing properties *Acta Biomaterialia* **8**(3) pp 1101–08
[3] Wu X, Jiang P, Chen L, Zhang J, Yuan F and Zhu Y 2014 Synergetic strengthening by gradient structure *Mater. Res. Lett.* **2**(4) pp 185–91
[4] Mahamood R and Akinlabi E 2017 *Functionally Graded Materials* (Springer) p 103
[5] Erdogan F 1995 Fracture mechanics of functionally graded materials *Comp. Eng.* **5**(7) 753–70
[6] Rizov V 2018 Lengthwise fracture analyses of functionally graded beams by the Ramberg-Osgood equation *Engineering Review* **38**(3) pp 309–20
[7] Rizov V 2018 Delamination of Multilayered Functionally Graded Beams with Material Nonlinearity *Int. J. of Structural Stability and Dynamics* **18**(4) 1850051
[8] Rice J 1968 A Path Independent Integral and the Approximate Analysis of Strain Concentration by Notches and Cracks *J. Appl. Mech.* **35**(2) pp 379–86
[9] Lukash P 1978 *Fundamentals of nonlinear structural mechanics* (Stroizdat, Moscow) p 204 (In Russian)

IOP Conf. Series: Materials Science and Engineering 1002 (2020) 012006 doi:10.1088/1757-899X/1002/1/012006

Theoretical analysis of fracture in non-linear elastic functionally graded beam of linearly changing thickness with three longitudinal cracks

V Rizov

Department of Technical Mechanics, University of Architecture, Civil Engineering and Geodesy, 1046 Sofia, Bulgaria

E-mail: v_rizov_fhe@uacg.bg

Abstract. The present paper is focussed on fracture analysis of a functionally graded non-linear elastic cantilever beam structure with three longitudinal cracks. The cross-section of the beam is a rectangle. The beam height increases linearly towards the clamped end. The material is functionally graded along the thickness of the beam. The *J*-integral approach is applied in the fracture analysis. The solutions to the *J*-integral for the three cracks are verified by deriving of the strain energy release rate. For this purpose, the complementary strain energy cumulated in the beam is differentiated with respect to the areas of the three cracks. A parametric investigation of the longitudinal fracture behaviour of the beam is carried-out by applying the solutions to the *J*-integral. The effects of the beam geometry, the crack length and the material gradient along the beam thickness on the longitudinal fracture behaviour are evaluated and discussed.

1. Introduction

The functionally graded materials whose properties vary continuously in the solid are employed in a wide range of engineering applications [1-4]. The great interest which has been shown recently towards the functionally graded materials is due to their intensive use in more and more sophisticated areas in aeronautics and automotive engineering [5]. The mechanical reliability of functionally graded structural members and components depends largely on their fracture behaviour [6]. It should be mentioned that functionally graded materials can be built-up layer by layer [5] which is a premise for appearance of longitudinal cracks between layers. Therefore, analyzing of longitudinal fracture is of paramount importance for guarantying the integrity and reliability of functionally graded structural members and components.

The present paper deals with fracture analysis of a functionally graded cantilever beam configuration with three longitudinal parallel cracks in contrast to previous papers which consider fracture behaviour of functionally graded non-linear elastic beams with one longitudinal crack [7-9]. The present study is motivated by the fact that the layered structure of the beams and material gradient in the thickness direction create a high risk of appearance of more parallel longitudinal cracks. The three longitudinal cracks are analyzed by applying the *J*-integral approach in the present paper. The fracture is analyzed also in terms of the strain energy release rate for verification.

2. Analysis of functionally graded non-linear elastic beam with three longitudinal cracks

The present paper studies fracture behaviour of the functionally graded non-linear elastic beam with three parallel longitudinal cracks shown in figure 1. The beam is clamped in its right-hand end. The loading consists of a bending moment M, applied at the free end of the beam. The length of the beam is l. The beam cross-section is a rectangle of width b and thickness h.

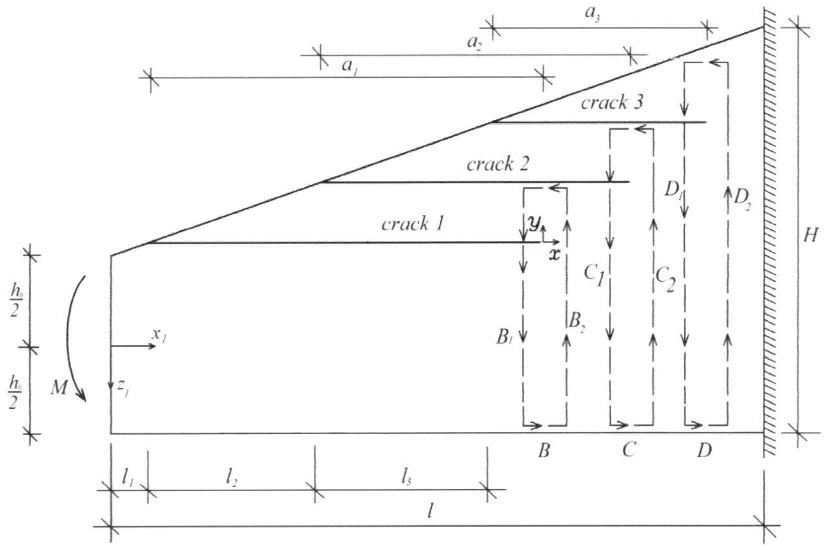

Figure 1. Geometry and loading of a functionally graded cantilever beam configuration of linearly changing thickness with three longitudinal cracks.

The thickness varies linearly from h_0 at the free end of the beam to H at the clamping (figure 1)

$$h = h_0 + \frac{H - h_0}{l} x_1. \tag{1}$$

Three parallel longitudinal cracks are located in the beam as shown in figure 1. The lengths of cracks 1, 2 and 3 are a_1, a_2 and a_3, respectively. The beam is functionally graded in the thickness direction. The longitudinal fracture behaviour of the beam is analyzed by applying the J-integral approach. First, crack 1 is studied. The J-integral is solved along the integration contour B, shown by a dashed line in figure 1. It should be mentioned that contours are chosen to facilitate the integration (for example, the stress along the upper and lower horizontal segments of the contours are zero). The solution of the J-integral is written as

$$J = J_{B_1} + J_{B_2}, \tag{2}$$

where J_{B_1} and J_{B_2} are the J-integral values in segments B_1 and B_2, of the integration contour (segments B_1 and B_2 coincide with the cross-sections of the lower arms of cracks 1 and 2, respectively). The J-integral in segment B_1 is written as

$$J_{B_1} = \int_{B_1} \left[u_{01} \cos\alpha - \left(p_x \frac{\partial u}{\partial x} + p_y \frac{\partial v}{\partial x} \right) \right] ds, \tag{3}$$

where u_{01} is the strain energy density. The quantities involved in (3) are obtained as

$$p_x = -\sigma;\ p_y = 0;\ ds = dz_3;\ \cos\alpha = -1;\ \frac{\partial u}{\partial x} = \varepsilon, \tag{4}$$

where σ is the stress, z_3 is the vertical centroidal axis of the cross-section of the lower arm of crack 1, ε is the strain. The stress is related to strain ε, by the following non-linear constitutive law [10]:

$$\sigma = \frac{E\varepsilon}{\sqrt{1+\varepsilon^2}}, \tag{5}$$

In (5), E is a material property that is distributed exponentially along the beam thickness

$$E = E_0 e^{m\frac{\frac{h}{2}+z_2}{h}}, \tag{6}$$

where E_0 is the value of E at the upper surface of the beam, m is a parameter that controls the material gradient in the thickness direction, z_2 is the vertical centroidal axis of the beam. It should be noted that the stress σ in (4) is obtained by using (5).

The distribution of the strains along the thickness of the lower arm of crack 1 is expressed as

$$\varepsilon = \kappa_1(z_3 - z_{3n}), \tag{7}$$

where κ_1 is the curvature, z_{3n} is the coordinate of the neutral axis. The following equations for equilibrium of the elementary forces in the cross-section of the lower arm of crack 1 are used to determine κ_1 and z_{3n}:

$$N_1 = \iint\limits_{(A_1)} \sigma da; \; M_1 = \iint\limits_{(A_1)} \sigma z_3 \, da, \tag{8}$$

where N_1 is the axial force, M_1 is the bending moment, A_1 is the area of cross-section. It is obvious that $N_1 = 0$, $M_1 = M$ (figure 1). After substituting of (5) and (7) in (8), the equations for equilibrium are solved with respect to κ_1 and z_{3n} by the MATLAB computer program.

The strain energy density is found by integrating of (5). The result is

$$u_{01} = E\left(\sqrt{1+\varepsilon^2} - 1\right). \tag{9}$$

The J-integral in segment B_2 is expressed as

$$J_{B_2} = \int\limits_{B_2} \left[u_{02}\cos\alpha_{B_2} - \left(p_{x_{B_2}} \frac{\partial u}{\partial x_{B_2}} + p_{y_{B_2}} \frac{\partial v}{\partial x_{B_2}} \right) \right] ds_{B_2}, \tag{10}$$

where the strain energy density u_{02} is obtained by replacing of ε with ε_{B_2} in (9). The strain ε_{B_2} is found by replacing of κ_1, z_3 and z_{3n} with κ_2, z_4 and z_{4n} in (7). Here κ_2 is the curvature, z_4 is the vertical centroidal axis, z_{4n} is the coordinate of the neutral axis of the cross-section of the lower arm of crack 2. Equations (8) are used to determine κ_2 and z_{4n}. For this purpose σ, z_3 and A_1 are replaced with σ_{B_2}, z_4 and A_2, respectively. Here σ_{B_2} is the stress, A_2 is the area of the cross-section of the lower arm of crack 2. The stress σ_{B_2} is obtained by replacing of ε with ε_{B_2} in (5). The other components of (10) are found as

$$p_{x_{B_2}} = \sigma_{B_2}; \; p_{y_{B_2}} = 0; \; ds = -dz_4; \; \cos\alpha_{B_2} = 1; \; \frac{\partial u}{\partial x_{B_2}} = \varepsilon_{B_2}. \tag{11}$$

After substituting of (3) and (10) in (2), the integration is performed by the MATLAB computer program.

The J-integral is used also to investigate crack 2. The integral is solved along the integration contour C, shown by dashed line in figure 1.

Finally, the J-integral is applied to analyze crack 3. The integration is carried-out along the integration contour D, shown by a dashed line in figure 1.

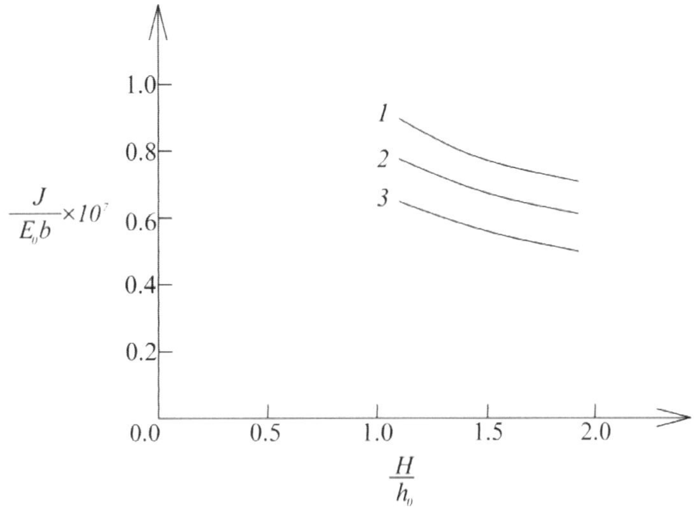

Figure 2. The J-integral value in non-dimensional form plotted against H/h_0 ratio (curve 1 – for crack 1, curve 2 – for crack 2 and curve 3 – for crack 3).

In order to verify the solutions to the J-integral, the tree cracks are studied also in terms of the strain energy release rate G. First, crack 1 is analyzed. The strain energy release rate is derived by applying the following dependence [11]:

$$G = \frac{dU^*}{bda_1}, \tag{12}$$

where U^* is the complementary strain energy, da_1 is an elementary increase of the length of crack 1. The complementary strain energy is obtained by integrating the complementary strain energy density in the volume of the beam. The complementary strain energy density u_{01}^*, in the lower arm of crack 1 is written as [9]

$$u_{01}^* = \sigma\varepsilon - u_{01}. \tag{13}$$

By combining of (5), (9) and (13), one obtains

$$u_{01}^* = E\left(1 - \frac{1}{\sqrt{1+\varepsilon^2}}\right). \tag{14}$$

The complementary strain energy densities in the other parts of the beam are found by replacing of ε with the corresponding strain in (14).

Cracks 2 and 3 are also studied in terms of the strain energy release rate. For this purpose,da_1 is replaced, respectively, with da_2 and da_3 in (12). It should be noted that the strain energy release rates found by (12) are exact matches of J-integral values which is a verification of the fracture analysis of the functionally graded non-linear elastic cantilever beam with three parallel longitudinal cracks developed in the present paper.

3. Numerical results

The numerical results reported here are obtained by applying the solutions to the J-integral derived in section 2 of the present paper. It is assumed that $b = 0.010$ m, $h_0 = 0.005$ m and $l = 0.300$ m.

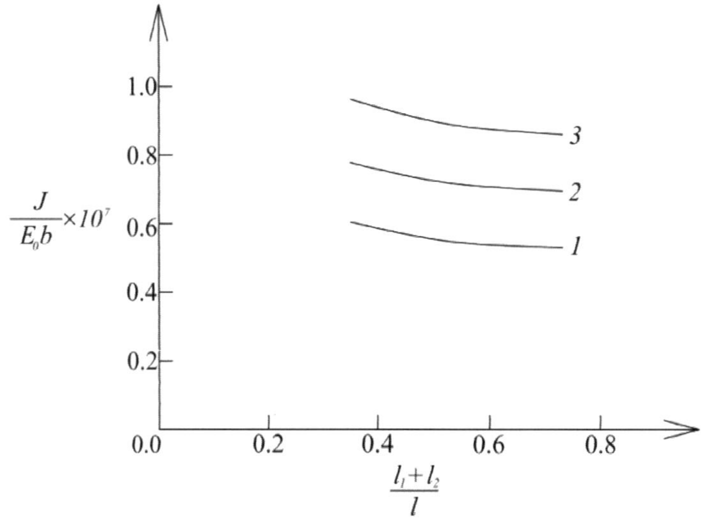

Figure 3. The *J*-integral value in non-dimensional form plotted against $(l_1 + l_2)/l$ ratio (curve 1 – at M = 4 Nm, curve 2 – at M = 5 Nm and curve 3 – at M = 6 Nm).

The *J*-integral value is presented in non-dimensional form by using the formula $J_N = J/(E_0 b)$.

The values of the *J*-integral in non-dimensional form are plotted against H/h_0 ratio in figure 2. One can observe in figure 2 that the *J*-integral value decreases with increasing of H/h_0 ratio. It can be observed also in figure 2 that the *J*-integral has highest, intermediate and lowest value for cracks 1, 2 and 3, respectively.

The location of crack 2 is characterized by $(l_1 + l_2)/l$ ratio. The *J*-integral value in non-dimensional form is plotted against $(l_1 + l_2)/l$ ratio in figure 3 at three values of the bending moment M, by using the *J*-integral solution for crack 2. The curves in figure 3 indicate that the *J*-integral value decreases with increasing of $(l_1 + l_2)/l$ ratio.

The *J*-integral value in non-dimensional form is plotted against parameter m, in figure 4 at three a_3/l ratios by applying the *J*-integral solution for crack 3. It can be observed in figure 4 that the *J*-integral value decreases with increasing of a_3/l ratio and m.

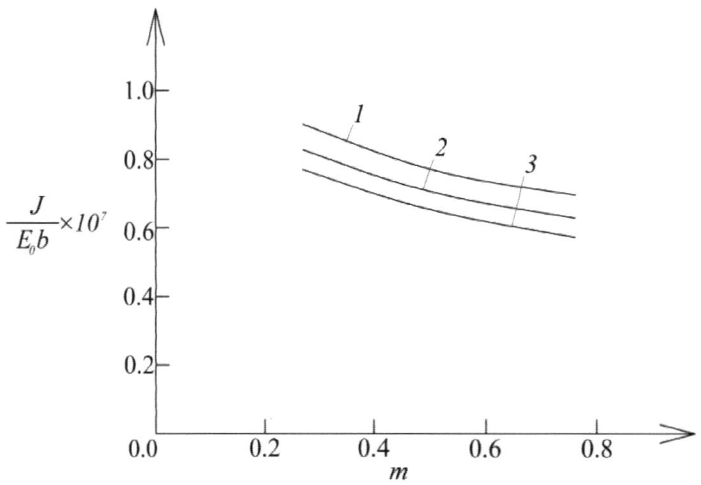

Figure 4. The *J*-integral value in non-dimensional form plotted against m (curve 1 – at a_3/l = 0.1, curve 2 – at a_3/l = 0.2 and curve 3 – at a_3/l = 0.3).

4. Conclusions

Longitudinal fracture of a functionally graded non-linear elastic cantilever beam is analyzed. The basic novelty of the present study is in the fact that three parallel longitudinal cracks are analyzed in contrast to previous studies which deal with one crack only [7-9]. The appearance of more parallel cracks is due to the fact that functionally graded structural members and components can be built-up layer by layer. The thickness of the beam increases linearly from the free end towards the clamping. The beam is functionally graded in the thickness direction. The fracture is studied by applying the *J*-integral approach. The fracture is analyzed also in terms of the strain energy release rate for verification of the solutions to the *J*-integral. Numerical results are obtained by using the solutions to the *J*-integral. It is found that the *J*-integral value decreases with increasing of H/h_0 ratio. It is found also that the *J*-integral value for cracks 1, 2 and 3 is higher, intermediate and lower, respectively. The analysis reveals that the *J*-integral value for crack 2 decreases with increasing of $(l_1 + l_2)/l$ ratio. The *J*-integral value for crack 3 decreases with increasing of a_3/l ratio and m. Form viewpoint of practical engineering, the results obtained in the present paper can be used in design of functionally graded non-linear elastic beam structures of continuously varying thickness in cases when the longitudinal fracture behavior has to be addressed. For example, the present paper clarifies the question for the influences of varying thickness, locations of and lengths of the longitudinal cracks on the fracture behavior.

References

[1] Hirai T and Chen L 1999 Recent and prospective development of functionally graded materials in Japan *Materials Science Forum* **308-311** pp 509–14
[2] Miyamoto Y, Kaysser W, Rabin B, Kawasaki A and Ford R 1999 *Functionally Graded Materials: Design, Processing and Applications* (Kluwer Academic Publishers, Dordrecht/London/Boston)
[3] Gasik M 2010 Functionally graded materials: bulk processing techniques *International Journal of Materials and Product Technology* **39** pp 20–9
[4] Nemat-Alla M, Ata M, Bayoumi M and Khair-Eldeen W 2011 Powder metallurgical fabrication and microstructural investigations of Aluminum/Steel functionally graded material *Materials Sciences and Applications* **2** pp 1708–18
[5] Mahamood R and Akinlabi E 2017 *Functionally Graded Materials* (Springer)
[6] Tilbrook M, Moon R and Hoffman M 2005 Crack propagation in graded composites *Composite Science and Technology* **65** pp 201–20
[7] Rizov V 2017 Analysis of longitudinal cracked two-dimensional functionally graded beams exhibiting material non-linearity *Frattura ed Integrità Strutturale* **41** pp 498–510
[8] Rizov V 2018 Non-linear longitudinal fracture in a functionally graded beam *Coupled Systems Mechanics* **7**(4) pp 441–53
[9] Rizov V 2018 Non-linear fracture in bi-directional graded shafts in torsion *Multidiscipline Modeling in Materials and Structures* **15** pp 156–69
[10] Lukash P 1978 *Fundamentals of nonlinear structural mechanics* (Stroizdat, Moscow) p 204 (In Russian)
[11] Rice J 1968 A path independent integral and the approximate analysis of strain concentration by notches and cracks *J. Appl. Mech.* **35**(2) pp 379–86

Evaluation of the load-bearing capacity of variously shaped steel-concrete slabs under short term loading

G L Vatulia[1,4], N V Smolyanyuk[2], A A Shevchenko[3], Ye F Orel[3], and M O Kovalov[1]

[1] Structural Mechanics and Hydraulics Department, Ukrainian State University of Railway Transport, Feuerbach sq. 7, 61050 Kharkiv, Ukraine
[2] Bridges, Structures and Structural Mechanics Department, Kharkiv National Automobile and Highway University, Yaroslava Mudrogo st. 25, 61002 Kharkiv, Ukraine
[3] Railway Survey and Design Department, Ukrainian State University of Railway Transport, Feuerbach sq. 7, 61050 Kharkiv, Ukraine

[4] E-mail: glebvatulya@gmail.com

Abstract. The article presents the evaluation results of load-bearing capacity exhaustion of steel-concrete slabs under concentrated and distributed short-term loading. The authors investigate the performance peculiarities of square, circular, and octagonal samples. The results of the experimental research were verified by comparison with the results of finite element modeling in the LIRA-SAPR software package. The obtained results proved the theoretical equations that allow to define the load-bearing capacity of steel-concrete slabs during their destruction over the cross-section and along the contact of sheet reinforcement with concrete. It is recommended to find the required intensity of sheet reinforcement anchoring from the condition of equality of ultimate loads. The authors provide further FEM analysis of square slabs to evaluate the impact of anchor spacing, the thickness of the steel sheet, and supporting conditions of slabs on their load-bearing capacity.

1. Introduction

Today, steel-concrete and composite structures, which are a combination of sheet steel or profiled steel, reinforcing bars, and concrete that work together, are commonly used worldwide [1-6], and their design is regulated by various standards [7-9]. The experience of the past three decades has proved the effectiveness of external reinforcement in bending structures. Extensive studies in this area are conducted to increase concrete strength. The existing research has two main directions: the constructive and the technological. The technological one deals with changing the concrete mixture composition. The constructive one deals with the rational bonding of concrete to reinforcement [10-13] or application of a certain type of reinforcement [14-16].

Solid-cast reinforced concrete slabs on profiled steel flooring, which simultaneously functions both as main reinforcement and as formwork, is becoming ever more commonly used [17-19]. This flooring design can significantly reduce labor costs, shorten construction time, and reduce the dead load of the floor on 30% vs. conventional reinforced concrete. With the reduced dead load of the floor, the consumption of materials for the casing and foundations of buildings can be also reduced, seismic resistance can be increased and costs of the construction of high-rise buildings can be cut significantly. Successful examples include the utilization of structures with external reinforcement in frames KUB2.5, IMET, ARCOS [20].

2. Description of experimental samples

The purpose of the study was the evaluation of stress-strain and ultimate states of steel-concrete slabs, the nature of their cracking and load-bearing capacity exhaustion. The work peculiarities of steel-concrete flooring slabs were determined during laboratory testing of the experimental samples that reflect the performance of a real structure [21]. The investigation examines square, circular and octagonal slabs, in which concrete and steel sheet external reinforcement were connected with the help of loop anchors, triangular keys, and U-shaped perforations.

Twenty samples were manufactured for the experimental studies, including:

- twelve samples (SqSCS 1-6 and SqSCSU 1-6) – square steel-concrete slabs, with dimensions in plan 1000×1000 mm and height h = 50 mm (figure 1a,d,e);
- four samples (OcSCS 1-4) – octagonal steel-concrete slabs, symmetrical in plan, with dimensions of the circumscribed circle R = 541 mm, inscribed circle r = 500 mm, side b = 414 mm and height h = 50 mm (figure 1b);
- four samples (CiSCS 1-4) – circular steel-concrete slabs, with diameter r = 1000 mm and height h = 50 mm (figure 1c).

Steel sheets with a thickness of t = 1 mm (SqSCS, SqSCSU, OcSCS) and 2 mm (CiSCS) were used as main external reinforcement.

In SqSCS 1-3 samples (figure 1a), the steel sheet was bonded to the concrete layer by means of the inclined anchors made from plain reinforcing steel (class Vr-1) with a diameter of d = 4 mm. In samples SqSCS 1-6, the steel sheet was bonded to the concrete with similar anchors with a diameter of d = 5 mm. The anchor spacing in SqSCS 1-6 samples was 50 mm.

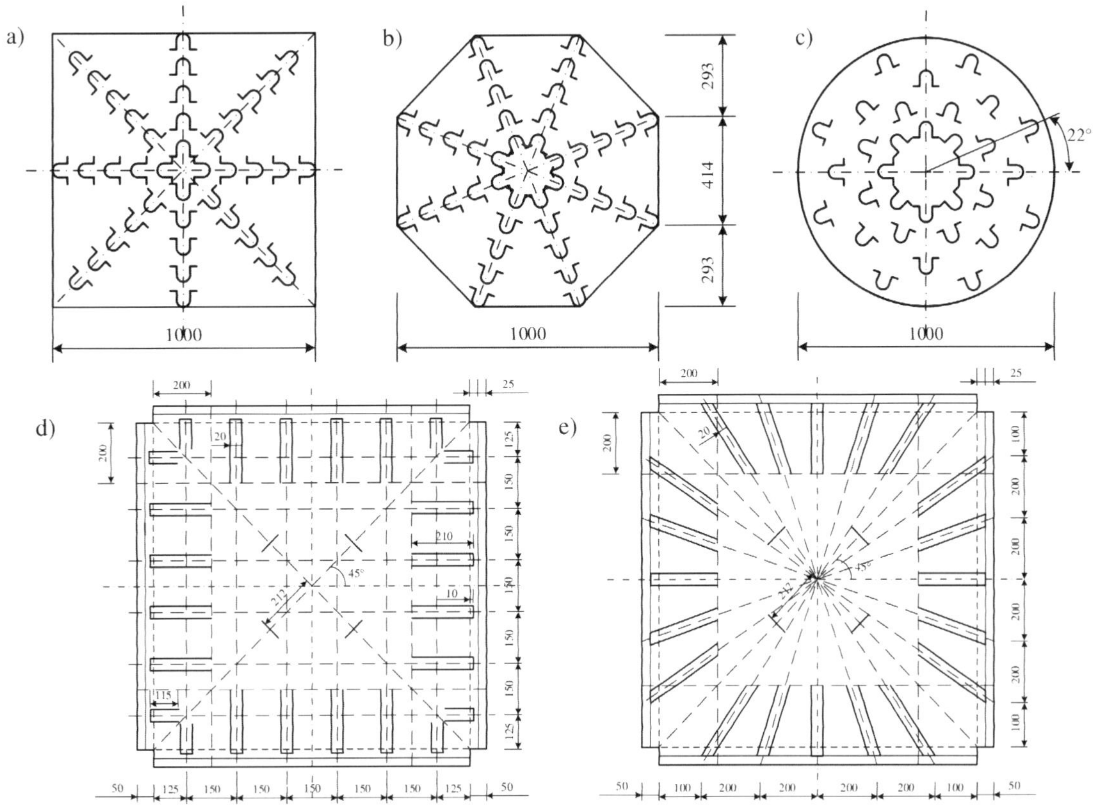

Figure 1. Slab samples: a) SqSCS, b) OcSCS, c) CiSCS, d) SqSCSU 1-2, e) SqSCSU 3-4.

In SqSCSU 1-6 samples (figure 1d, e), the steel sheet was made with flange and bonded with concrete using U-shaped perforations with a width of 20 mm and length from 80 to 260 mm, made in steel sheet. The perforations crossed the bending line of the flanges. The base of each perforation was placed on the steel sheet, while its bent edge was on the sheet flange. The following perforation layout variants were used: mutually perpendicular (at an angle of 90° relative to the sheet edge) in SqSCSU 1-2 samples (figure 1d), fanlike, at angles from 54° to 90° relative to the sheet edge, in SqSCSU 3-4 samples (figure 1e) and on corners – in SqSCSU 5-6 samples. For extra adhesion of the steel sheet with concrete in samples without U-shaped perforation, the loop anchors from the plain wire (class Vr-1) with diameter d = 2 mm, were bonded to the steel sheet. Anchors were located on diagonals of the steel sheet for each slab model.

In the OcSCS 1-4 samples, the steel sheet was bonded to the concrete core with anchors made of corrugated reinforcing wire (class Vr-1) with a diameter d = 4 mm.

At the same time, to obtain data on the physical and mechanical properties of concrete and steel [21], concrete cubes with dimensions of 100×100×100 mm were made, as well as prisms with dimensions of 150×150×600 mm. The steel strips with dimensions of 250×25×1 mm (2 mm) were cut from the external reinforcement of experimental models. Concrete and steel samples were tested following the requirements of the national standards [22, 23].

3. Tests progress and experimental results analysis

The SqSCS, OcSCS samples were tested for the concentrated load applied in the center of the slab through the rigid stamp with dimensions of 200×200×20 mm.

SqSCSU slabs were tested under the action of distributed load applied in the center of the slab through a rigid stamp with external dimensions of 400×400 mm and internal dimensions of 200×200 mm, which provided load transfer in a closed strip with an area of 0.12 m^2.

CiSCS slabs were tested under the action of a concentrated load applied in the center of the slab, which was transmitted through a rigid stamp with a radius of 10 mm. The test specimens were placed on a rigid support circuit of the test bench. To prevent the break-off of the slab from the support contour, brackets were installed in the corners of the experimental samples of the SqSCS and SqSCSU series, which allowed for free angular displacement.

The slabs were loaded in steps of approximately 0.1 of the destructive load value (F_{ul}). The readings of the measuring equipment without load were taken as zero reading. After each step, the load was maintained for 10 minutes to take the measuring equipment readings.

The vertical displacements of slabs during the loading were recorded along the slab symmetry axis by clock-type indicators with a division value of 0.01 mm. The indicators were located at a distance of 150, 300, 500 mm (i.e. in the center of the slab) from the slab edge (figure 2c).

Relative strains in all samples were measured using strain gauges BF350-2EB, which were glued both to the concrete and steel sheet. Strain gauge station VNP-8 was used to record sensor readings. The layout of the sensors and their quantity depends on the characteristics of each sample. On the concrete compressed zone the sensors were located equally in all samples: in pairs, perpendicular to each other in the transverse and longitudinal directions, in the middle and diagonally of the slab (figure 2a,b,d).

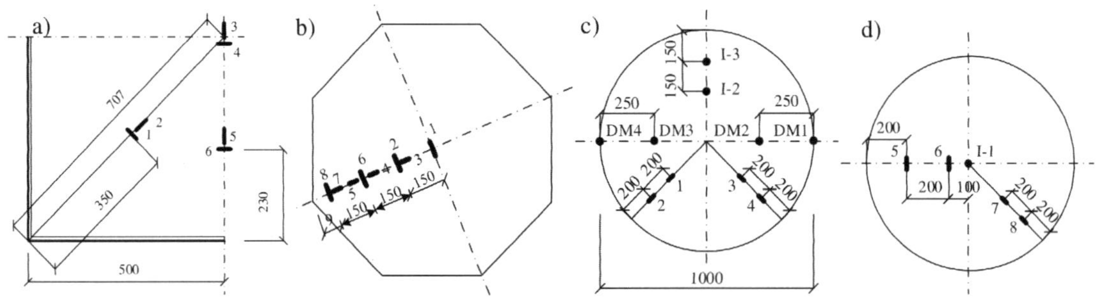

Figure 2. Sensors, indicators and deflection meters layout.

Slab deflections were measured in the middle (below the load point) by a clock-type indicator with the division value of 0.01 mm. Figures 3-6 show the "load-deflection" curves for samples of all series in the middle of the slab.

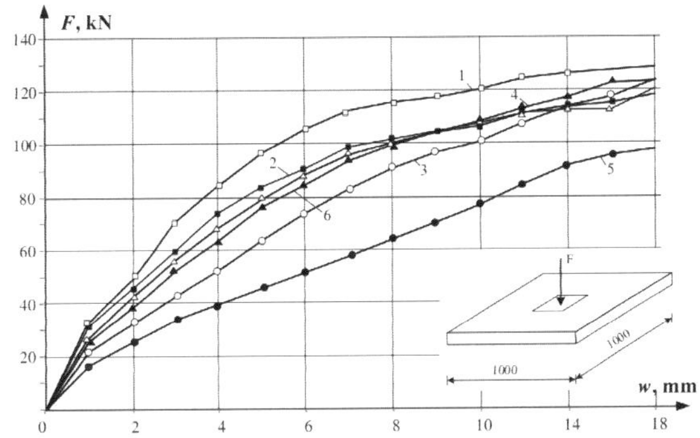

Figure 3. "Load-deflection" curves for samples SqSCS.

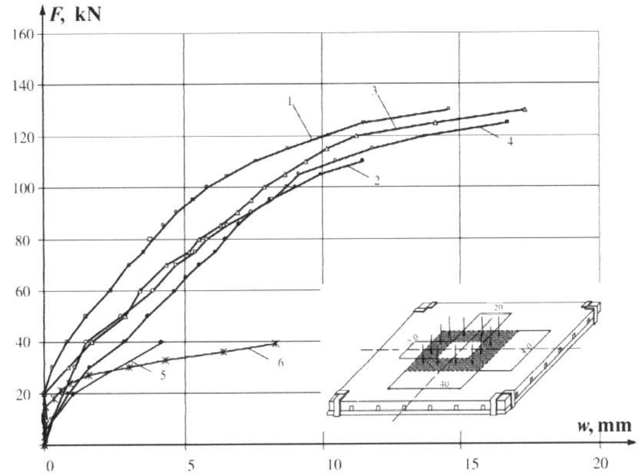

Figure 4. "Load-deflection" curves for samples SqSCSU.

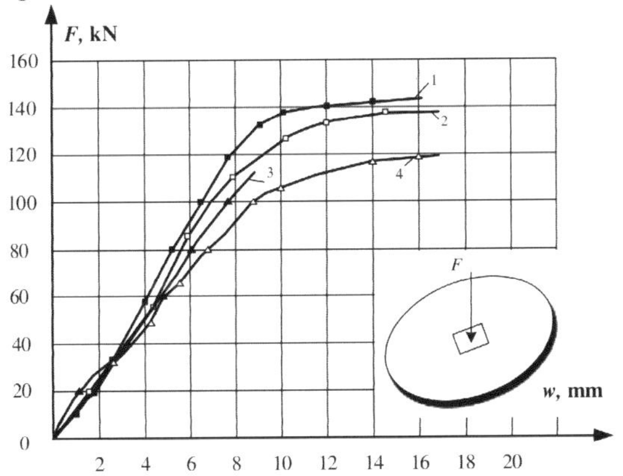

Figure 5. "Load-deflection" curves for samples CiSCS.

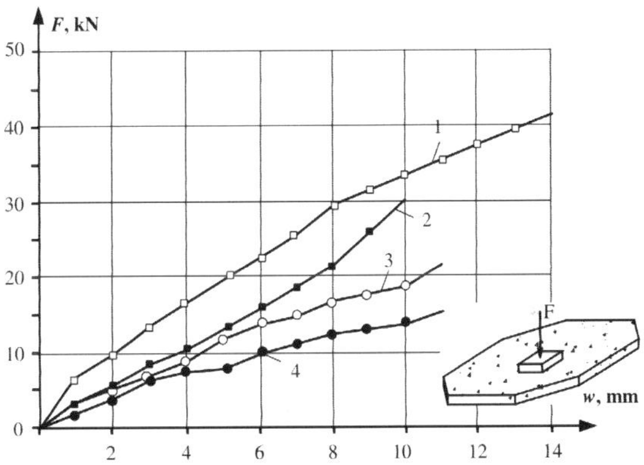

Figure 6. "Load-deflection" curves for samples OcSCS.

It can be noted that the "load-deflection" diagrams for all samples of steel-concrete slabs were nonlinear, due to the crack formation in the concrete stretched zone and the development of plastic deformations in the cross-sectional components.

The results of the deformation measurements in the steel sheet and concrete of SqSCS samples showed that plastic deformations in the sheet appear at a load of 40 kN, while at a load of 50-60 kN the destruction of the upper concrete fiber occurs in the middle of the slab.

Meanwhile, analysis of the nature of slabs deformation and the development of plastic deformations in the steel sheet during destruction, suggests that the load-bearing capacity exhaustion in the test samples occurred due to the strength loss of normal sections [5, 11, 19].

Analysis of experimental test results of CiSCS samples suggests that plastic deformations in the steel sheet appeared at a load of about 60 kN, while at a load of 80-90 kN the destruction of the upper concrete fiber occurs in the middle of the slab.

A specific feature of slabs SqRCSU with triangular keys was that the bearing capacity exhaustion was accompanied by the break-off of the edges of the perforations from the flanges of the steel sheet at the welding sites. When all samples were tested, this phenomenon was present only in the slabs where the perforations were welded perpendicular to the edge of the sample (SqSCSU 1-2).

As for steel-concrete slabs (OcSCS 1-4), the results of the deformation measurements of external reinforcement and concrete show that plastic deformations in the sheet appears at a load of 12 kN, while at a load of 25 kN the destruction of the upper concrete fiber occurs in the middle of the slab.

The dismantling of the slab steel sheet was carried out to examine the character of crack formation in the stretched zone of concrete.

4. Results verification and implementation

LIRA-SAPR software package [25], which implements the finite element method (FEM), was used during the numerical modeling of steel-concrete slabs. The following finite elements were used to create the calculation mathematical model: FE 241 (shell physically non-linear FE for slabs' concrete simulation), FE 341 (shell geometrically non-linear FE for steel sheet simulation), and FE 55 (FE for anchors' linear and angular compliance simulation). Exponential dependence was used to take into account the actual physical and mechanical characteristics of concrete at different stress levels.

The theoretical results (F_{FDM}) were obtained utilizing the equations which allow to define the load-bearing capacity of steel-concrete slabs during their destruction over the cross-section and along the contact of sheet reinforcement with concrete [3, 24]. The finite difference method (FDM) was used during the theoretical investigations.

The results of the experimental (F_{exp}) and theoretical (F_{FDM}) data comparison with numerical (F_{FEM}) calculations are presented in table 1. The additional values mentioned in table 1 describe the mechanical characteristics of the implemented materials, as follows [9]: f_y – nominal value of the yield strength of structural steel and $f_{c,prism}$ – value of the prismatic compressive strength of concrete.

Table 1. Results of the comparison of experimental and theoretical data with numerical calculations.

Samples	f_y, MPa	$f_{c,prism}$, MPa	F_{exp}, kN	F_{FDM}, kN	F_{FEM}, kN	F_{exp}/F_{FDM}	F_{exp}/F_{FEM}
SqRCS-1	220	25.68	128	115	118	1.11	1.08
SqRCS-2	220	25.68	119	115	118	1.03	1.01
SqRCS-3	220	25.68	121	115	118	1.05	1.03
SqRCS-4	220	25.68	123	115	118	1.07	1.04
SqRCS-5	220	25.68	100	115	118	0.87	0.85
SqRCS-6	220	25.68	123	115	118	1.07	1.04
SqRCSU-1	220	31.54	130	118	122	1.10	1.07
SqRCSU-2	220	31.54	120	118	122	1.02	0.98
SqRCSU-3	220	31.54	135	123	125	1.10	1.08
SqRCSU-4	220	31.54	130	123	125	1.06	1.04
OcRCS-1	190	25.68	40	32.3	33	1.24	1.21
OcRCS-2	190	25.68	34	32.3	33	1.05	1.03
OcRCS-3	190	25.68	30	29.7	29	1.01	1.03
OcRCS-4	190	25.68	22	20.7	21	1.06	1.05
CiRCS-1	221	16.8	143	139	138	1.03	1.04
CiRCS-2	221	16.8	149	139	138	1.07	1.08
CiRCS-3	221	24.3	122	115	114.6	1.06	1.06
CiRCS-4	221	24.3	118	115	114.6	1.03	1.03

The maximum load-bearing capacity was obtained for a slab with a fanlike perforation layout (SqSCSU 3-4). A comparison of the experimental values of the destructive loads (F_{exp}) with the loads for the SqSCS slabs 1-6 (see table 1) shows that the proposed method of combining (using perforations) helps to increase the slab load-bearing capacity to 10%.

Due to the adequacy of the obtained calculation results during the finite element modeling, the load-bearing capacity exhaustion of square steel-concrete slabs (SqSCS) was further investigated under different support conditions, anchors spacing, and thickness of steel sheet.

On the basis of the analysis of the experimental diagrams [11, 19] and results of FEM investigation, when spacing of loop anchors increases from 50 mm to 100 mm, under the same conditions of hinged support along the contour, the load-bearing capacity reduces to 17% in the SqSCS slabs, and to 15% in the CiSCS slabs.

If steel sheet thickness changes from 2 mm to 4 mm the load-bearing capacity of steel-concrete slabs increases from 17 to 38% when the hinged support along the edge is changed to rigid restraint.

The authors additionally examine five types of slab supporting commonly used in frames KUB2,5, IMET, ARCOS [20], namely: hinged slab; restrained slab; slabs with hinged and restrained edges; slabs with two restrained edges, one hinged edge and one unrestrained edge; and slabs with one restrained edge, one hinged edge and two unrestrained edges (figures 7-11).

The theoretical values of slab deflection (FD method) were obtained utilizing the methodology described in [3]. The comparison of the calculation (FEM) results with the theoretical (FDM) ones of square slab deflection for different supporting conditions are shown in figures 7-11. As a result, the proposed approaches for finite-element modeling of the frames with external reinforcement structures can be utilized during their design and reconstruction.

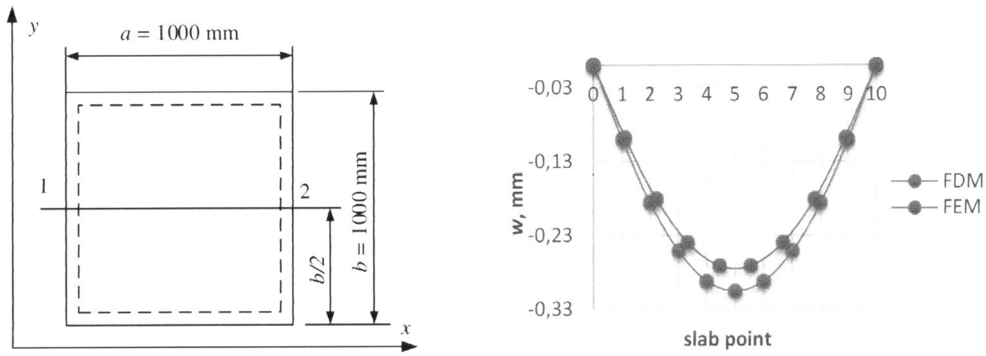

Figure 7. Deflection curves for hinged slab.

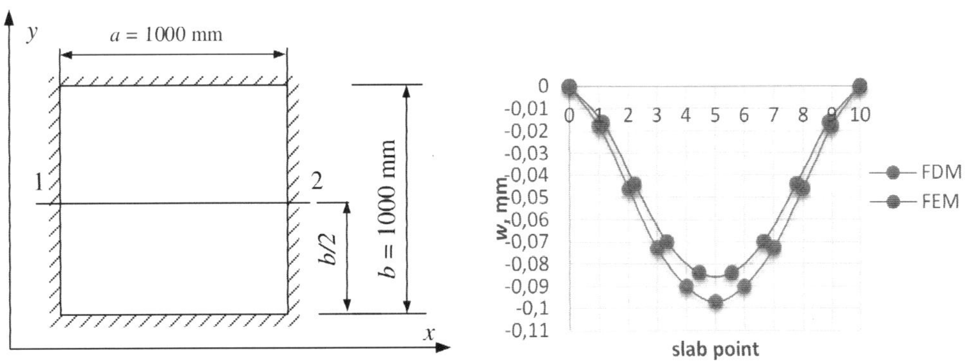

Figure 8. Deflection curves for the restrained slab.

IOP Conf. Series: Materials Science and Engineering 1002 (2020) 012007 doi:10.1088/1757-899X/1002/1/012007

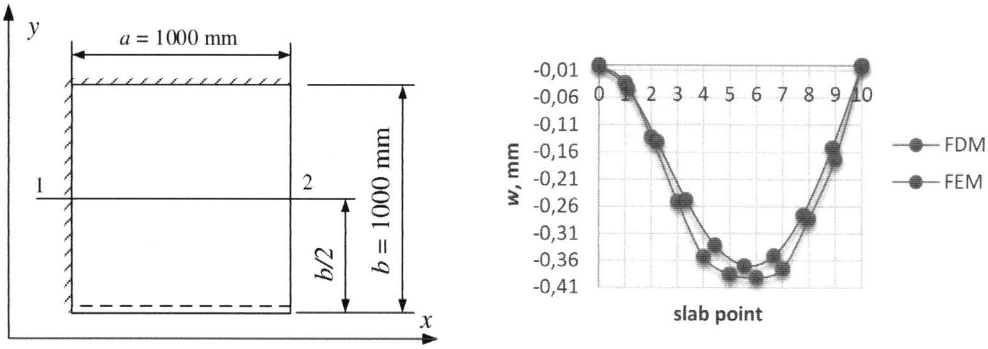

Figure 9. Deflection curves for a slab with hinged and restrained edges.

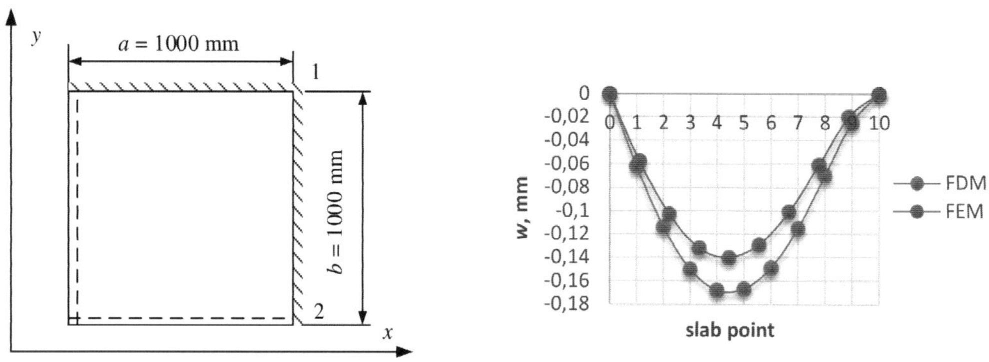

Figure 10. Deflection curves for a slab with two restrained edges, one hinged edge, and one unrestrained edge.

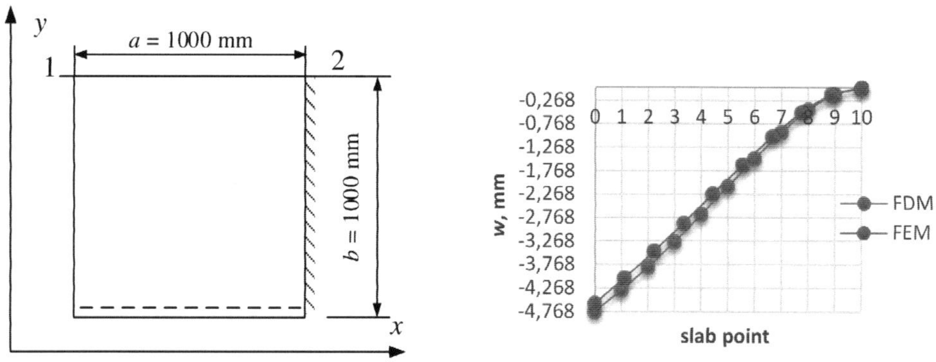

Figure 11. Deflection curves for a slab with one restrained edge, one hinged edge, and two unrestrained edges.

5. Conclusion

On the basis of the conducted research, conclusions were made about the load-bearing capacity exhaustion of steel-concrete slabs, the pliability of shear connections, and the impact of the support and loading conditions, as well as the thickness of a steel sheet on the load-bearing capacity of structures with external reinforcement:

- if the anchor spacing increases from 50 mm to 100 mm, under the same conditions of hinged support along the contour, the load-bearing capacity reduces to 17% in the SqSCS slabs, and to 15% in the CiSCS slabs;
- the method of concrete and steel sheet external reinforcement connection (using perforations) helps to increase the SqSCSU slab load-bearing capacity to10 % in comparison with loop anchors used in SqSCS slabs;
- change of the steel sheet thickness from 2 to 4 mm allows for increasing the load-bearing capacity of steel-concrete slabs from 17 to 38% if hinged support along the edge is changed to rigid restraint;
- the comparison of calculated (F_{FEM}), theoretical (F_{FDM}) and experimental (F_{exp}) results demonstrate the adequacy of the finite-element modeling of steel-concrete slabs. The stresses, deformations, and crack distribution fields, which were obtained during the FE modeling, correspond to the physics of the process that occurs during the mechanical testing of bending structures.

References

[1] Nie J and Zhao J 2009 Flexural Behavior of Steel-Concrete Composite Beams *Key Engineering Materials* **400** pp 37–42
[2] Ong K C G, Mays G C and Cusens A R 1982 Flexural tests of steel-concrete open sandwiches *Magazine of Concrete Research* **34**(120) pp 130–138
[3] Chikhladze E D,Vatulia G L and Kitov Y P 2006 *Basis for calculation and design of composite and steel-concrete constructions* (Transport of Ukraine) p 104
[4] Vatulia G, Rezunenko M, Petrenko D and Rezunenko S 2018 Evaluation of the load-bearing capacity of rectangular steel-concrete columns *Civil and Environmental Eng.* **14** pp 76–83
[5] Smolyanyuk N V 2004 *Stress-strained and ultimate state of steel-concrete plates* (Ph.D. thesis) (Ukrainian State University of Railway Transport) p 184
[6] Krainskyi P, Blikharskyy Y, Khmil R and Blikharskyy Z 2018 Experimental study of the strengthening effect of reinforced concrete columns jacketed under service load level *MATEC Web Conf.* **183** 02008
[7] DBN V.2.6-160 2010 Composite structures: General rules (Kyiv: Min. bud-va, arkhit. ta zhytlovo-kom. hosp) 2011
[8] ANSI/AISC 360-16 2016 Load and resistance factor design specification for structural steel buildings (Chicago, Illinois) 2016
[9] CEN 2004 EN 1992-1-1: Eurocode 2: Design of concrete structures - Part 1-1: General rules and rules for buildings (Brussels: BSI) 2004
[10] Aboobucker M A M, Wang T Y and Richard Liew J Y 2009 An experimental investigation on shear bond strength between steel and fresh cast concrete using epoxy *IES J. Part A: Civil Struct. Eng.* **2**(2) pp 107–115
[11] Shevchenko A A Stress-strained state of circular steel-concrete plates *Collected scientific works of Ukrainian State University of Railway Transport* **130** pp 113–120
[12] Kwon G, Engelhardt M D and Klingner R E 2010 Behavior of post-installed shear connectors under static and fatigue loading *J. Constr. Steel Res.* **66**(4) pp 532–541
[13] Smitha M S and Kumar S R S 2013 Steel-concrete composite flange plate connections finite element modeling and parametric studies *J. Constr. Steel Res.* **82** pp 164–176
[14] Bobalo T, Blikharskyy Y, Kopiika N and Volynets M 2020 Serviceability of RC Beams Reinforced with High Strength Rebar's and Steel Plate *Lecture Notes in Civil Eng.* **47** pp 25–33
[15] Saravanan M, Marimuthu V, Prabha P, Arul Jayachandran S and Datta D 2012 Experimental investigations on composite slabs to evaluate longitudinal shear strength *Steel Compos. Struct.* **13**(5) pp 489–500
[16] Lobodanov M, Vegera P and Blikharskyy Z 2020 Planning Experiment for Researching Reinforced Concrete Beams with Damages *Lecture Notes in Civil Engineering* **47** pp 243–250

[17] Ong K C G and Mansur M A Punching Shear of Steel-concrete Open Sandwich Slabs *Magazine of Concrete Research* **37**(133) pp 216–226
[18] Nie J, Fan J and Cai C 2008 Experimental study of partially shear-connected composite beams with profiled sheeting *Eng. Struct.* **30**(1) pp 1–12
[19] Chikhladze E D and Vatulya G L 1999 Experimental Researches of Steel-Concrete Plates *Proc. of the IASS 40th Anniversary Congress* **1** pp 13–18
[20] Shmukler V S, Klimov Y A and Burak N P 2008 Lightweight Frame Systems (Kharkiv) p 336
[21] Vatulia G L, Orel Y F and Kovalchuk A M 2011 Steel-concrete slabs experimental studies, *Collected scientific works of Ukrainian State University of Railway Transport* **125** pp 314–318
[22] DSTU B V.2.7-217:2009 Construction Materials. Concrete. The methods of the prismatic strength, modulus of elasticity and Poisson's ratio evaluation (Kyiv: Min. region bud) 2011
[23] GOST 12004-81 The reinforcement steel. The tensile test methods (Moskow) 1995
[24] Chikhladze E D and Vatulya G L 2011 Stress-strained state of steel-concrete structures under force and temperature effect *Proc. of the 5th International Conference on Dynamics of Civil Engineering and Transport Structures and Wind Engineering* pp 181–184
[25] Gorodetsky A and Evzerov I 2007 *Computer models of structures* (Kyiv: Fact)

Simulation of steel rod atmospheric corrosion by alternate immersion in salt solution

Veselin Tsonev[1], Nikolay Nikolov, Kamen Penkov and Yordanka Marcheva

Technical University – Sofia, 8 Kliment Ohridski Blvd., 1000 Sofia, Bulgaria

[1] E-mail: tzonev@tu-sofia.bg

Abstract. Corroded B235 reinforcing steel rods, which were exposed to an open air environment for 25 years, with a certain specific mass loss, were studied. Test pieces were prepared from them by lathe machining. The test pieces were subjected to accelerated corrosion through an alternate immersion in NaCl solution. The same specific mass loss as on the initial corroded rods was achieved for 52 days. Further the test pieces were cleaned, observed, photographed and tensile tested. The cleaned surfaces and the stress-strain curves of the test pieces were compared with results, obtained from the initial corroded rods. It was found out about the tested steel rods that alternate immersion in salt solution for 52 days is a good simulation of twenty-five-year atmospheric corrosion.

1. Introduction

Atmospheric corrosion is a common phenomenon that reduces the mechanical properties of steel parts and structures. It is caused by moisture in the air and depends on many factors – humidity, temperature, winds, presence of salts or other chemicals, pollution, pH, climate, location. Atmospheric corrosion is usually uneven and creates spots, ulcers and pits.

Atmospheric corrosion in steel is a slow process. Obtaining corroded test pieces by exposing them to natural conditions takes a long time. For example, [1] describes a 10-years experiment in which unprotected steel specimens are attached to the exposed surfaces of 8 steel bridges. Reinforcing rods after 30 and 40 years of service as steel-concrete structure in natural environment are studied in [2] searching for a relation between natural and accelerated artificial corrosion. The study [3] uses test pieces whose preparation, in a combination of artificial and natural environment, took 26 years. These examples show that the field tests duration are not always acceptable for the laboratory practice. It is usually appropriate to use accelerated corrosion methods for test pieces preparation. In [4, 5] it is summarized that the corrosion influence on the mechanical properties of steel is mainly a geometric effect. The shape, depth, and density of the pits are crucial. And the corroded surface relief strongly depends on the process conditions. The distribution of pits, obtained in two different environments – air exposure and immersion in sea water – is given in [5]. They are quite different, as the stay in sea water gives a smoother surface. Thus, the impact of the shapes of the defects on the mechanical properties is studied in [4]. The defects in [4] are not produced by a corrosion process, but by machining. The test pieces preparation is maximally facilitated and accelerated in this way. Computer simulations can also take advantage of the available accurate information about machined test piece geometry. Despite these obvious advantages, the surface relief obtained by machining is very different from that caused by corrosion and is therefore not preferred by researchers.

One of the most common accelerated corrosion methods is the Salt Spray Test. Salt mist (5% NaCl, 35° C, 10-60 days) is used in [2], causing about 11% mass loss on reinforcing rods with a diameter of 10 mm. The tensile test results are compared with those from naturally corroded rods with similar mass loss. It is found out that there is no direct relationship between the two types of results. The reason is the big difference in the surface relief caused by the two corrosion methods. The same problem is observed with other two easy to implement and frequently used methods – immersion in a corrosive solution and the galvanostatic method. The most common corrosive solution is sea water. A study of its impact on the mechanical properties of steel test pieces is made in [6]. In addition to causing a smoother surface than atmospheric corrosion, this method often has a long duration – the preparation of test pieces in [6] takes 1 year. To accelerate the corrosion process, electric current is often used – the galvanostatic method. Such an experiment is described in [7], where steel fibers are tested. It is found out that uniform corrosion is obtained, which leads to significantly better mechanical properties than in case of pitting caused by cyclic wetting and drying.

Cyclic Corrosion Testing (CCT) is the best way to obtain in a short time artificial pitting corrosion similar to that in natural atmosphere [8]. The test pieces are periodically exposed to a corrosive solution, while many important parameters are controlled – temperature, humidity, ultraviolet radiation, and others. A simplified and standardized version of CCT is the Alternate Immersion Test (AIT) [9]. This method is easier to implement, and it is preferred for the presented study. The relevant equipment and technology are described in [10, 11].

The aim of the present work is to determine whether the atmospheric corrosion effects on B235 steel rods can be simulated and investigated by periodically immersing the same rods in salt solution, using a standard method [9] and an available technique [10]. The studied rods have been stood outdoors for 25 years (corrosion in natural atmosphere), in vertical position, in a temperate continental climate zone, without anti-corrosion protection. Their specific mass loss and mechanical properties was determined in [11]. Here, new test pieces are made from these rods and then subjected to AIT until the same specific mass loss is obtained. The surfaces and stress-strain curves obtained after both methods of corrosion, are compared and analyzed.

2. Test pieces

Figure 1 (a) shows a photo of the initial corroded material – B235 steel rod with a nominal diameter of 6.5 mm [12, 13]. Two types of test pieces are made of it. Figure 1 (b) shows 3 test pieces intended for tensile testing. They have a diameter of 5 mm in gauge length. Figure 1 (c) shows specimens prepared to determine the specific mass loss at different exposure times during the AIT. They have a diameter of 5 mm and a surface area of 244.73 mm^2.

(a) (b) (c)

Figure 1. Test pieces: (a) Initial corroded rod; (b) Test pieces for tensile testing; (c) Specimens for specific mass loss determination during AIT.

3. Alternate Immersion Test

The accelerate corrosion method AIT used here is described in detail in [10, 11]. A water solution of 3.5% NaCl is used; the wetting/drying cycle includes 10 min of immersion in the solution followed by 50 min of air-drying at a temperature of 25 °C.

B235 steel can be assumed to be an analogue of the widespread brand of structural steel S235JR, whose corrosion behavior was studied in [10]. The blue line in figure 2 shows the specific mass loss of S235JR steel rod with diameter of 6 mm. The target specific mass loss is 1462.25 g/m^2, as obtained in [11] for the initial B235 rod. It can be assumed that the needed AIT duration is approximately 50 days.

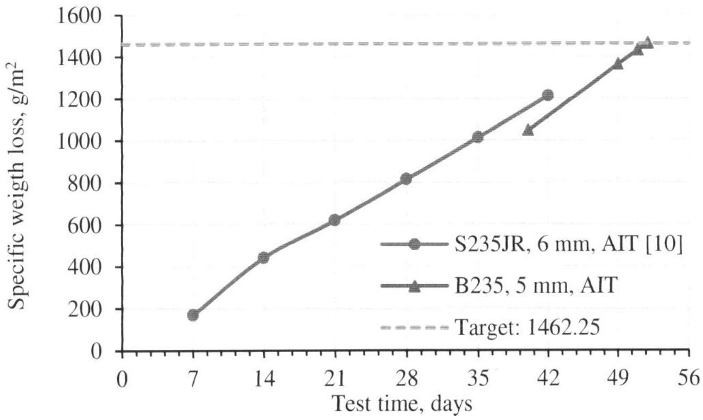

Figure 2. Specific weight loss of steel rods.

For more precise determination of the necessary AIT duration, which will lead to the target specific mass loss, the specimens shown in figure 1 (c) were used. They were pre-weighed with a precision of ±0.001 g, and divided into groups of 3 pieces. All groups were subjected to AIT simultaneously, together with the test pieces for tensile testing, and one group was removed every few days, starting on day 40. After cleaning, re-weighing and averaging the results, the specific mass loss for the respective days of exposure was established. The results are given in table 1, and are plotted as a red line in figure 2.

Table 1. Specific mass loss of B235 steel specimens after 40-52 days AIT.

AIT duration, days	40	49	51	52
Specific mass loss, g/m^2	1046.05	1362.73	1430.15	1462.84

Due to the achievement of desired specific mass loss, the AIT exposure was discontinued on day 52.

4. Surface observations

Pictures of the cleaned initial steel rod, after 25 years exposure to atmospheric environment are shown in figure 3 (a). Pictures of the cleaned test piece after 52 days of AIT are shown in figure 3 (b). The photos on the left were taken with a camera, and the photos on the right – with a microscope at magnification of ×15. The two types of test pieces from the same material have practically equal specific mass loss (approximately 1462 g/m^2) and their properties can be compared.

As can be seen in figure 3, the shape and the size of the pits are approximately the same. The depth of the surface defects is also similar. There is a difference in the pit distribution – the number of pits caused by natural corrosion is higher and they are more evenly distributed than the ones caused by

AIT. Due to the similarity in the pit shape and depth, similarity in mechanical properties can also be expected during the tensile test.

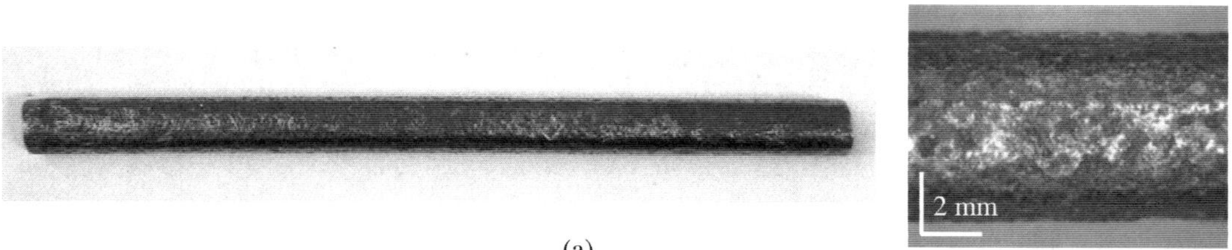

(a)

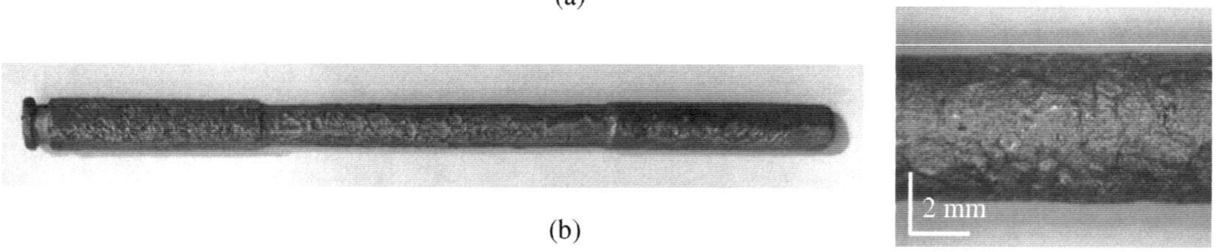

(b)

Figure 3. General appearance and close look at the cleaned corroded test pieces:
(a) Initial steel rod after 25-year atmospheric corrosion; (b) Test piece after 52 days of AIT.

The two types of test pieces differ in appearance – the cleaned initial rod is dark in color and glossy, while the AIT-subjected one is matte and yellowish. This difference is caused by the cleaning methods. The oxide layer mechanical strength and density are very high after a long exposure to natural atmosphere and very low after the short period of accelerated corrosion impact. Corrosion products were very difficult to remove from the initial rod. Chemical and mechanical methods (scrubbing with a soft brush) were used, which smooths the protrusions, and minimal residues of hard and dark-colored oxides remained in the depressions. In the case of an artificially corroded test piece, the cleaning was easy and entirely chemical, according to a standard procedure [14], which completely removes the corrosion products and creates a more contrasting texture, shown in figure 3 (b). These visual differences do not affect the mechanical properties determined by the tensile test.

5. Tensile test

The experiments were performed in accordance with EN ISO 6892-1:2019 [15]. A testing machine with a screw load mechanism and precise digital control was used [16]. Longitudinal strain was measured with a 25 mm gauge length extensometer. The loading rate was 120 N/s.

The three test pieces shown in figure 1 (b) were tensile tested. The tensile force and longitudinal strain were continuously recorded during the test. The stress-strain curves were drawn, the upper yield strength R_{eH} and the tensile strength R_m were determined. The values of these mechanical properties are shown in table 2.

Table 2. Mechanical properties of corroded B235 steel rods.

Corrosion method:	Atmospheric corrosion, 25 years			Alternate immersion test, 52 days		
Test piece	Nominal diameter, mm	R_{eH}, MPa	R_m, MPa	Nominal diameter, mm	R_{eH}, MPa	R_m, MPa
1		272	346		259	331
2	6.5	266	340	5	254	327
3		268	345		261	327
Average:		*269*	*344*		*258*	*328*

Figure 4 shows the stress-strain curves of the two types of test pieces as well as the curve of the virgin B235 steel.

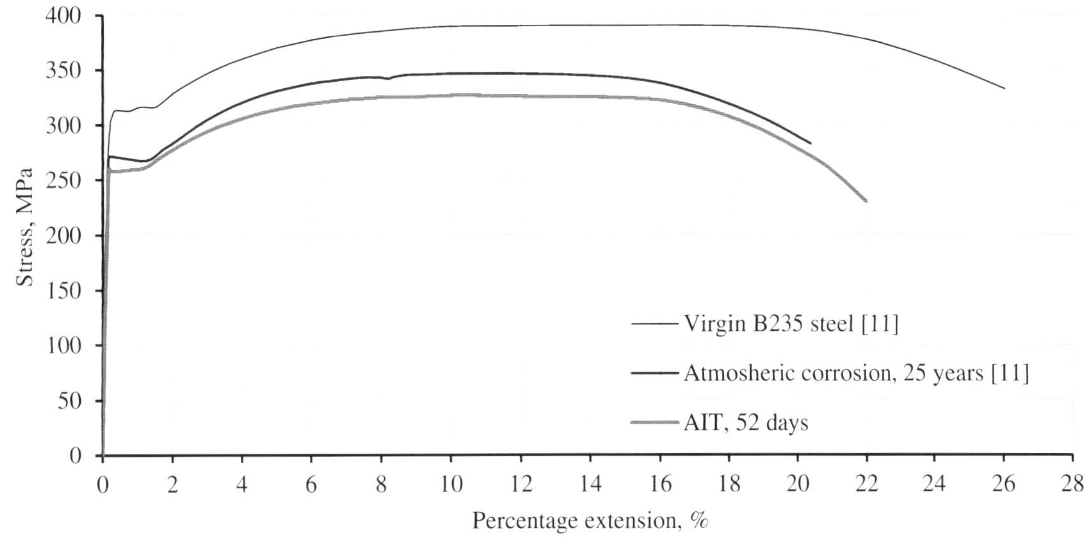

Figure 4. Stress-strain curves – test piece No 1 of each type.

6. Conclusions

AIT artificial corrosion disrupts the circular cross-sections of the test pieces. Pits are formed on the surface, similar in shape and size to those on naturally corroded specimens. The stress-strain curves obtained from test pieces with equal specific mass loss achieved by exposure to natural atmosphere or AIT are also close. The difference in the yield strength R_{eH} and tensile strength R_m is between –4 and –5%. The difference in maximum percentage extension is slightly larger: +7.8%, but there is generally a significant scattering in the experimental results according to this parameter.

The smaller diameter test pieces (5 mm, AIT) have greater percentage cross-section area losses than the larger diameter test pieces (6.5 mm, natural atmosphere). This explains the slightly lower strength of the AIT-exposed test pieces. An additional study is planned to determine the impact of the initial diameter on strength properties of the corroded rods.

The similar relief of the corroded surfaces, as well as the close strength and deformation parameters of B235 steel rods show that 25 years of atmospheric corrosion can be successfully simulated by 52 days accelerated corrosion test with periodic immersion in salt solution.

Acknowledgements

The authors would like to thank the Research and Development Sector at the Technical University – Sofia for the financial support.

References

[1] Křivý V, Urban V and Kubzová M 2016 Thickness of corrosion layers on typical surfaces of weathering steel bridges *Procedia engineering* **142** pp 56–62
[2] Apostolopoulos C A and Papadakis V G 2008 Consequences of steel corrosion on the ductility properties of reinforcement bar *Construction and Building Materials* **22**(12) pp 2316–24
[3] Zhu W and François R 2013 Effect of corrosion pattern on the ductility of tensile reinforcement extracted from a 26-year-old corroded beam *Advances in concrete construction* **1**(2) 121–36
[4] Islam M R and Sumi Y 2011 Geometrical effects of pitting corrosion on strength and deformability of steel rectangular plates subjected to uniaxial tension and pure bending *J. Japan Society of Naval Architects and Ocean Engineers* **14** pp 9–17

[5] Oka M, Kitada H and Watanabe T 1990 Experimental study on statical strength of corrosive mild steel *J. Society of Naval Architects of Japan* **167** pp 229–36
[6] Angel F C 2014 *Fractal effect of corrosion on mechanical behavior of unprotected structural steel* (In: Developments in Corrosion Protection, InTech) pp 551–65
[7] Chen G, Hadi M N, Gao D and Zhao L 2015 Experimental study on the properties of corroded steel fibres *Construction and Building Materials* **79** pp 165–72
[8] Nikolov N, Marcheva Y and Tsonev V 2019 Methods for preparation of corroded steel specimens for mechanical testing *MATEC Web of Conferences* **287** 07004
[9] EN ISO 11130:2018 Corrosion of metals and alloys - Alternate immersion test in salt solution
[10] Nikolov N, Tsonev V, Penkov K, Kuzmanov N and Borisov B 2019 Machine for accelerated cyclic corrosion tests through alternate immersion in salt solution *IOP Conf. Ser.: Mater. Sci. Eng.* **664** 012016
[11] Tsonev V, Nikolov N and Penkov K 2020 Impact of atmospheric corrosion on the mechanical properties of B235 steel rods *IOP Conf. Ser.: Mater. Sci. Eng. (TechSys-2020)*
[12] BDS 4758:2008 Steel for reinforcement of concrete. Weldable reinforcing steels B235 and B420
[13] BDS EN 10080:2005 Steel for reinforcement of concrete – Weldable reinforcing steel – General
[14] EN ISO 8407:2014 Corrosion of metals and alloys – Removal of corrosion products from corrosion test specimens
[15] EN ISO 6892-1:2019 Metallic materials – Tensile testing – Part 1: Method of test at room temperature
[16] Tsonev V, Kuzmanov N, Borisov B and Penkov K 2019 System for materials testing at static loading *IOP Conf. Ser.: Mater. Sci. Eng.* **618** 012048

Numerical investigation on the effects of the aerodynamic shading between the tower and the blade of wind turbine generator with horizontal axis

J A Genov and I A Angelov[1]

Department of Mechanics, Faculty of Transport, Technical University – Sofia, 1000 Sofia, Bulgaria

[1] E-mail: ivvoangelov@gmail.com

Abstract. The purpose of the current study is to evaluate the effects of the aerodynamic shading between the tower and the blade of NREL 5MW wind turbine generator. 3D models of the turbine and the tower are designed. They are reduced to plane models in order to simplify the convoluted, spatial aerodynamic interaction to a sequence of plane-2D problems. The plane models are numerically studied in Ansys Fluent, using dynamic-deformable mesh method. Visual and numerical results of the aerodynamic shading interaction are obtained. Force distributions on the blade and on the tower of NREL 5MW wind turbine generator are derived, for the exact moment when the aerodynamic shading occurs.

1. Introduction

In fluid dynamics, when two or more bodies are in a fluid dynamic interaction, and the bodies are close to one another, or passing close to one another, a phenomenon known as shading, occurs [1]. This phenomenon is characterized with deviations in pressure and velocity fields around the bodies, as well as a change in the aerodynamic forces, exerted on the bodies. Effects of this nature occurs between wind turbine generators in wind turbine farms, when a turbine`s aerodynamic wake shadows another turbine, resulting in less power of the second [2]. Similar aerodynamic shading can be observed in urban environment, building close to each other shadow one another [3].

While operating, the aerodynamic interaction between the wind and the wind turbine generator creates curtain pressure fields, close to the solid bodies (figure 1). They result in aerodynamic forces. When a blade of the turbine passes close to the tower of the structure, the pressure fields around the two bodies interact with each other, which result in deviation of the aerodynamic forces exerted on the bodies (figure 2). This interaction is associated with aerodynamic shading [4, 5].

Most of the traditional methods for analyzing the aerodynamic shading between tower and blade of wind turbine, such as Powles` method, Blevins' method are based on semi-analytical and experimental techniques [6, 7]. These methods are an approximation of the real physical medium and cannot guarantee adequacy of the numerical results. On the other hand the modern CFD softwares are capable of analyzing the aerodynamics of complex structures such as wind turbine generators, providing valuable and reliable numerical data [8, 9].

The purpose of the current study is a numerical evaluation of the effects the aerodynamic shading between the tower and blade of wind turbine generator with horizontal axis. All calculations are executed in Ansys-Fluent.

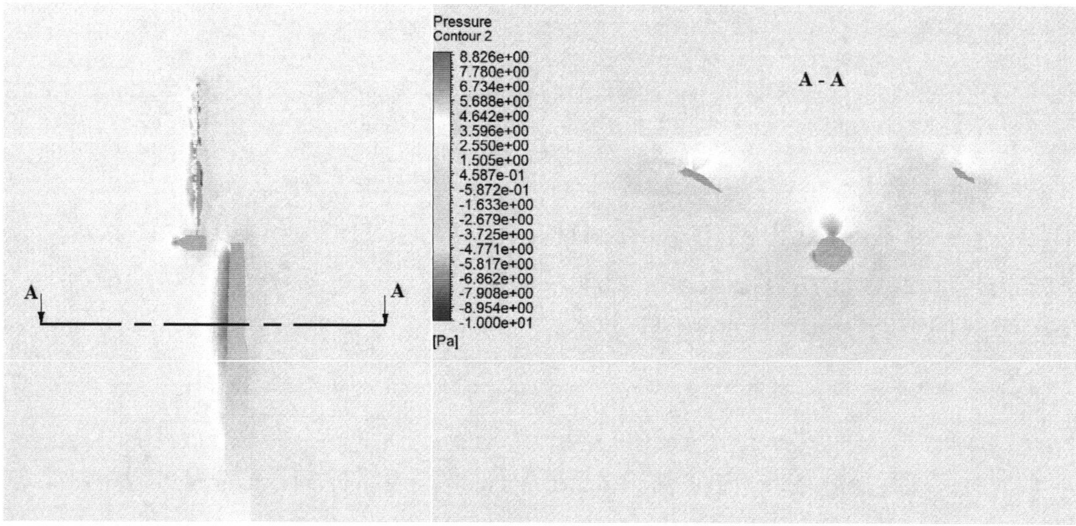

Figure 1. Pressure field distribution around the blade and the tower of an arbitrary wind turbine generator with horizontal axis.

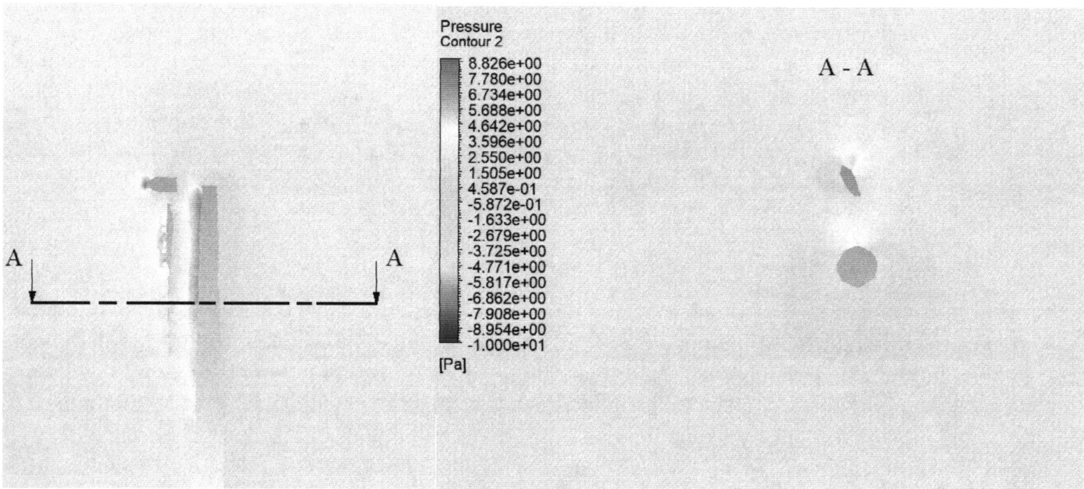

Figure 2. Pressure field deviation between the blade and the tower of an arbitrary wind turbine generator with horizontal axis caused by shading.

2. Model description

2.1 Wind model

In [10, 11] was shown that accounting for vertical wind speed gradient leads to significant fluctuations in the aerodynamic forces, especially for wind turbine generators of high class, like NREL 5MW [12]. In the current study, a logarithmic vertical wind speed distribution is chosen [13, 14] (figure 3).

$$u_0(z) = \frac{u^*}{k}\left[ln\left(\frac{z-d}{z_0}\right) + \psi(z, z_0, L)\right]. \tag{1}$$

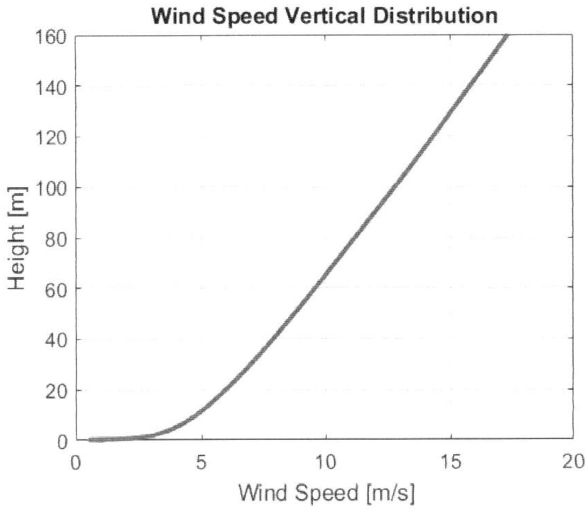

Figure 3. Wind speed vertical distribution.

2.2 Aerodynamics of the model

Considering the geometric dimensions of the generator [12], turbine diameter of 126 m and tower height of 87.6 m, a 3D transient numerical simulation of the turbine generator will cost a lot of computational time and may lead to inaccurate numerical values. In order to reduce the computational time and to improve the result accuracy, a 2D plane simulation is chosen. Similar to Blade Element Momentum (BEM) theory [15], the blade of the turbine is divided in finite cross sections. Each of the cross sections has 2 wind velocity components figure 4.

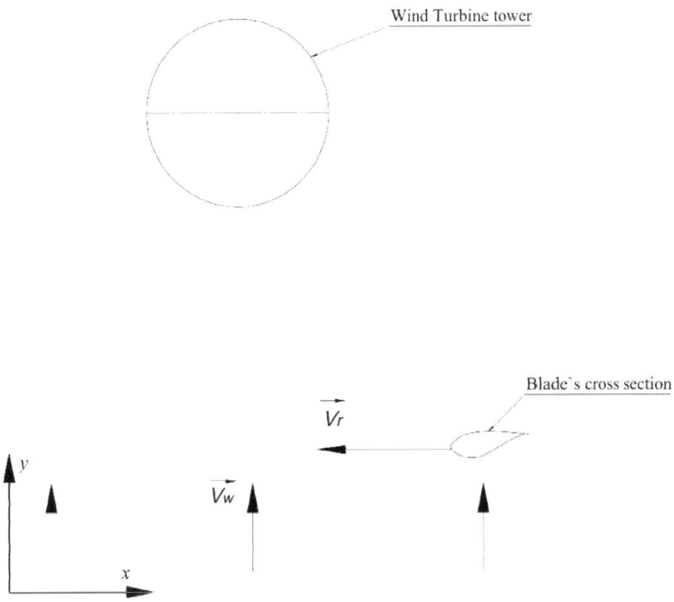

Figure 4. Aerodynamics of the model.

One component V_r corresponding to the rotation of the turbine blade, depending on the distance r from the cross section to the axis of rotation:

$$V_r = \omega r, \tag{2}$$

In the 2D case V_r is a translational velocity of the cross section. The second component V_w is the wind speed velocity itself. Its magnitude depends on wind speed vertical distribution (figure 3) and the cross sectional height from the ground. The frame of reference shown in figure 4 will be used to define the direction of the acting forces.

2.3 Numerical model
The described aerodynamic model is implemented in Fluent Ansys. The software solves numerically, the governing equations of fluid-dynamics: the Continuity equation (3), the Navier-Stokes` equation (4) and the Energy equation (5) [16], using Finite Volume Method (FVM) [17]. In the most general form, the equations are:

$$\frac{\partial \rho}{\partial t} + \nabla.(\rho \vec{V}) = 0\,; \tag{3}$$

$$\rho \frac{\mathrm{D}\vec{V}}{\mathrm{D}t} = -\nabla p + \rho \vec{g} - \rho \vec{\omega} \times \vec{V} + g\rho T \vec{j} + \nabla\left(\mu \nabla.\vec{V}\right); \tag{4}$$

$$\frac{\partial}{\partial t}\left[\rho\left(e + \frac{V^2}{2}\right)V\right] + \nabla.\left[\rho\left(e + \frac{V^2}{2}\right)V\right] = \rho\dot{q} - \nabla.(p\vec{V}) + \rho(\vec{f}.\vec{V}) + Q`_{\text{viscous}} + W'_{\text{viscous}}\,. \tag{5}$$

Depending on the assumptions of the real physical system, the equations are modified in a more specific form. For the current study, the following assumptions are made:

- the air is assumed incompressible - with a constant density;
- the fluid is considered with isotropic viscosity;
- k-ω SST turbulence model is used [18]. The turbulence model adds two additional equations to the system of Partial Differential Equations (PDE);
- no heat transfer is considered.

The assumptions made, result in the following system of PDE

$$\left|\begin{array}{l} \nabla.\vec{V} = 0 \\ \dfrac{D\vec{V}}{Dt} = \vec{f}' - \nabla\left(\dfrac{p}{\rho}\right) + \nu\Delta\vec{V} \\ \dfrac{\partial k}{\partial t} + U_j \dfrac{\partial k}{\partial x_i} = P_k - \beta^* k\omega + \dfrac{\partial}{\partial x_j} + [(\nu + \sigma_k \nu_T)\dfrac{\partial k}{\partial x_j}] \\ \dfrac{\partial \omega}{\partial t} + U_j \dfrac{\partial \omega}{\partial x_i} = \alpha S_2 - \beta\omega^2 + \dfrac{\partial}{\partial x_j}[(\nu + \sigma_k \nu_T)\dfrac{\partial k}{\partial x_j}] + 2(1 - F_1)\sigma_\omega \dfrac{1}{\omega}\dfrac{\partial k}{\partial x_i}\dfrac{\partial \omega}{\partial x_i} \end{array}\right. , \tag{6}$$

where σ_ω=0.5, σ_k=0.85, η_0=4.38; β=3/40; β*=0.009, α=0.55 are coefficients of k-ω SST model [18].

Transient analysis is executed for physical time of 3.5 seconds with a time step of 0.0025 seconds. Dynamic mesh is used, in order to set linear velocity V_r of the blade`s cross sections. For this purpose, a UDF is compiled from a source file, using Microsoft Visual Studio [19]. The dynamic remeshing for different time steps of the DU40 airfoil calculation is shown in figure 5.

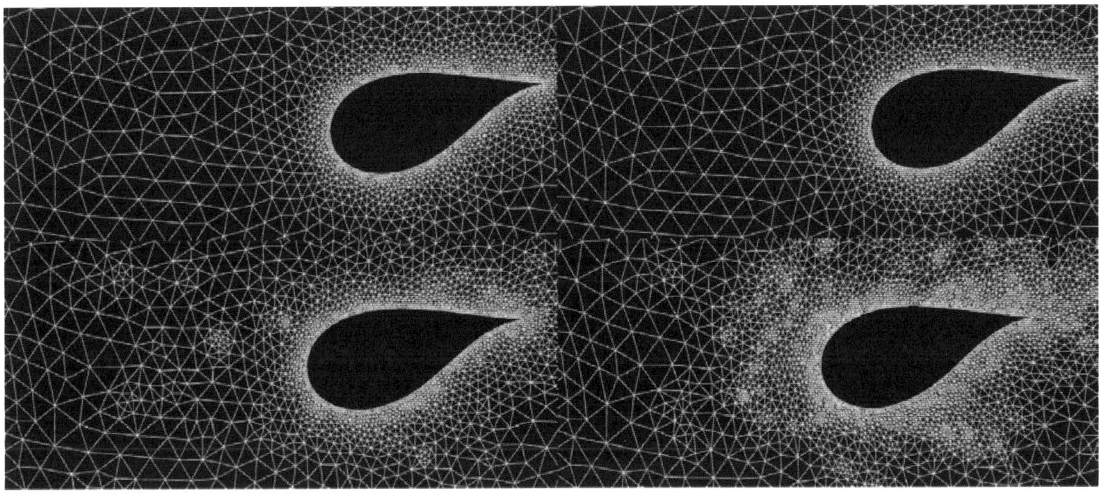

Figure 5. Dynamic mesh of the DU40 airfoil simulation for 1st, 3rd, 5th and 10th time steps of the calculation.

It is interesting to note the remeshed grid of finite elements, after the calculation finish (figure 6).

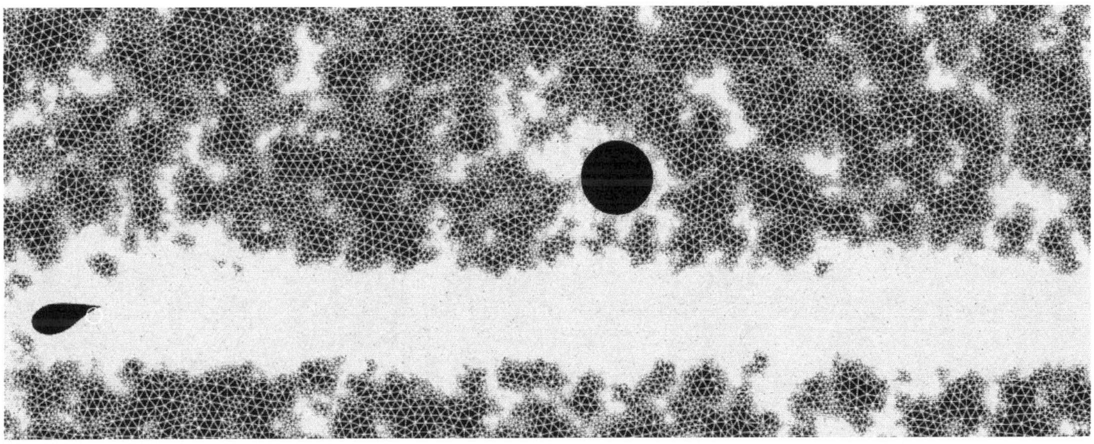

Figure 6. Finite element mesh at the end of the calculation.

The wind velocity V_w is set as inlet boundary condition. Multiple number of simulations are calculated for different cross sections of the blade. The purpose of the calculations is to define the behaviour of the pressure and velocity fields while the blade passes in front of the tower. Numerical values of the acting aerodynamic forces are obtain.

3. Results

As mentioned, calculations are made for multiple number of cross sections. The results show a certain pattern in the deviation of the aerodynamic parameters. In order to specify this pattern, the results for the DU30 airfoil are illustrated and analysed in details. The DU30 airfoil cross section is at distance of 24.05 m from the center of rotation. The two velocity components are V_w=10.5 m/s and V_r=30 m/s. The velocity and pressure distribution in the moment of a blade passing in front of the tower is shown in figure 7 and figure 8.

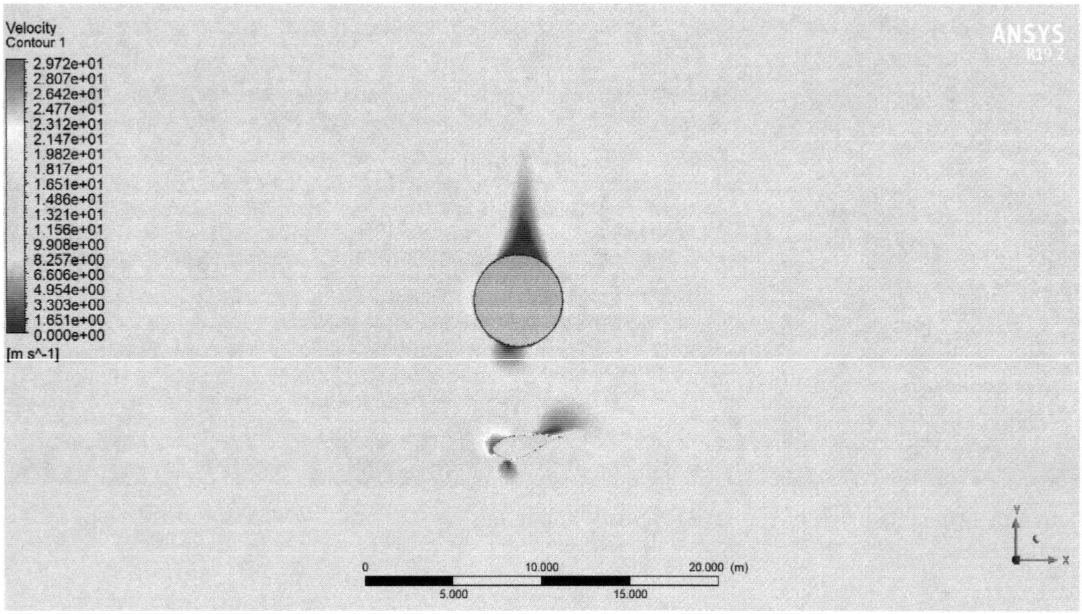

Figure 7. Velocity field of DU30 passing in front of the tower.

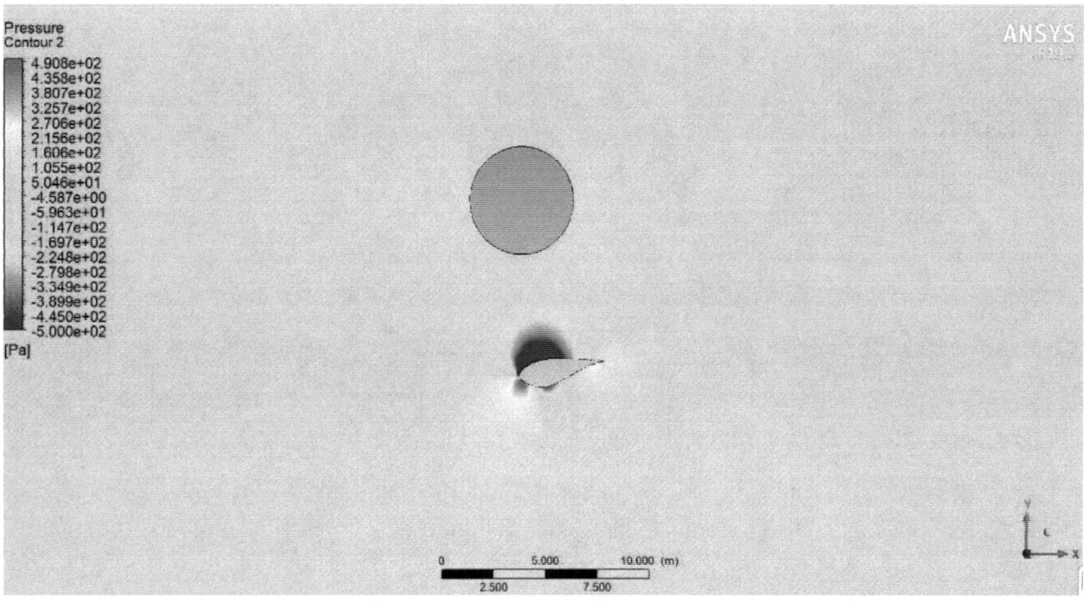

Figure 8. Pressure field of DU30 passing in front of the tower.

As expected, regions with relatively high speed occur close to the leading edge of the airfoil. This results in a low pressure close to the leading edge. The low pressure disturbs the pressure distribution around the tower, creating a region with low pressure in front of the tower. The effect of low pressure in front of the tower changes the direction of the drag force, exerted on the tower, from y positive direction (downstream) to y negative direction (upstream). The low pressure field close to the leading edge of the blade generates a lift force exerted on the tower. While the blade passes in front of the tower, the lift force changes its direction from x positive, to x negative. The numerical values of the tower drag and lift forces as a function of time are shown in figure 9 and figure 10.

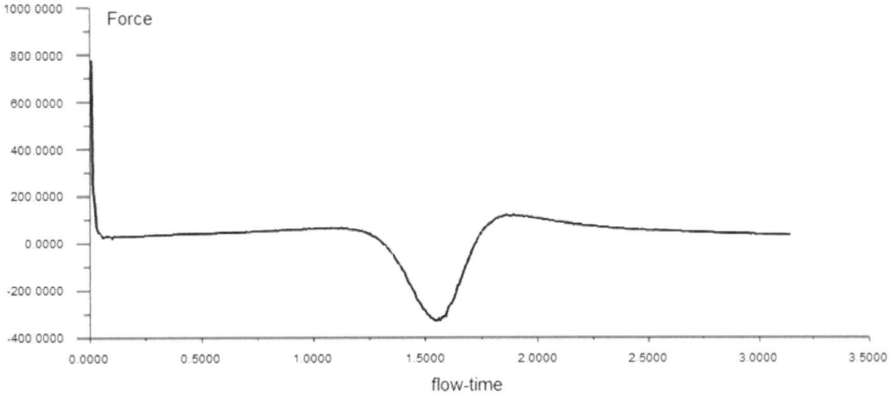

Figure 9. Tower drag force as a function of time.

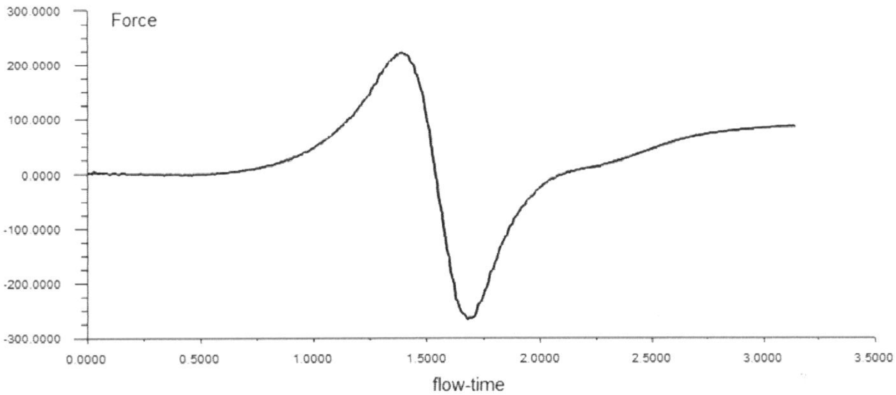

Figure 10. Tower lift force as a function of time.

The exact moment of time, when a blade is passing in front of the tower is 1.5 seconds. The aerodynamic shading influences the thrust (positive y direction) and the torque (negative x direction) forces exerted on the blade`s cross section. For the considered case, the drop in the thrust force is 15% and the drop in the torque force is close to 25%. The change in the magnitude of two forces is shown in figure 11 and figure 12.

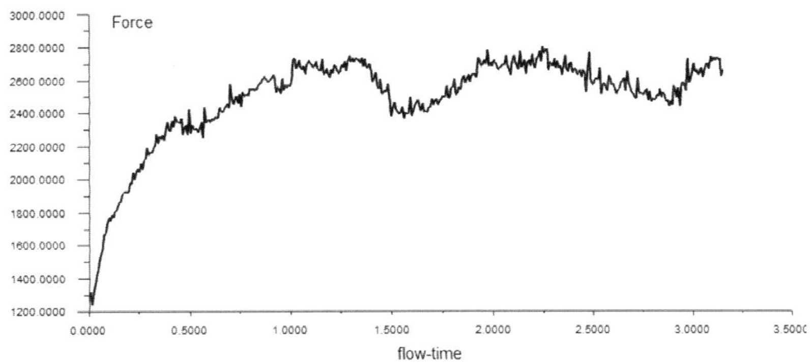

Figure 11. DU30 thrust force change in time.

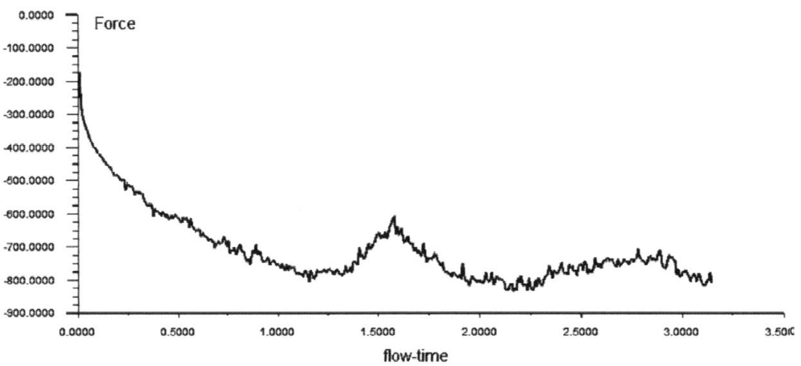

Figure 12. DU30 torque force change in time.

The described behaviour of the pressure and velocity fields, the drag and lift forces exerted on the tower, as well as the thrust and torque forces exerted on the blade is similar for all numerical calculations executed in the study. The accounting of all calculations and the coupling of all results, give us the following results. Figure 13 shows the drag force distribution on the tower of NREL 5MW wind turbine generator, without any consideration of aerodynamic shading. Figure 12 shows the deviation in the force, for the exact moment, when the blades 'shadows' the tower.

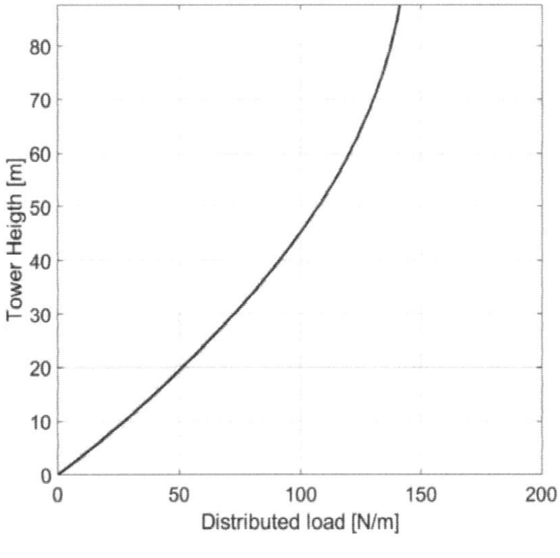

Figure 13. Tower drag force distribution without shading effects.

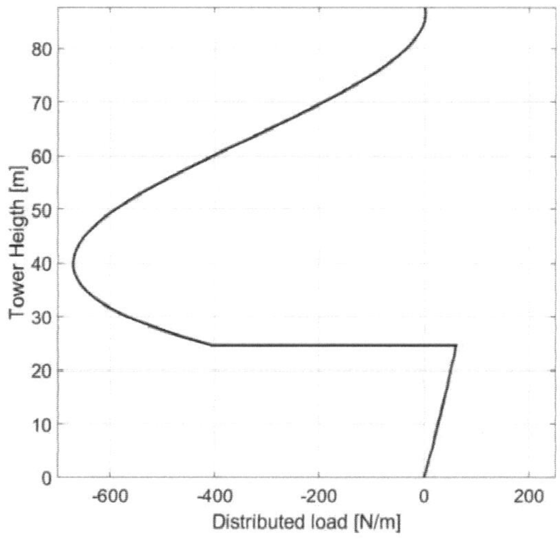

Figure 14. Tower drag force distribution with shading.

The rapid change in the distributed loads at a height of 25 m appears (figure 14), because no blade tip vortices are considered. In future researches, Glauert correction factor [21], associated with the tip loss, will be taken into account.

Figure 15 shows the thrust force distribution, exerted on the blade of the wind turbine for the vertical bottom position of the blade, without considering the aerodynamic interaction with the tower. Figure 16 shows the drop in the thrust force (normal to the plane of the wind turbine), when the shading is taken into account.

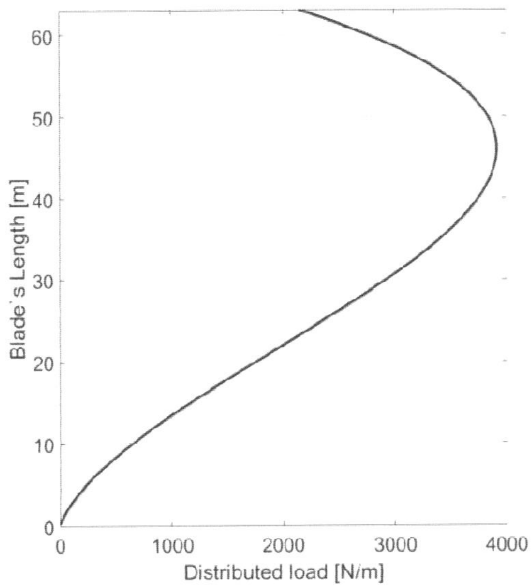

Figure 15. Blade thrust force distribution without shading effects.

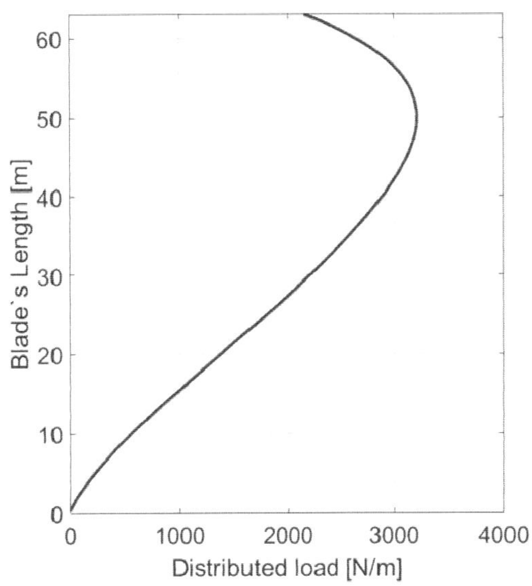

Figure 16. Blade thrust force distribution with shading.

As a confirmation of the numerical results validity, it is appropriate to mention that the aerodynamic forces, exerted on the blade`s cross sections, before the shading, correspond with lift and drag coefficients published in [20]. Before the shading manifests, the lift force on the tower is zero (figure 10).

4. Conclusion

The numerical studies made in this paper show the influence of the aerodynamic shading, which occurs between the blade and the tower of NREL 5MW wind turbine generator. Deviation in pressure and velocity fields is established, as well as change in the aerodynamic forces, exerted on the blade and the tower of the turbine generator. The drag and lift forces exerted on the tower change both their magnitudes and directions as shown in figure 9, figure 10 and figure 14. The analysis shows a drop in the blade thrust and torque forces between 15% and 25% compared to the primary values in figure 11, figure 12 and figure 16.

The paper suggests that the effect of aerodynamic shading is significant and should be analysed for every wind turbine design independently.

The obtained results could be used as inputs in a coupled dynamic model of the turbine and the tower in order to evaluate vibrations and fatigue of the constructive components. The model could be useful for the synthesis of a control system of the wind turbine to reduce loads and stresses on the components and to achieve better uniformity of torque and hence the quality of energy generated. The results could be used for verification of an analytical model of the aerodynamic shading, derived as a function of wind speed, distance turbine – tower etc.

Acknowledgments

This work was supported by the Bulgarian Ministry of Education and Science under the National Research Program E+ „Low carbon energy for transport and live", approved by DCM №577/17.08.2018.

References

[1] Cummin R, Mason W, Marton S and McDoniel D 2015 *Applied Computational Aerodynamics* (Cambridge University Press) p 253

[2] Sørensen J, Mikkelsen R, Troldborg N, Okulov V and Shen W 2013 *The aerodynamics of wind turbines* (Springer)

[3] Michallef D and Bussel Gerard 2018 A review of urban wind energy research: aerodynamics and other challenges *Energies MDPI* **11**(9) 2204

[4] Leishman J2011 Aerodynamics of horizontal axis wind turbines *Advances in wind energy conversion technology* (Springer) pp 1–70

[5] Wang Y, Wang J Zhang J 2019 Effects of wind rotor tilt angle on aerodynamic power of wind turbine under typical periodic disturbances *Advances in Mechanism and Machine Science* (Springer) pp 3459–68

[6] Reiso M 2013 *The tower shadow effect in downwind wind turbines* (NTNU – Trondheim)

[7] Das S, Karnik N and Santoso S 2011 Time-domain modeling of tower shadow and wind shear in wind turbines *ISRN*

[8] Zahle F, Madsen A and Sørensen N. 2009 Evaluation of tower shadow effects on various wind turbine concepts *EFP-2007-II*

[9] Pinting Z, Shuhong H, Tao Y, and Jianlan L 2014 3D potential flow model of tower interference for upwind wind turbine *Advances in Mechanical Engineering* 612453

[10] Genov J and Angelov I 2016 Influence of vertical wind speed distribution on the dynamic loads in large wind turbines *Proc. Int. Sci. Conf. BulTrans-2016 (Sozopol, Bulgaria)* pp 136–40 (In Bulgarian)

[11] Genov J, Kralov Iv and Angelov I 2018 Dynamic loads of the turbine blades in high power class wind generators caused by the vertical wind turbulence *1st Inter. Conf. of Marine Eng. (Varna, Bulgaria)*

[12] Jonkman J, Butterfield S, Musial W and Scott G 2009 Definition of a 5-MW reference wind turbine for offshore system development, *Tech. Report NREL/TP-500-38060*

[13] Manwell J, McGowan J and Rogers A 2009 *Wind energy explained: theory design and application* (DNV – Global Energy Concepts, Washington, USA)

[14] Landberg L 2016 *Meteorology for Wind Energy - an Introduction* (John Wiley & Sons Ltd.) p 215

[15] Grant I 2011 *Wind Turbine Blade Analysis Using the Blade Element Momentum method*

[16] John D and Anderson J 1991 *Fundament of Aerodynamic* (University of Maryland) p 792

[17] Moukalled F, Mangani L and Darwish M 2016 *Finite Volume Method in Computational Fluid Dynamics* (Springer)

[18] SST k-omega model https://www.cfd-online.com/Wiki/SST_k-omega_model Accessed on: 15.02.2020

[19] *ANSYS Fluent Tutorial Guide 18* 2017 (ANSYS, Inc.)

[20] Moriaty P and Hansen A 2005 *AeroDyn Theory Manual* (NREL/TP-500-36881) p 47

IOP Conf. Series: Materials Science and Engineering 1002 (2020) 012010 doi:10.1088/1757-899X/1002/1/012010

Crack resistance of reinforced-concrete sleepers with elastic rail fastening systems without base-plate

A A Plugin[1], S V Miroshnichenko[1], O V Lobiak[2,3], OA Kalinin[1] and DA Plugin[1]

[1] Department of Building Materials and Structures, Ukrainian State University of Railway Transport, Feuerbach sq. 7, 61050 Kharkiv, Ukraine
[2] Department of Structural Mechanics and Hydraulics, Ukrainian State University of Railway Transport, Feuerbach sq.7, 61050 Kharkiv, Ukraine

[3] Email: lobiak@ukr.net

Abstract. The article presents the main results of complex experimental and theoretical research into the crack resistance and the limit state of reinforced-concrete sleepers with elastic rail fastening systems without base-plate. The objective of this study is to uncover reasons for the cracks and other breakages during operation, demonstrate their importance, and produce a set of recommendations for prevention of the most serious defects. The research also involved statistical analysis of the defect factor for sleepers, and determined the number of defects by type according to track layout, traffic density, tonnage transported, concrete grade and other characteristics based on the results of full-scale research of track sections and surveys from maintenance specialists. The full-scale research was made on rail sections under standard operational modes and on a test track section of straight and curved elements with sleepers of concrete grades C32/40, C35/45 and C40/50.The study was based on numerical methods, modelling of the stress-strain behaviour of the track, and estimation of the performance of sleepers with cracks.The method used incremental-iterative algorithms to consider the elastic plastic properties of materials, solve contact problems in operation of rail fastening systems, and also reveal the structural nonlinearity of joint action for track elements. The method was realized in LIRA-SAPR software on the basis of Building Information Modelling (BIM) and the Finite Element Method (FEM).The results were verifiedby experimental research into the sleepers under static and dynamic loads with estimation of crack development in terms of concrete grades, deformation modulus, rigidities of the ballast layer and the sub-grade, position of the reinforcement and other characteristics.

1. Introduction

The most popular type of reinforced-concrete sleepers on Ukrainian Railwaysis Sh1-type sleepers reinforced with high-tensile steel wire and pre-stressed with KB-type terminal-bolted rail fasteners [1].This structure has a steel base-plate with a rubber pad between the rail base and rail seat of a sleeper, which re-distributes the loads across a large concrete area. Since the 1990s Ukrainian Railways have introducedthe SB3-type sleeper (figure 1) for elastic rail fasteners KPP-5 without base-plate; they are popular as they have a few details, low metal capacity and workingintensity.A prototype of these fastening systems was the SB-3-typefastener (Poland). The SB3-type sleeper differs from the Sh1-type sleeperbythe presence of embedded anchors instead of holes for the insert bolts, and by a higher rail seat.

The sleepers are made of concrete grade C32/40; they comply with the maximum loads 232kN/axle from the rolling stock [2], and undergo final inspection in accordance with [1]. However, in operation they suffer from crack development and other defects, which can be estimated through the tonnage carried [3]. Since the 2000s railways have witnessed new types of defects, e.g. a crack network emerging after 1-3 years of operation, which develops and results in damage. Track facility engineers have noticed that SB3-type sleepers are less reliable than Sh1-type sleepers, and their damage rate depends on the traffic intensity of the rail section. Therefore, Ukrainian Railways use elastic rail fasteners without base-plate only for sections with traffic capacity of less than 30 million gross tons per year. However, the authors did not manage to find any substantiated quantitative analysis proving this view.

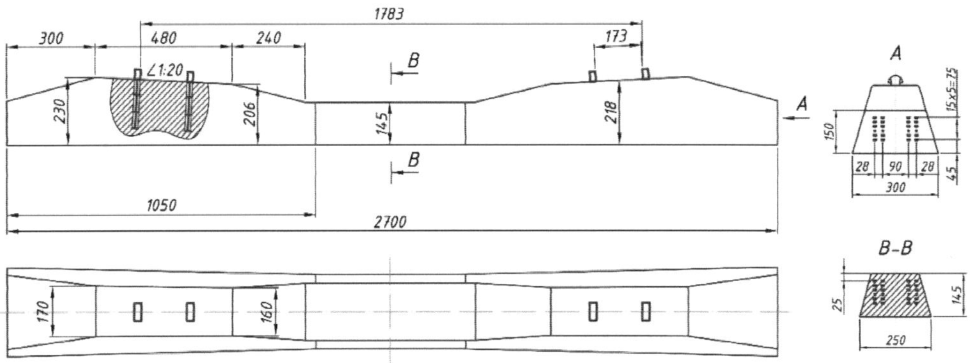

Figure 1. Structure of SB3-type sleeper.

2. Critical analysis of the results of earlier research

Practically, cracks in reinforced-concrete sleepers begin to develop when they are installed [4-7]. Study [4] substantiates the conditions under which force cracks do not open, and presents the most crack resistant sleeper structure; studies [5-7] propose an improved production technology aimed at greatercrack resistance and durability of rail sleepers. Studies [8-12] explain the connection between crack network development andconcrete corrosion due to the interaction of alkali in concrete with reactive aggregates [13-16]. However, the authors in [4-7] do not study sleepers with elastic fasteners without base-plate, and in [8-12] they consider only alkaline corrosion, irrespective of other factors of defects. The authors did not manage to find data on the dependencies between the damage factor of sleepers and the concrete grade, however, as known, many countries use higher concrete grades, unlike it is in Ukraine.

3. Statistical analysis of defects in sleepers

The full-scale research of track sections and statistical analysis of the results revealed typical defects in sleepers (table 1).

Table 1. Defects in sleepers with elastic rail fasteners without base-plate, percentage of sleepers with defects of all types.

№	Damage (defect)	%	Reason for defect, note
1	Abrasive wearof sleeper end by ballast (figure 2a)	41.4	Ballast settlement under sleeper ends and rail seats, vibrations of sleeper ends under vertical dynamic loads (development of transverse cracks near the anchors and in mid-span of sleeper)
2	Transverse cracks in mid-span of sleeper (figure 2b)	24.1	Ballast settlement under rail seats, deflections of sleepers under vertical loads (may result in fracture of sleeper)
3	Transverse cracks near anchors (figure2c)	20.0	Crack development under horizontal loading component on anchor. Defect is intensified by ballast settlement under rail seats (results in complete damage of sleeper near anchor)
4	Longitudinal cracks at sleeper ends (figure 2d)	19.4	Irregular ballast settlement under rails, formation of shear stresses, cracks and spalls due to uneven stiffness of foundation (may result in spalls in sleeper)

5	Crack network on sleeper ends, sometimes along sleeper (figure 2e)	18.0	Crack network development due to corrosion of aggregates caused by the alkali of cement, alternate freezing and de-freezing (can lead to concrete degradation along cracks)
6	Transverse cracks under rail seats	4.5	Ballast settlement under rail seats, crack development under vertical loads
7	Longitudinal cracks across anchor	2.4	Crack development under horizontal load on the anchor (can result in splits of sleeper)
8	Longitudinal cracks in mid-span (figure 2f)	2.1	Corrosion of aggregates caused by the alkali of cement, crack development under initial tension of reinforcement due tolower concrete strength (results in splitsin sleeper or complete concrete degradation)

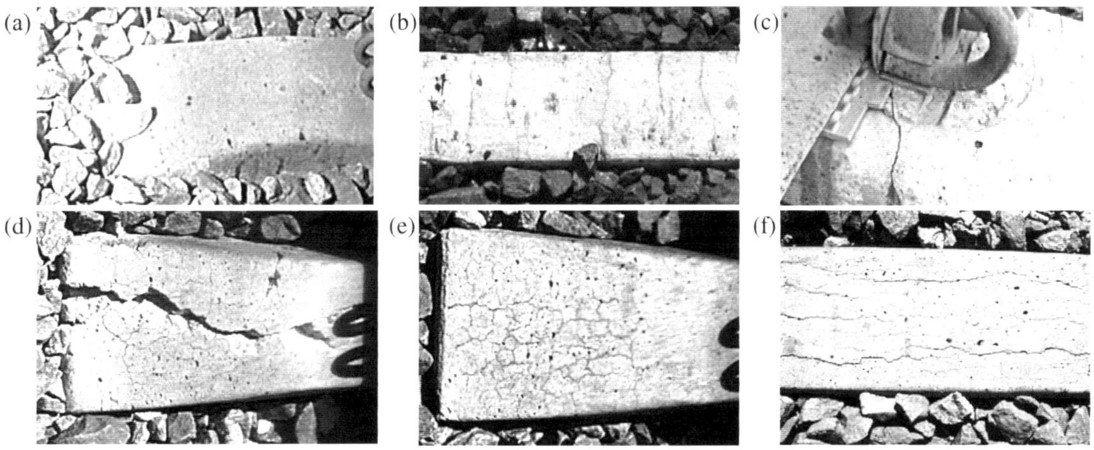

Figure 2. Defects in sleepers with elastic rail fasteners without base-plate.

Dependency of the defect rate of a sleeper on the tonnage carried is given in figures 3 and 4 (a) and on the traffic volume – in figure 4 (b). It demonstrates a slight correlation by Chaddock's scale (the squared correlation coefficient R^2 does not exceed 0.3) between the tonnage carried and the number of sleepers with defects, which to the utmost corresponds to the linear dependency. However, dependencies between sleepers with defectsin pcs/km×million tons on the traffic density were not observed.

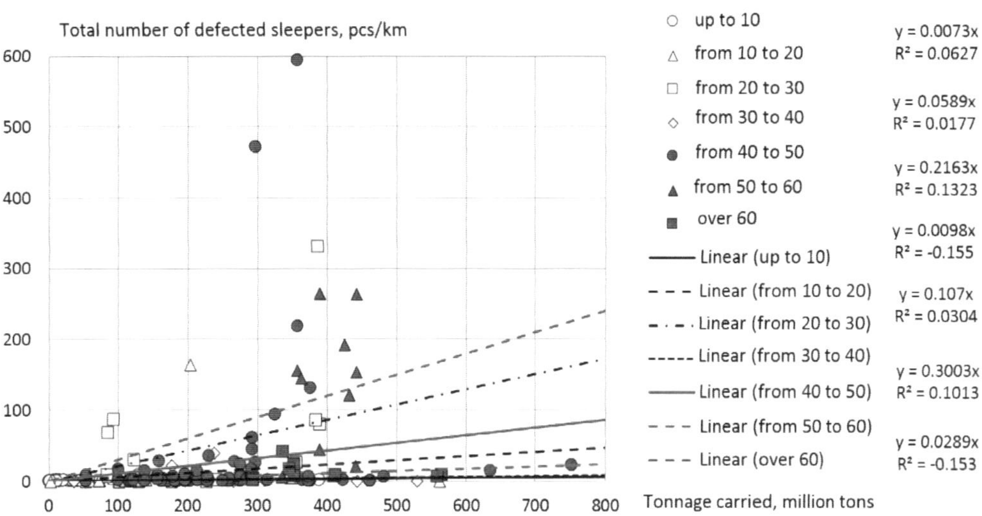

Figure 3. Dependency of the number of sleeper with defects (pcs/km) on tonnage carried (million gross tons) along sections with traffic density of up to 10, from 10 to 20, from 20 to 30, from 30 to 40, from 40 to 50, from 50 to 60, and over 60 million tons per year.

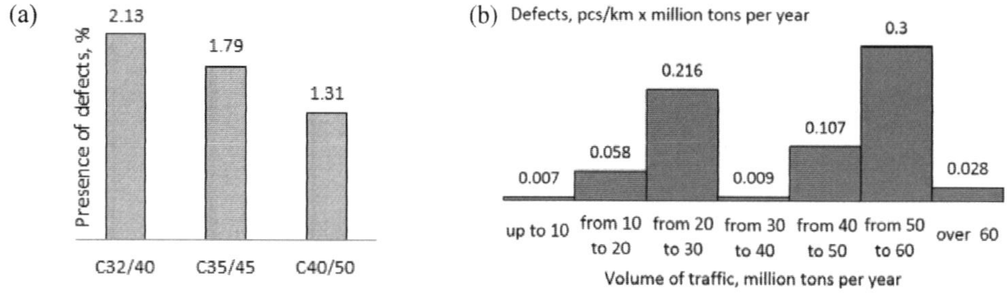

Figure 4. Dependency of the number of sleepers, with transverse cracks near anchors after 107 million gross tons transported,on concrete grade (a), dependency of defect rate on traffic density (b).

4. Analysis of stress-strain behaviour of sleepers by finite element method

4.1. Design diagram

Generally, the design engineering of today uses industrial software as the basic tool for computer modelling. An improved calculation of a track with rail fasteners without base-plate can be fully realized in LIRA-SAPR software [17] which earlier was proved to be an effective tool to study transport structures [18].

The following hypotheses were assumed in designing the finite element model (figure 5).

1. Steel elements, rail pads and inserts were designed with universal Solid finite elements of homogeneous isotropic linear elastic material in a three-dimension elasticity problem.

2. The strength and deformability of concrete sleepers were characterized by the parabolic rectangular deformation diagrams $\sigma_c - \varepsilon_c$ in accordance with EN 1992-1-1 (Eurocode 2), the pre-tension forces were set by decreasing the temperature of the axial finite elements installed in the axes of reinforcement of a sleeper. The research considered concrete grades C32/40, C35/45 and C40/50.

3. The models considered structural non-linearity of an assembly unit guaranteed by special bi-nodal finite elements of one-way connection (certain contacts were intended only for compression or compression with friction). These finite elements were inlaid between the adjacent nodes of the contacting elements and described the work of these contacts with the tensors of the axial rigidities. Generally, six contact problems were considered: rail pad – rail (S-R); anchor – insert (A-B); insert – rail (B-R); rail clip – insert (C-B); rail clip – anchor (C-A); anchor – sleeper.

4. A track fragment restricted by two sleepers was selected out of the track structure through the method of fragmentation. The boundary conditions of the fragment considered repeated the corresponding boundary conditions of a simplified full structure model. Thus, the fragment was calculated with a more detailed design diagram.

5. The rigidity of the sub-sleeper foundation was determined according to a three-dimensional model of the soil foundation and the ballast layer. According to this model the authors determined the moduli of sub-grade reaction C_1, C_2 across the whole sleeper sole (these moduli depended on loads on the sleeper), and also the compressed depth and compaction. Therefore, the boundary conditions were set by the rigidity values of the rail seat and the loads in the rail sections on the boundaries of a fragment.

6. The load transfer from wheels to rail was performed through elliptical flanges of the finite element of the rail, the total area of which was 100mm^2.

By applying series of loads such factors as dead weight impact, pre-tension of the reinforcement, and contemporary impacts were considered. The model assumed a possible loading on the rail with the axial load in any position within a span between the sleepers, which can be important for the analysis of other track elements. The analysis of the maximum loads in the sleepers was based on the position of the plate of loading, which laterally coincided with the sleeper axle and admitted displacement in the transverse section was conditioned by entering a curve. The operational loading was taken 232 kN/axle in accordance with [2]; it corresponded to the impact from rail passenger vehicles running at speed exceeding 160 km/h. The study considered loads for straight and curved track sections (table 2). For a curved line the horizontal loading was applied at 13mm lower than the rolling surface of the high rail.

Table 2. Temporary loads on rail.

Layout element	Forces, kN				
	Vertical				Horizontal
	Left rail		Right rail		
	P_{left}	Eccentricity	P_{right}	Eccentricity	H_{right}
Straight	116	8 mm in the middle of track	116	8 mm in the middle of track	-
Curved	114	23 mm in the middle of track	118	7 mm high rails	120

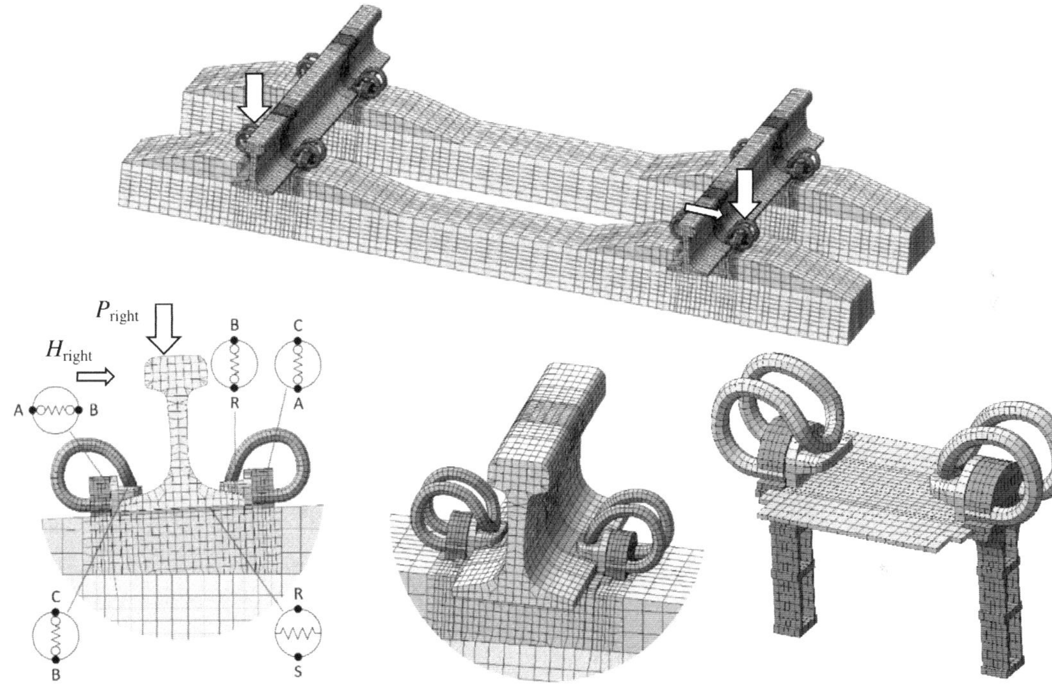

Figure 5. Design diagram of track with elastic rail fasteners without base-plate.

Five technicaltrack stateswere taken for the research: 1 – working (the deformation modulus of the ballast 120-145MPain the rail seat, 40-90MPa in the mid-span); 2-5 – the deformation modulus decreases (figure 6). Changes in the deformation modulus did not influence the average rigidity for all epures; a decrease in the rigidity characteristics was determined by re-distribution of the deformation modulus values from the rail seat to the mid-span of a sleeper.

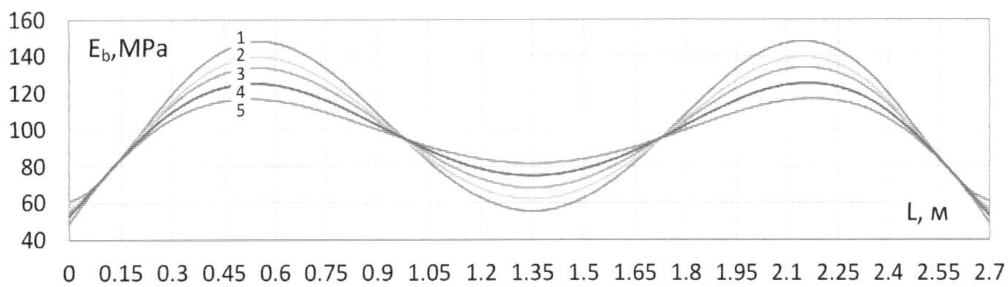

Figure 6. Changes in deformation modulus of ballast along sleeper according to the technical state of track.

A displacement of the pre-stressed reinforcement package of a sleeper from the design position was considered.The calculation was made for six levels of deviation in reinforcement: 1 – design position, 2-6 – geometric centers of reinforcement package displaced to sleeper bottomby 5–25mm, respectively.

4.2. Analysis of stress-strain behaviour of sleepers

Stresses and deformations in a sleeper under the operational condition of a track are given in table 3. Figure 7 presents an example of stress fields for concrete grade C32/40 in acurved track section.

Table 3. Calculation results

Layout element	Concrete grade	Maximum stresses near anchor, MPa	Stresses in mid-span of sleeper, MPa	Vertical deformations, mm	Rail deflection, mm
Straight	C32/40	18.3	9.6	3.4	0.36
	C35/45	18.7	8.5	3.4	0.29
	C40/50	18.9	7.2	3.3	0.25
Curved	C32/40	30.3	11.2	4.28	0.72
	C35/45	31.4	10.4	4.28	0.65
	C40/50	32.2	8.7	4.23	0.51

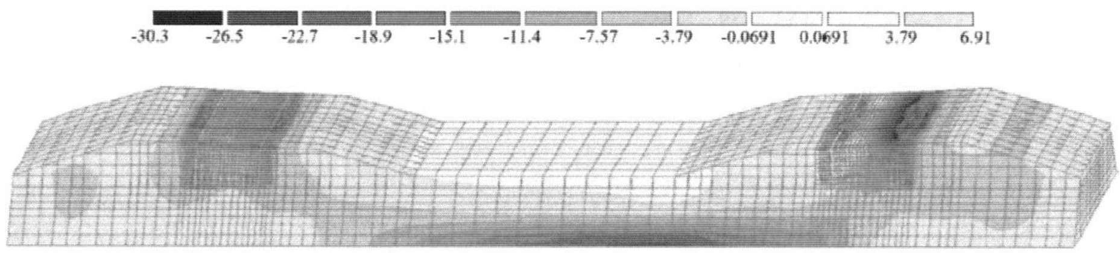

Figure 7. Main stresses in C32/40 sleeper for curved track section, MPa.

From the results of the calculation for different technical states of a track the research establishedchanges in the stress-deformed state and the moment of crack development (the load corresponding to the emergence of the first crack) in the upperflange of a sleeper in the mid-span (figure 8). Deflections of the pre-stressed reinforcement package of a sleeper from the design position were illustrated in figure 9.

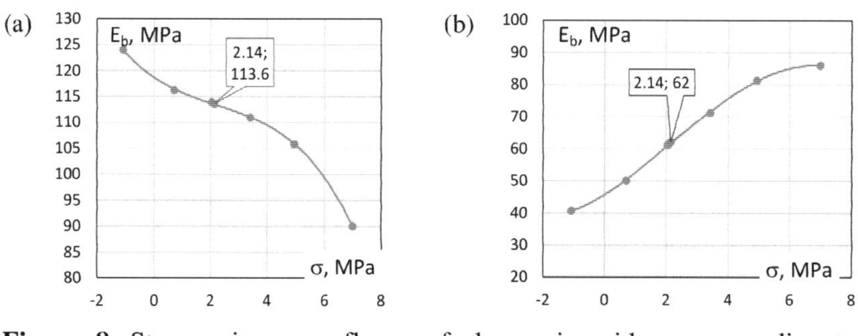

Figure 8. Stresses in upper flange of sleeper in mid-span according to deformation modulus of ballast in rail seat (a) and in mid-span (b).

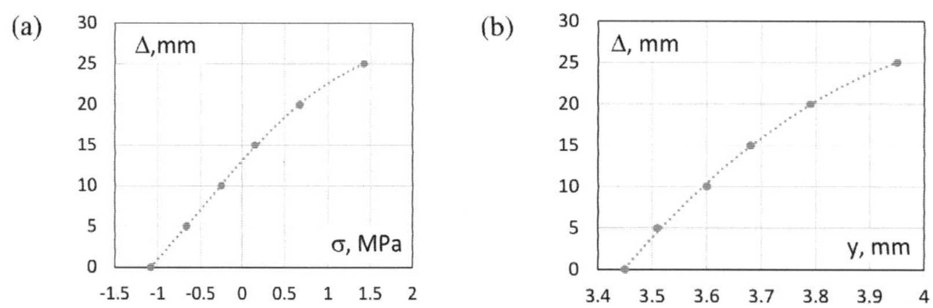

Figure 9. Changes in stresses σ in upper flange of sleeper in mid-span (a) and maximum vertical deformations *y* (b) according to deflections of reinforcement package from design position.

4.3. Main conclusions from the results of the calculation

Thus, the calculation by means of the finite element method established an effect of ballast settlement under rail seat, deflections of areinforcement package and concrete grade on the stress-strain behaviour.

Ballast settlement under the rail seats results in the development of transverse cracks in the middle of a sleeper. The moment of crack development is reached under settlement conditioned by a simultaneous decrease of the average deformation modulus of the ballast of up to 113.6MPa in the rail seat and an increase to 62MPa in the middle of a sleeper corresponding to the total rigidity decrease of the foundation by 22%.

Deflections of a reinforcement package from the design position in the sleeper production do not considerably affect crack development. A downshift of 25mm leads to a sign change of stresses in the middle of a sleeper reaching +1.4MPa, meanwhile the moment of crack development is not reached.

A change in the concrete grade from C32/40 to C40/50 results in:

- (for a straight line) an increase of the maximum loads in the anchor zone by 3%, a decrease of the loads in the midspan of the sleeper by 25%, a decrease in the vertical deformations by 3%, a decrease of the sleeper deflection by 31%;
- (for a curved line) an increase of the maximum loads in the anchor zone by 6%, a decrease of the loads in the midspan of a sleeper by 22%, a decrease in the vertical deformations by 2%, a decrease of the sleeper deflection by 29%.

Despite an increase of the maximum loads in the anchor zone by 3-6%, change of the concrete grade from C32/40 to C40/50, considering an increase in the compressive resistance of concrete by 25%, conditions a higher reserve of strength and rigidity of no less than 19%.These conclusions are proved by the coincidence with the results of the analysis of the on-site observation of the sections. Particularly, one of the most popular defects in sleepers is transverse cracks in the anchor zone (20%, table 7, table 3), and damage factor of sleepers by this type decreases with a higher concrete grade (figure 4a).

5. Experimental research

The study deals with static research into 17 sleepers specially manufactured with artificially formed deflectionsof a reinforcement package from the design position and lower depth of cross-sections. The sleepers were tested by static loading with a hydraulic press according tothe diagrams presented in figure 10.

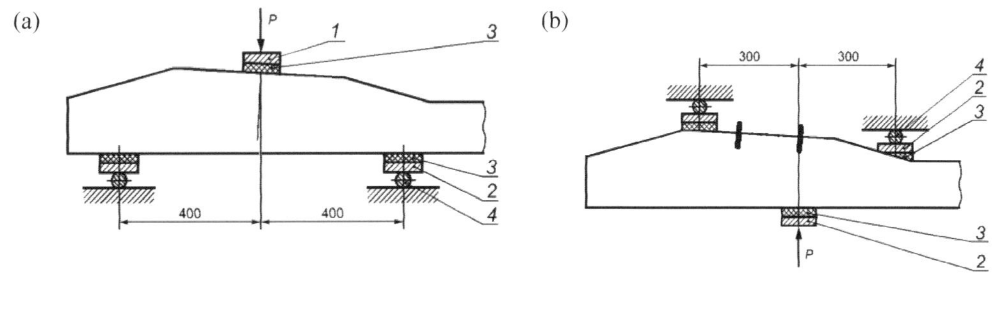

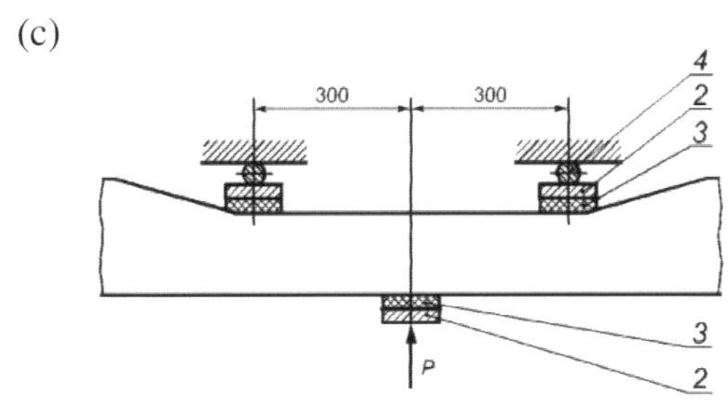

Figure 10. Test diagrams of sleeper for crack resistance and strength:(a) – rail seat in standard position; (b) – rail seat in reverse position (force is applied under internal anchor); (c) – midsection; *1* – steel plate with deflection of lower base 1:20, with a size of 250×100 mm and average thickness of 25 mm; *2* – steel plate 250×100×25 mm; *3* – rubber pad 250×100×10 mm; *4* – steel roller with a diameter of 40 mm and length of 250mm.

The results of the research into crack resistance and strength of the experimental sleepers under static loading are given in figure 11.

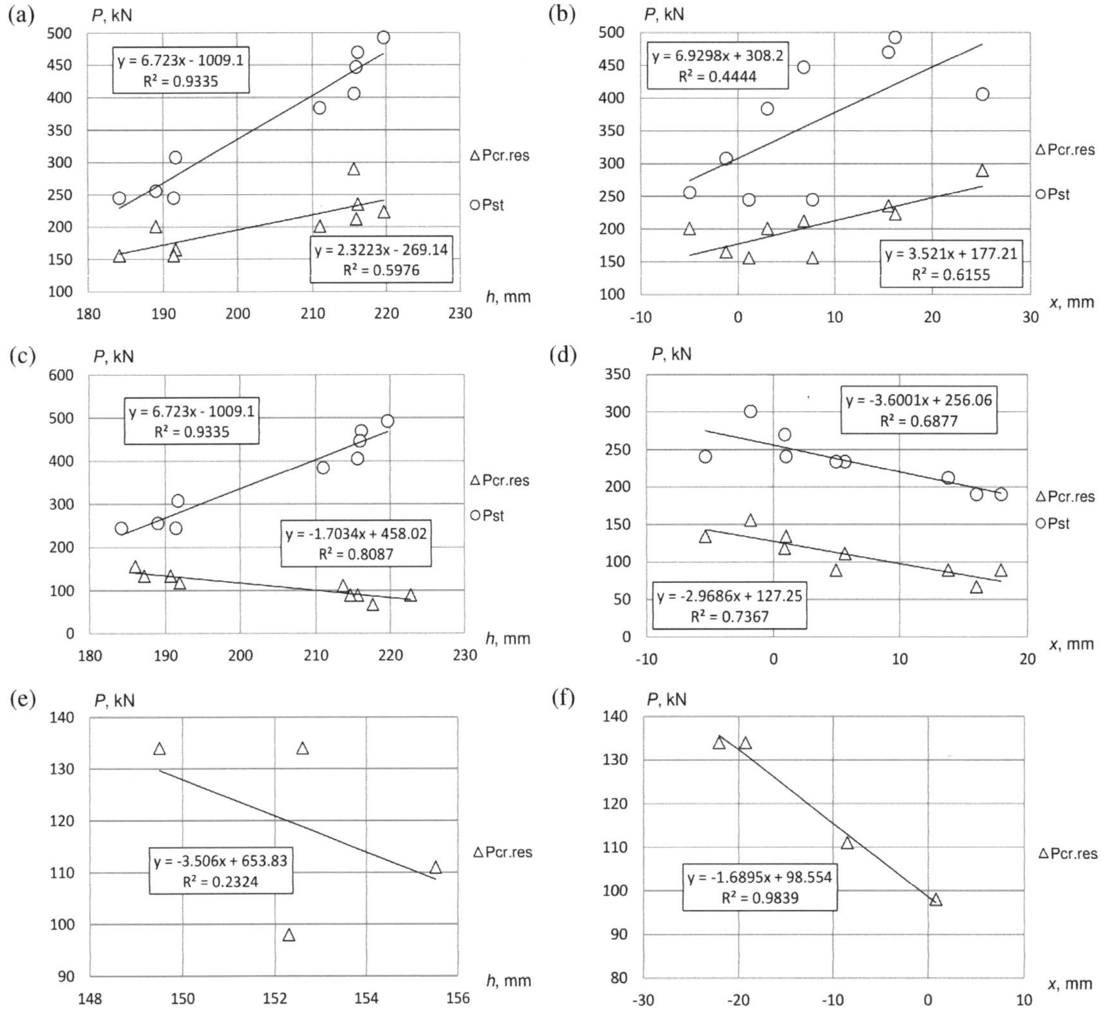

Figure11. Dependency of crack resistance boundary $P_{cr.res.}$, kN and strength boundary P_{st}, kN of sleeper under loading:(a), (b) – rail seat section on top at height *h* (a) and deflection of reinforcement package *x* (b); (c), (d) – rail seat section at the bottom at height *h* (c) and deflection of reinforcement package *x* (d); (e), (f) – midsection at the bottom at height *h* (d) and deflection of reinforcement package *x*(f).

Therefore, the static tests established that due to the loading on the rail seat of sleepers fromthe top, which corresponded to the vertical loading component from the rolling stock, a decrease of the rail seatdepth from 218 to 185mm conditioned a fall of the crack resistance boundary by 33%, and the strength boundary – by 48%. Deflection of a reinforcement package upward by 5mm conditioned a fall of the crack resistance boundary by 9%, and the strength boundary – by 13%.Deflection of a reinforcement package downward conditioned an increase of the crack strength boundary and the strength boundary.

When the sleeper section was loaded from the bottom under the internal anchor, which modeled an effect of the horizontal loading component from the rolling stock on the anchor, owing to the development of a similar moment in the cross section, a decrease in the rail seatdepth from 218 to 185mm conditioned an increase of the crack resistance boundary by 75%, and a decrease of the strength boundary – by 104%. An upward deflection of thereinforcement package conditioned an increase of thecrack resistanceboundary and the strength boundary. A downward deflection of the reinforcement package by 25mm conditioned a decrease of the crack resistance boundary by 40%, and of the strength boundary – by 38%.

When the middle section of a sleeper was loaded from the bottom, which modeled ballast settlement under the rail seats, an increase of the middle section depth from 145mm to 155mm conditioned a decrease of the crack resistance boundary by 16%. An upward deflection of the reinforcement package conditioned an increase of the crack resistance boundary.

The authors tested sleepers of concrete grades C20/25, C32/40, C35/45, C40/50 with dynamic (impact) loading.The research was conducted with a shock testing machine by multiple drops of 10-kg weight from a height of 0.4 m (figure 12a). The tests used cubic models of 150-mm ribs with anchors installed in production. Some grooves were made in the concrete under the anchors (figure 12 b-f). The breaking knife impacted the anchor simulating the horizontal transverse force in the track. Three models were produced of each concrete grade. Each model was used to determine energy E spent for the development of the defect.

$$E = Nmgh/1000, \quad (1)$$

where N – number of impacts generating defect; m – falling-free weight, 0.1 kN; g – free fall acceleration, 9.81 m/s^2; h – drop height, 0.4 m; 1000 – J to kJ conversion factor.

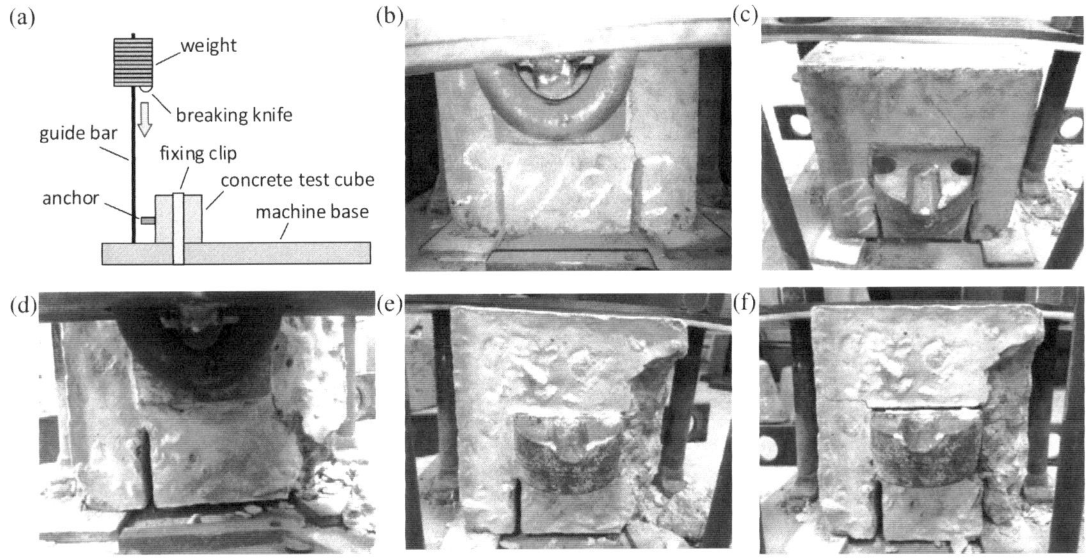

Figure12. Tests on sleeper models through dynamic (impact) loading: (a) – model of shock testing machine; (b) – (f) – sleeper models with defects: initial cracks (b, c, concrete grades C35/45 and C40/50, respectively); initial spall, larges palls, complete damage with separation of anchor from concrete (d, e, f, concrete grade C20/25, respectively).

The criterion, which characterizes a longer service life of sleepers due to higher concrete grade, is an increase in fracture energy in percentage to the fracture energy of the reference grade (C32/40):

$$\Delta P_{C35/45} = 100\times(E_{C35/45}-E_{C32/40})/E_{C32/40} \quad (2)$$
$$\Delta P_{C40/50} = 100\times(E_{C40/50}-E_{C32/40})/E_{C32/40}. \quad (3)$$

The results of the research are given in figure 13. The dependencies were approximated: for initial defects (cracks and spalls) – by linear dependencies; for large spalls and complete damage – by exponential ones.

The analysis of the dependencies demonstrated that for initial defects the correlation between concrete strength and the number of impacts (energy) by Chaddock's scale was very slight, for large spalls (R^2 = 0.57; R = 0.75) it was noticeable, and for complete damage (R^2 = 0.92; R = 0.96) – very high.

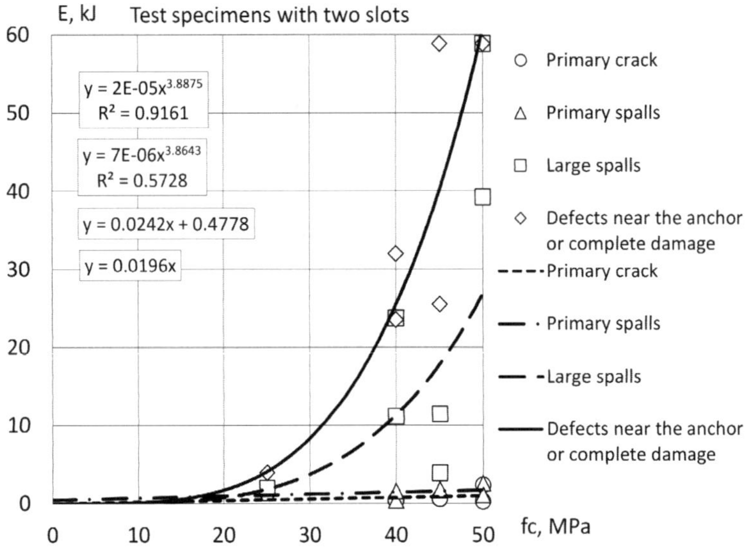

Figure 13. Dependency of energy E spent for development of defects (initial crack, initialspalls, large spalls, complete damage), on concrete compressive strength f_c.

The evaluation of a life gain is given in figure 14; the average sleeper life before development of defects due to higher concrete grade from C32/40 to C35/45 and C40/50 can increase by 33% and 187%, respectively.

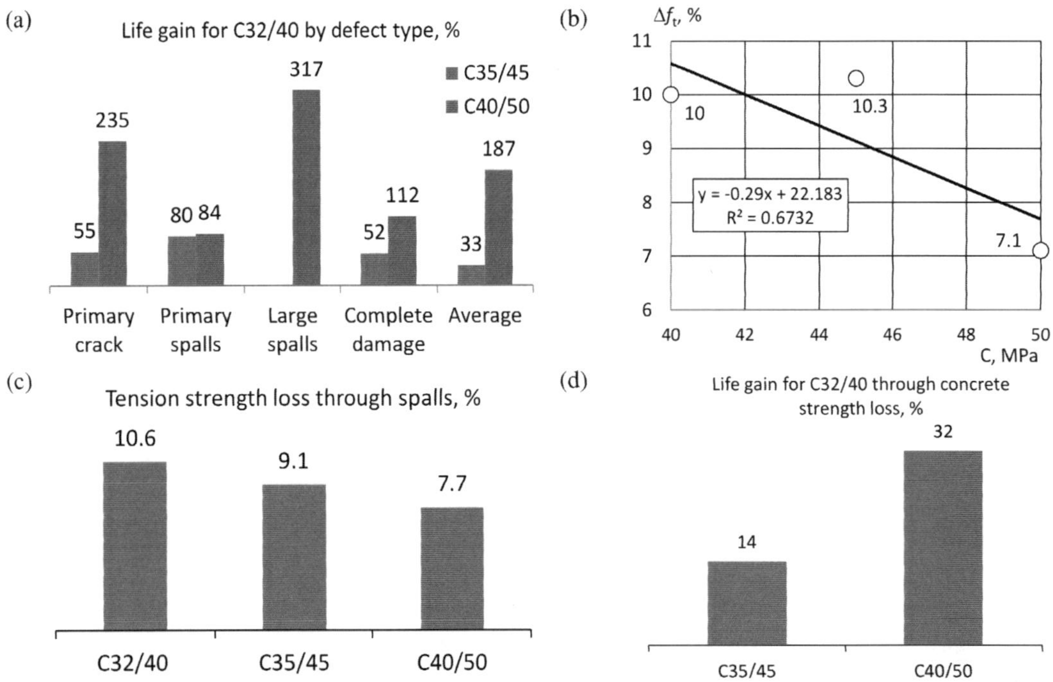

Figure 14. Results of tests on sleeper models by dynamic loading: (a) – by impact loading – better resistanceto defects due to higher concrete grade from C32/40 to C35/45 and C40/50; (b) – (d) – by vibration loading: dependency between decrease in torsion strength at splitting Δf_t and concrete grade C (by strength factor of cubic samples, b, c); better resistance to spills due to higher concrete grade from C32/40 to C35/45 and C40/50.

The tests were conducted on sleeper models from concrete grades C20/25, C32/40, C35/45, C40/50 with dynamic (vibration) loading.The models were half-sleepers witharail fragment and two assembled fasteners, in which a polymeric insert was replaced with a steel one; it provided a direct transfer of the horizontal loading component to the anchor and through it – to concrete (figure 15 a, b). The tests on half-sleepers with a rail and a pair of fasteners were made with the vertical force 22.5 ton-force and the load rate 9 Hz with the dynamic machine МУП-50, equipped with a base (figure 15a); it allowed for installation of the half-sleeper at an angle of 15° to the horizon and appropriate de-composition of the force relative to the sleeper into normal and transverse components (similar to vertical and horizontal forces in the track).

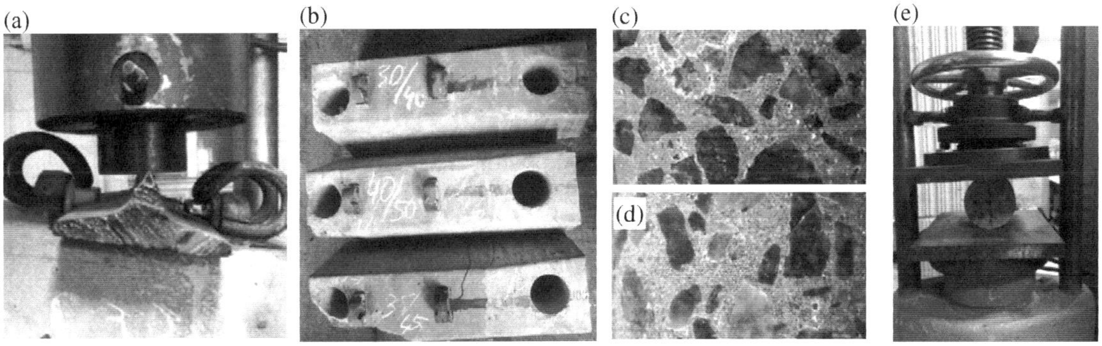

Figure 15. Dynamic tests on sleeper model with fastening assembly: (a) – model installed on МУП-50 at15° (with right steel insert); (b) – models after selection of test samples; (c), (d) – test samples of concrete grades C32/40 (c) and C40/50 (d) treated with organic phosphor solution in Vaseline oil, in ultraviolet light; (e) – determination of concrete tensile strength under splitting according to [19].

After 1.5 million cycles of vibration impact,the 85-mm-diameter test samples were selected out of the modelswith circular bit: near the anchor where dynamic impact was maximum (left, figure 15c) and in the minimum dynamic impact zones (right, figure 15b).The test samples were studied with fluorescent non-destructive testing, they were used for manufacturing of sample-cylinders of 77–113mm in height; these cylinders were used to determine the characteristic which is most sensitive to micro-cracks, i.e. cylinder-splitting tensile strength (figure 15d). Besides, a decrease (in percentage) of this characteristic after vibration impact in the maximum impact zone in comparison with that of the minimum impact zone was also determined.

Using the results of the fluorescent non-destructive testing the authors established changes in the structure of concrete grade C32/40 near the anchor (in the left hand part of the photo, figure 14c) as a wavy crack network testifying the beginning of concrete segregation. Concrete grades C35/45, C40/50 did not demonstrate any structural change (figure 14d). A strength decrease in the maximum dynamic impact zone, obtained by the regression equation $\Delta f = -0.29C+22.18$ (figure 13b), was 10.6% for concrete grade C32/40, 9.1% for concrete grade C35/45, and 7.7% for concrete grade C40/50. Consequently, a life gain of sleepers due to higher concrete grade can be estimated as 14% for C35/45 and 32% for C40/45.

6. Conclusions

1. The stresses in sleepers, which cause defects, do not depend on the traffic density of a section which indicates only intra-repair terms due to the tonnage carried. These stresses are determined by axle vertical loads, lateral horizontal loads (curve radii and traffic speeds), and dynamic impacts (maintenance conditions).

2. The stresses in the sleepers near the anchors are on the boundary of possible local concrete deterioration and crack development, thus, when concrete does not comply with the requirements for strength and homogeneity, such factors as track deterioration, additional influence of watering and continuous leakage currents cause defects and damage in sleepers.

3. The authors believe that better resource of sleepers with fasteners without base-plate for the sections of any traffic density can be reached with the following recommendations:

- to use concrete grade C35/45 and higher for sleepers;

- for manufacturers of reinforced sleepers: to carefully follow requirements for quality of aggregates, particularly, reactive silica, and requirements for strength, freeze thaw and crack resistance by improving all types of control;

- not to apply fasteners without base-plate for sleepers in the track sections of epures less than 1840 sleepers/km; and

- to decrease tonnage norms for sections with fasteners without base-plate, and fill in ballast under the sleepers with worn-out ends.

References

[1] DSTU B V.2.6-209:2016 *Pre-stressed reinforced concrete sleepers for 1520 and 1435-mm gauge railways: Technical regulations* (Kyiv: Ministry for Communities and Territories Development of Ukraine)
[2] DSTU 7571:2014 *Railway Rolling Stock Permissible exposure norms to the railway track width 1520mm* (Kyiv: The Ministry of Economic Development, Trade and Agriculture of Ukraine)
[3] TsP-0150 2006 *Technical regulations for application of used materials for the permanent track on Ukrainian Railways* (Kyiv: NKTB TsP Ukrzaliznytsia) p 56
[4] Zolotarskiy A, Yevdokimov B, Isaiev N, Krysanov L, Serebrenikov V and Fedulov V 1980 *Reinforced sleepers for rail track* (Moscow: Transport) p 270
[5] Mchedlov-Petrosian O 1966 *Reinforced sleepers (manufacture and maintenance)* (Kharkov: Papers of KhIIT) **86** p 64
[6] Mchedlov-Petrosian O 1969 *Reinforced sleepers (strength and durability)* (Kharkov: Papers of KhIIT) **109** p 64
[7] Mchedlov-Petrosian O 1971 *Technological support in terms of longer life of reinforced sleepers* (Kharkov: Papers of KhIIT) **122** p 64
[8] Rybkin V, Zaiats Yu, Lysniak V and Kovalenko V 2011 *Research into reasons for damage of reinforced sleepers and elaboration of recommendations for higher concrete strength: Research report* (Dnepropetrovsk: DNURT) p 164
[9] Kovalenko V, Kovalenko S and Zaiats Yu 2015 *Research into reasons for premature fraction of reinforced Sh3-type sleepers produced at Reinforced sleeper production plant (Korosten): Research report* (Dnipropetrovsk: DNURT) p 33
[10] Pshinko P 2012 *Higher reliability of reinforced sleepers: Synopsis of Thesis... Candidate of Engineering Sciences. 05.23.01* (Dnipro: DNURT) p 20
[11] Sheinich L, Mykolayets M, Krylov Ye and Mudrik M 2019 *Conduct physical and chemical research of reinforced sleepers and make a conclusion on reasons for corrosion: Research report* (Kiev: SRIBC) p 129
[12] Alekseev S, Ivanov F, Modry S and Shissl P 1990 *Durability of reinforced concrete in aggressive mediums* (Moscow: Stroiizdat) p 320
[13] Stark J and Wicht B 2001 *Dauerhaftigkeit von Beton der Baustoffals Werkstoff* (Berlin: Springer) p 338
[14] Stark J *Alkali-Kieselsäure-Reaktion* 2008 (Veimar: F.A.Finqer Institute fürBaustoffkunde) p 139
[15] Rozental N and Liubarskaya G 2012 *Concrete corrosion under interaction of alkalis with silicon dioxide of aggregate* (Concrete and reinforced concrete) **1**(6) pp 50–60
[16] Plugin A N, Kalinin O A, Plugin A A, Litvinov M I and others 1998 *Research into basic production factors responsible for cement consumption under production of reinforced structures: Research report* (Kharkiv: Kharkiv State Academy of Railway Transport) p 36
[17] Gorodetsky A and Evzerov I 2007 *Computer models of structures* (Kyiv: Fact) p 394
[18] Lobiak A, Plugin A, Kravtsiv L, Kovalova O 2018 Modelling of motorway bridge spans under modernization with consideration of rheological properties of the materials *MATEC Web of Conf.* **234** 04004
[19] DSTU B V.2.7-214:2009 *Construction Materials. Concretes. Methods for determination of strength with reference samples* (Kyiv: Ministry for Communities and Territories Development of Ukraine)

Increasing the safety operation of railway tanks in emergency situations by improving the design of boiler bottom protective elements

Pavol Šťastniak, Juraj Gerlici and Kateryna Kravchenko[1]

University of Žilina, Department of transport and handling machines, 01026, Univerzitná 8215/1, Žilina, Slovak Republic

[1] E-mail: kateryna.kravchenko@fstroj.uniza.sk

Abstract. This article presents an analysis of the problem of reliability of tank boilers in an emergency event. The designs of the boiler bottom protection and the prevention of wagon crawls on top of each other in collisions are evaluated. An analysis of the requirements for the tank protective elements is performed. A technical solution to improve the protective shield of the bottom is developed. The proposed solution will increase rigidity and strength of the protective shield structure, reduce the risk of damage to the railway tank boiler bottom when the protruding elements of the adjacent wagon collide. The work presents the results of the stress-strain state of the proposed design protective shield in the case of rail car collision and, accordingly, the force impact of 150 kN according to the RID TE25 standard.

1. Introduction

Railway transport plays an important role in the social and economic development of countries, as the developed railway systems are a precondition for formational economic growth and improvement of its competitiveness [1]. Accidents occur in rail transport as well as in other modes of transport. Carrying dangerous goods involves the risk of an incident due to the fault of other traffic participants, climatic conditions, badly chosen packaging materials or lack of marking [2-5].

The transportation of dangerous materials is essential to the economy of the countries and to the well-being of its people. Dangerous materials are used in water purification, farming, manufacturing, and other industrial applications. As the need for dangerous substances increases, the flows of dangerous goods also grow.

Railroads carry bulk of dangerous materials annually, including millions of tons of explosive, poisonous, corrosive, flammable, and radioactive materials. The need for dangerous materials to support essential services means that the transportation of highly dangerous materials is unavoidable. Rail transportation of dangerous materials is a safe method for moving large quantities of dangerous materials over long distances. The vast majority of dangerous materials shipped by rail tank car every year arrive safely and without incident, and railroads generally have an outstanding record in moving shipments of dangerous materials safely. However, rail shipments of dangerous materials frequently move through densely populated or environmentally-sensitive areas where the consequence of one incident could be loss of life, serious injury, or significant environmental damage [6-8].

2. Analysis of statistical data on accidents in railway transport

Over the 2012-2016 period there have been, on average, just under 1950 significant accidents each year on the EU railways (figure 1). In these accidents, on average, just under 1050 persons are killed and 850 persons seriously injured each year [9]. According to the statistics of the Eurostat (EuroStat 2020) DATABASE: *Annual number of accidents involving the transport of dangerous goods* (2004-2015), open between 2006 and 2015, the average number of accidents involving the transport of dangerous goods was 44 (figure 2 and 3) [10].

Figure 1 displays a fall of 3.6% in the number of significant accidents between 2012 and 2016. Percentages are expressed as Compound Annual Growth Rates (CAGR). However, in 2016, not all of these Common Safety Indicators (CSIs) improved across Europe.

The estimated economic impacts of these accidents follow a similar pattern. Figure 1 shows that the rate FWSI/Significant Accident has been improving since 2010, although there is considerable variation between the years and an ascending trend is noticed for the period 2014-2016.

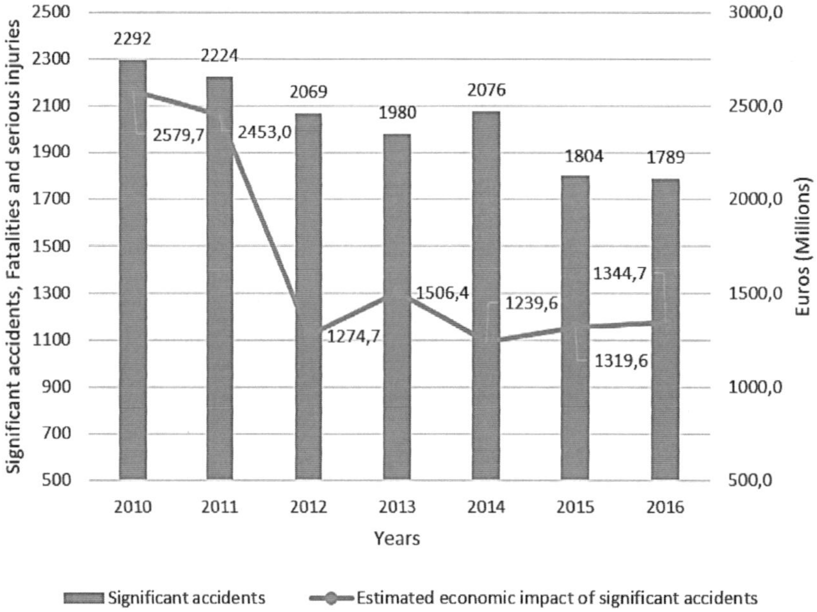

Figure 1. Significant accidents EU 28, 2010-2016.

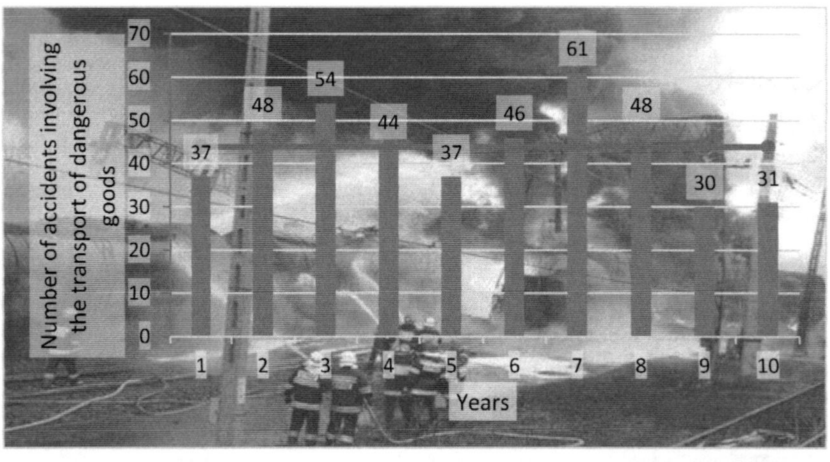

Figure 2. Annual number of accidents involving the transport of dangerous goods on railways EU 28, 2006-2015.

Figure 3. Annual number of accidents involving the transport of dangerous goods on railways, by country, 2004-2015.

Having estimated what threat dangerous substances pose to the environment, individuals and the whole infrastructure of transport, it is essential to guarantee the safety of these processes [11, 12].

In a collision, there may be situations where one of the railway vehicles climbs on top of the other and, through its buffers or automatic coupler, collides with the front surface of the tank cover of the other wagon. In such a collision scenario, the packaging may be torn (figure 4), and the dangerous substance may leak into the air or soil.

a) b)

Figure 4. Damage to the tank bottom when one wagon climbs on top of another a) tank wagons colliding with a track-ram shunting Mussalo Kotka 08.07.2016 [13], b) damage to the end of the tank container in the container wagon (Photo: OTKES) [14].

In the sector of freight carriage by rail, the problem of transporting dangerous goods is found to occur mainly in the area of the structure and the equipment of tank wagons and tanks. In order to mitigate the effects of adverse events, including damage to the bearing structure of vehicle platforms and the transported goods as a result of, for example, travelling at an excessive speed or derailment, the vehicles feature a range of structural solutions and sub-assemblies classified as passive safety mechanisms [16-20]. Tank wagons used for transport of dangerous materials can be made more resistant to damage in accidents through the use of a thicker steel tank and other protective features [15, 21].

3. Classification of structures to protect the end of the rail tank car boiler from collision damage

A common scenario of a rail accident involves one railway vehicle running into another at an excessive speed, which leads to their structures becoming deformed and – quite frequently – their buffers missing each other in the vertical plane, which results in neighbouring vehicles climb one on top of the other. To prevent this highly harmful phenomenon, leading to, among others, leakage of the tank of a wagon carrying dangerous goods, freight wagons are constructed in a way to feature special solutions preventing the occurrence of climbing [16]. They are mounted on the wagon front usually. These devices are generally called over-buffering protection devices. Another type of protection is known by the term protective shield, which serves as a barrier to the tank body.

Over-buffer protection and a protective shield serve to reduce damage in the event of a railway vehicle crash. The purpose of these devices is to eliminate, as far as possible, damage caused by a collision. Both devices are idle in normal operation, they are activated only when the railway vehicles move towards each other in the vertical direction.

The design of the protective elements of both the over-buffering protection device and the protective shield is very different. These elements work only in emergency situations, therefore, these structures must have a minimum weight and a sufficiently simple structure for easy disassembly.

The significant variety of designs can be divided into the following groups (figure 5):

a) Over-buffering protection:
- permanently installed:
  – made in the form of a cone [European patent EP 2033868A1];
  – rectangular shape [European patent EP2808222A1];
  – with grate or teeth [European patent EP 2727792B1, European patent EP 3530544A1];
- triggered in a collision [Anti-climbing device 2020 Innova Systems And Technologies].

b) Protective shield:
- made with repetition of the tank profile [Patent US4466356A];
- average height [Patent US3994239A];
- high [European patent EP0816197A2];
- with shockproof materials (e.g. elastomer) [Patent DE19833935A1].

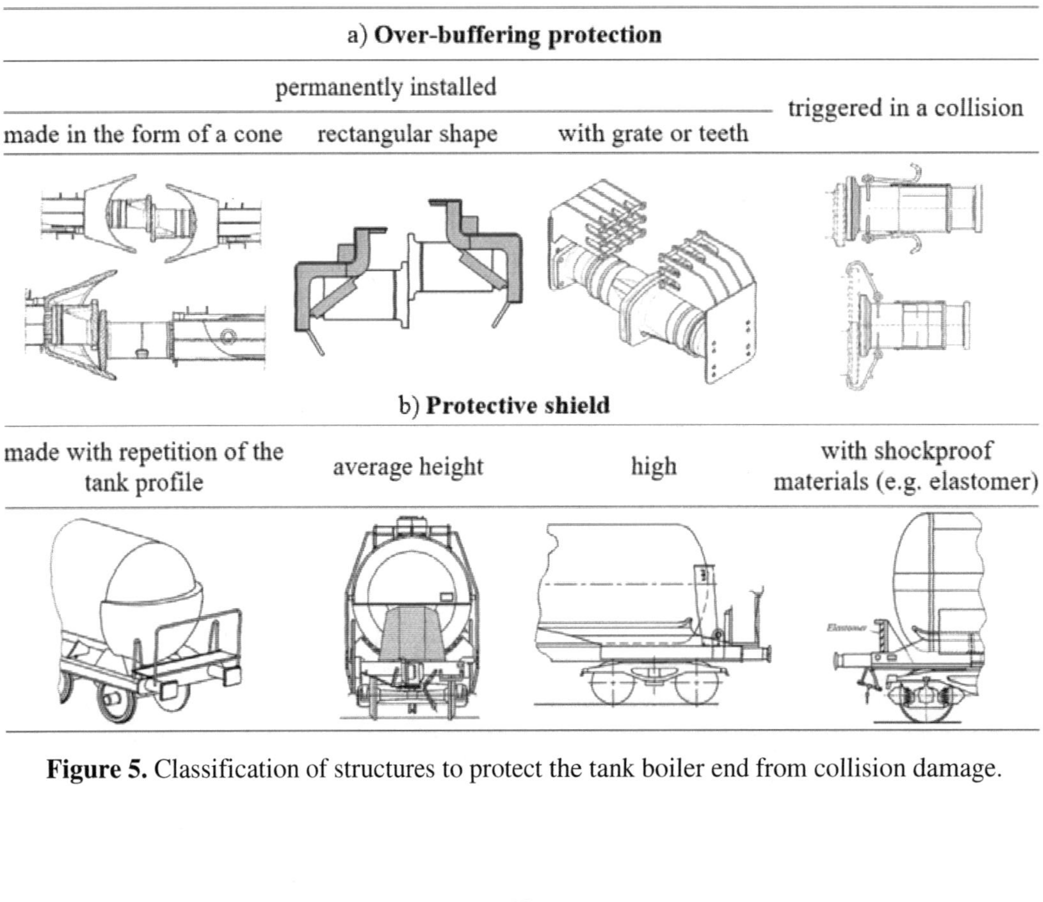

Figure 5. Classification of structures to protect the tank boiler end from collision damage.

A protective shield is a device used to protect the tank of a tank-wagon that transports dangerous goods. The shield is located at the ends of the railway vehicle above the main frame. It protects the tank bottom from punctures that could occur as a result of railway vehicles climbing on top of each other, while it is active only in the event of failure of the over-buffering protection, or in its absence. The tank ends are usually still perforated to increase safety.

4. Basic requirements for the design of the protective shield

The tanks of wagons must be protected against stacking and derailment in the event of an accident or at least limit damage to the tank from protruding structural elements (buffers) of an adjacent wagon. The requirements for the protective shield, a measure to limit damage when jumping over buffers are defined in the requirements of the Code for the International Carriage of Dangerous Goods by Rail (RID) in Chapter TE25 [15].

According to the requirements of RID in Chapter TE25, the measures to limit damage in the event of an accident are:

- Use of a protective shield at each wagon end. The protective shield must cover the relevant width of the tank in each case up to its appropriate height. In addition, the width of the protective shield must be at least as great along the entire height of the shield as the distance defined by the outer edge of the buffer plates. The protective shield must, at a height measured from the upper edge of the buffer support, cover either 2/3 of the tank diameter or at least 900 mm and must additionally be fitted with a device at the upper edge to catch the rising buffers. The wall thickness of the protective shield must be at least 6 mm. Lastly, the shield and its attachment points must be designed in such a way as to minimize the possibility of the bottom to be punctured by the shield.
- Protective shield at both ends of wagons equipped with automatic couplings. If a protective shield is used at both ends, it must cover the tank end to a height of at least 1100 mm, measured from the upper edge of the buffer plate, the couplings must be fitted with anti-slip devices to prevent unintentional disconnection and the protective shield must be at least 1200 mm wide. The protective shield must have walls at least 12 mm thick and its points of attachment must be such as to minimize the possibility of the tank ends piercing the protective shield itself.

5. Technical proposal for a protective shield design

on the basis of the analysis of the design of protective shields, a promising design of the protective shield has been developed. The design takes into account the requirements of the rail tank car manufacturer, as well as RID rules. An application for an invention (a2020 03605 from 16.06.2020 UKRPATENT) and utility model (u2020 03804 from 24.06.2020 UKRPATENT) has been filed for this technical solution.

The method of fixing this structure is based on the use of two consoles 13. To ensure the best strength properties of the consoles 13 and protective shields of the rail tank car, it is advisable to place the consoles 13 around the axis of the buffers. Console 13 is attached to the rail tank car by welding in two places, by means of one of the transverse plates 1 (figure 6) with the frame of the rail tank car and by means of the main vertical stiffening plate 2 with the bearing structure of the buffer. The second transverse plate 1 connects the main vertical stiffening plate 2 with the auxiliary vertical stiffening plate 3.

Depending on the design of the rail tank car, the thickness of the console elements may vary. For example, for tank truck Za (c) ns 98 m^3 (Tatravagónka Poprad) the best thickness is 10 mm.

The height of the protective shield of the rail tank car above the upper edge of the buffer must be not less than 900 mm. The width of the protective shield of the rail tank car, as well as the height are subject to the rules of RID TE25. Accordingly, the width should be at least such as the distance between the outer surfaces of the buffer plates.

The design of the protective shield of the rail tank car must have the lowest possible weight. Therefore, according to the strength calculations, the lowest weight of the structure for the rail tank car Za (c) ns 98 m^3, will be achieved when the thickness of the metal sheet 4 is 6 mm.

To ensure high stiffness of the structure in the longitudinal direction, metal sheet 4 is made of a peak shape and cut in the shape of a trapezoid. Several stiffeners are placed on the front side of the protective shield of the rail tank car.

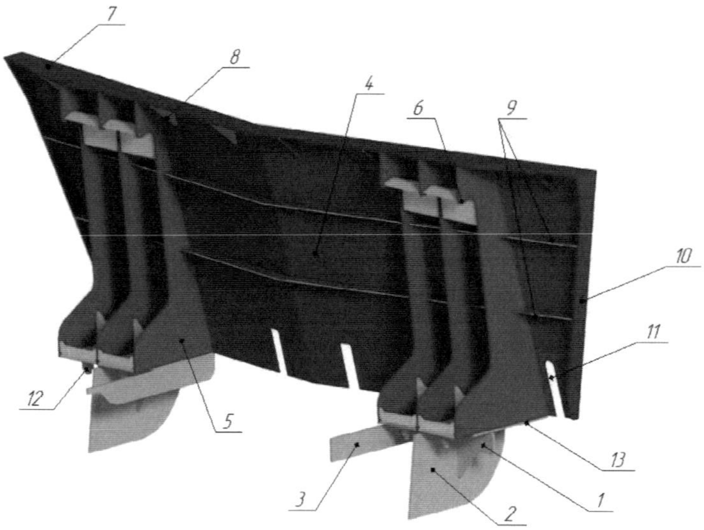

Figure 6. The protective shield design.

The main reinforcement of the structure are three vertical stiffeners 5, which are located opposite the buffers on the left and right hand sides of the trapezoid of metal sheet 4 and in the upper part have gripping elements 6. In addition, the vertical stiffeners 5 are arranged so that they are directly connected to the main vertical stiffening plate 2 of console 13 and the frame of the rail tank car. In the event of an accident, the buffers of the adjacent rail car rise above the frame of the rail tank car and must be captured by the gripping elements 6, as shown in figure 7.

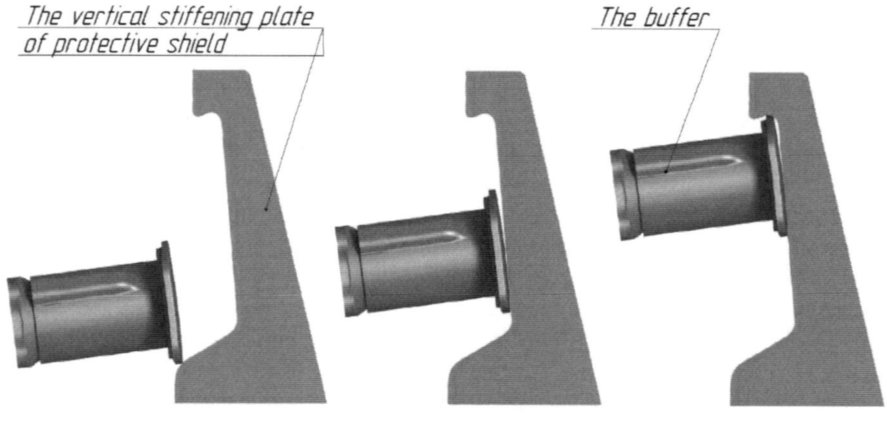

Figure 7. The process of capturing the buffers of adjacent rail car.

The connection of the vertical stiffeners 5 is realized, in addition to the metal sheet 4, by joints at three levels of height. The first connection is located on the upper edge of the protective shield of the rail tank car and is realized by means of a plate 7, which is placed on its entire upper part and fixed on a metal sheet 4 by means of triangles 8. The second connection is in the area of the rail car buffer capture. The last connection is at the bottom in the connection area with the consoles 13.

To strengthen the structure of the protective shield of the rail tank car on its outer surface horizontal ribs 9 in two rows and side vertical ribs 10 are fixed on the edge of the protective shield of the rail tank car along its entire height, in the lower part of metal sheet 4 there are holes 11 which are parts of the frame rail tank car.

When two cars collide (figure 8 a, b), one of them will begin to rise above the second (figure 8, c). This will lead to the capture of buffers of one of the rail cars by the elements of the second rail car protective shield (figure 8, c, d). In this situation, it is possible that there will be a plastic deformation of the protective shield. At the rear, the protective shield is pressed against the rail tank car bottom, so that the force from the buffer plates is decomposed into a larger plane. This will preserve the integrity of the rail tank car bottom.

When the rail cars collide, one of the rail cars will start approaching the other and the buffers of the oncoming rail car will be captured by the gripping elements 6 of the rail tank car protective shield, there will be forces that will be transmitted to the rail tank car frame through screws 12 in the holes.

Material EN S 355 can be used for the construction of the rail tank car protective shield withinsignificant economic costs and sufficient yield strength for this design. In addition, it should be borne in mind that in the event of a rail car collision, the protective screen may be destroyed and will interact with the rail tank car bottom.

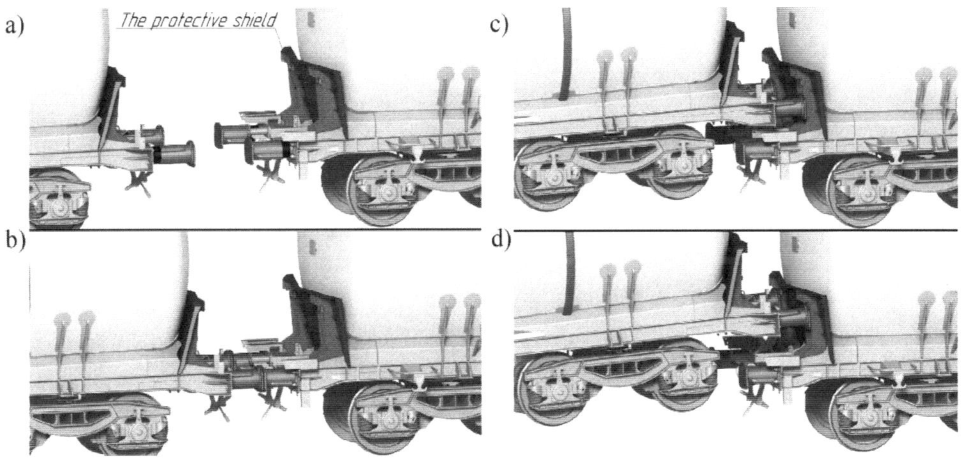

Figure 8. The collision.

6. Strength analysis of the protective shield

Strength analysis was performed in ANSYS 19.2. The shield geometry was inserted into a numerical program, which was created as an assembly in the Catia V5R20 program.

In accordance with the dimensions of the construction, the mesh size was 10 mm. Based on the requirements of the protective shield, the main stressed part is the vertical stiffening ribs. The load from the rising buffers is applied to these stiffening ribs and at the same time these reinforcements reach up to the places where the shield is attached to the rail car frame. For this reason, the mesh size is 6 mm on the given parts of the structure.

When selecting the boundary conditions, the shield was fixed behind the elements located around the screw holes (figure 9). A force of 150 kN was applied to the edges of the vertical stiffening ribs catching the buffers.

Stresses distribution in the construction is most expressive in the ribs of rigidity. Critical stress points are located in the upper parts of the stiffening ribs catching the buffers. The maximum stress value is 342.27 MPa (figure 10).

To determine the stresses at the welds, the analysis of the elements around the buffer catch is given. The maximum stress value in the welds between the stiffening ribs and their reinforcing plates is 242.2 MPa (figure 11).

In terms of capturing the buffers, the resulting deformations of the entire structure are the maximum displacement. The maximum analysed displacement is 2.15 mm. The place of occurrence of the largest displacement is on the upper edge of the structure. The displacement in the vertical direction did not exceed 1.19 mm and in the longitudinal direction of the rail car 1.80 mm. The distribution of the total deformation construction is shown in figure 12.

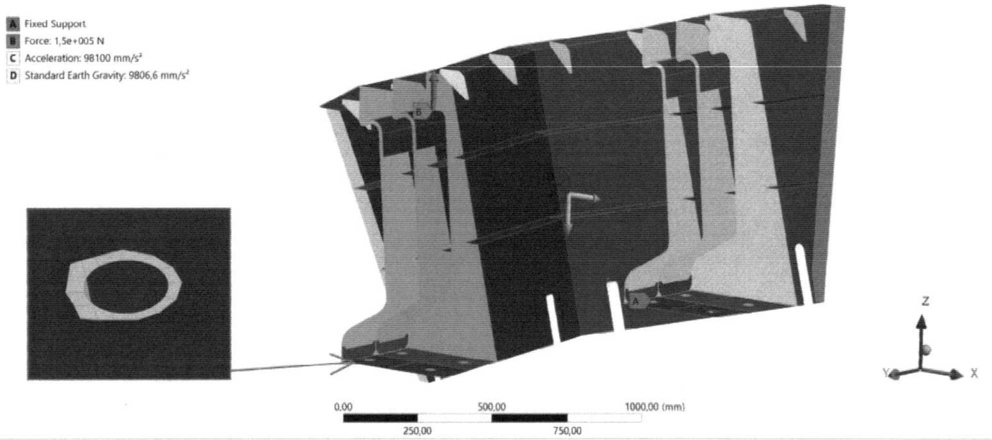

Figure 9. Boundary conditions of the protective shield.

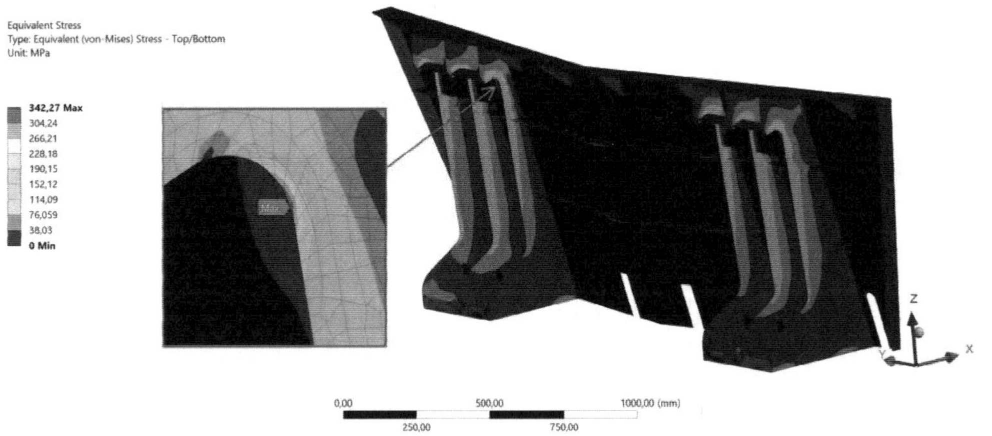

Figure 10. Protective shield stresses analysis.

The most important result from the analysis is the distribution of the stress. The calculation revealed that the maximum stress was 342.27 MPa. This value does not exceed the yield strength of the EN S355 material used. Due to the construction of welded joints, the stresses in the most stressed welds are verified. The most critical welded joints were the places where the main stiffening ribs were connected with their auxiliary bent plates, i.e. the place where the rising buffers were caught. The critical stress in the given welds is 242.2 MPa. This means that the yield strength of the welded structure for EN S355 was not exceeded in the welds.

Another important finding is the relatively low overall deformation of the structure. From the displacement results, a maximum of 2.15 mm was recorded at the upper edge of the protective device.

The largest displacement in the vertical direction is 1.19 mm. In the longitudinal direction of the rail vehicle, the maximum displacement is 1.80 mm. The given displacement results indicate the largest deviations of the whole structure. Due to its dimensions, the deformations are low enough.

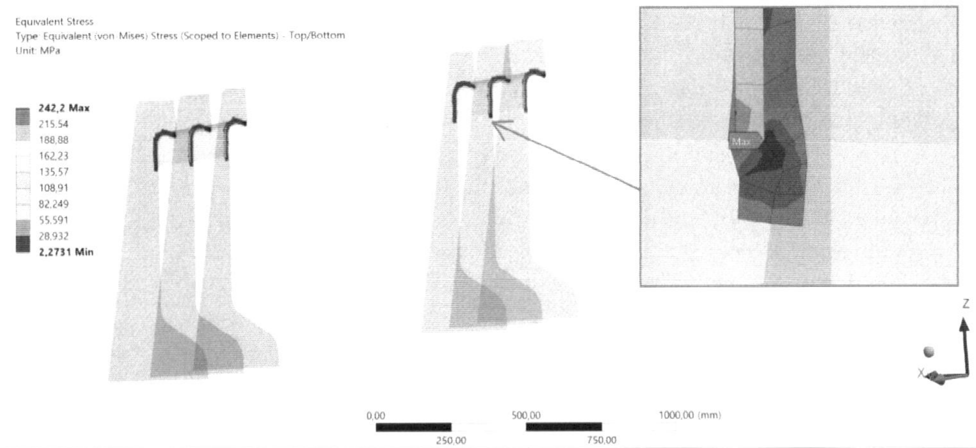

Figure 11. Stress in the area of the most stressed welds.

Under the loads specified by the manufacturer, the protective shield is sufficiently rigid and no plastic deformations occur in it. The deformation values of the structure are relatively small. The ability of the shield to catch rising buffers is not significantly affected by this load. The final weight of the protection shield is 234.95 kg.

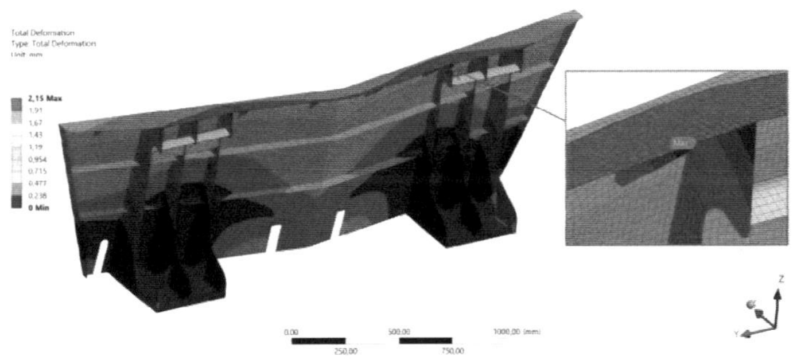

Figure 12. Deformation of the protective shield.

7. Conclusion

The analysis of emergency situations in railway transport showed the need to develop new designs that protect the rail tank body from damage during collisions. For this, existing designs for protecting the bottom of rail tanks from damage are evaluated. On the basis of the requirements of regulatory documents and manufacturers of railway cars, the design of a protective shield for the end part of railway tanks is proposed. The protective shield is made of separate pieces of sheet metal. Its construction is designed to protect the tank bottom as effectively as possible.

The calculations showed the reliability of the structure under the action of 150 kN forces, specified in requirements RID TE25. Critical stress points are located in the upper parts of the stiffening ribs catching the buffers. The maximum stress value is 342.27 MPa. The final weight of the protection shield is 234.95 kg.

The developed design of the protective screen makes it possible:

- to reduce the risk of damage to the rail tank car bottom in the event of collision of the protruding elements of the adjacent car, which may occur in accidents;
- to provide high rigidity of the protective shield in the longitudinal direction due to the use of the peak form of the rail tank car protective shield;
- to provide durability of the protective shield of the rail tank car due to use of stiffening ribs and strengthening elements of the design (triangles, plates);
- to be used for different designs of rail tank cars in connection with the use of special holes for placement of rail car elements (for example, vehicle frames);
- to ensure simple production of the protective shield and installation on the rail tank car, due to the fact that the shape of individual parts of the structure is relatively simple;
- to ensure the integrity of the rail tank car bottom due to the fact that all stiffeners and reinforcing structural elements are placed on the front side of the rail tank car protective shield, respectively, the rear side of the protective shield is made without irregularities that could damage the rail tank car in case of an accident.

Acknowledgments
This publication was made thanks to the support within Operational Program Research and Innovation for project: "New generation of freight railway wagons" (Project code in ITMS2014+: 313010P922) co-financed from the resources of the European Regional Development Fund.

References

[1] Fomin O V, Burlutsky O V and Fomina Yu V 2015 Development and application of cataloging in structural design of freight car building *Scientific and Technical J. Metallurgical and Mining Industry* **2** (Dnipro: Ukrmetallurginform STA Ltd) pp 250–6
[2] Kelrykh M et al 2019 Features of tank car testing for dangerous cargoes transportation *IOP Conf. Ser.: Mater. Sci. Eng.* **659** 012055
[3] Bagheri M, Saccomanno F, Chenouri S and Fu L 2011 Reducing the threat of in-transit derailments involving dangerous goods through effective placement along the train consist *Accident Analysis and Prevention* **43**(3) (Elsevier) pp 613–20
[4] Das A, Mazumdera T N and Gupta A K 2012 Pareto frontier analyses based decision making tool for transportation of hazardous waste *J. of Hazardous Materials* **227-228** (Elsevier) pp 341–52
[5] Huang W, Zhang R, Xu M, Yu Y, Xu Y and De Dieu G J 2020 Risk state changes analysis of railway dangerous goods transportation system: Based on the cusp catastrophe model *Reliability Engineering & System Safety* **202** (Elsevier) 107059
[6] Batarlienė N and Jarašūnienė A 2014 Analysis of the accidents and incidents occurring during the transportation of dangerous goods by railway transport *Transport* **29**(4) (Taylor & Francis) pp 395–400
[7] Gheorghe A V, Birchmeier J, Vamanu D, Papazoglou I and Kröger W 2005 Comprehensive risk assessment for rail transportation of dangerous goods: A validated platform for decision support *Reliability Engineering & System Safety* **88**(3) (Elsevier) pp 247–72
[8] Vaezi A and Verma M 2017 An analytics approach to dis-aggregate national freight data to estimate hazmat traffic on rail-links and at rail-yards in Canada *J. of Rail Transport Planning & Management* **7**(4) (Elsevier) pp 291–307
[9] *Report on Railway Safety and Interoperability in the EU 2018* (Luxembourg: Publications Office of the European Union) p 132
[10] Smit C, Katuin A and Pruis G 2014 Cost/benefit analysis for proposal on crash buffers for tank-wagons intended for the carriage of dangerous goods *3rd Session of the RID Committee of Experts' Standing Working Group Intergovernmental Organisation for Int. Carriage by Rail Inf* **6** p 38
[11] Kelrykh M and Fomin O 2014 Perspective directions of planning carrying systems of gondolas *Scientific and Technical J. Metallurgical and Mining Industry* **6** (Dnipro: Ukrmetallurginform STA Ltd) pp 64–7

[12] Fomin O V 2015 Increase of the freight wagons ideality degree and prognostication of their evolution stages *Naukovyi Visnyk Natsionalnoho Hirnychoho Universytetu* **2** (National Mining University of Ukraine) pp 68–76
[13] *Collision of tank wagons with a rail buffer during replacement work in Mussalo* Kotka 8 July 2016 Investigation ID: R2016-03 Investigation report 3/2017
[14] *Collision of a shift work unit with stationary wagons in Kouvola on 21 September 2017* Investigation ID: R2017-02 Investigation report 6/2018
[15] *Convention concerning international carriage by rail (COTIF)* 2019 Appendix C – Regulations concerning the international carriage of dangerous goods by rail (RID) p 1082
[16] Grzywna M and Rasiński T 2019 Passive safety mechanisms in freight wagons *Problemy Kolejnictwa* **183** (Instytut Kolejnictwa) pp 113–9
[17] Vatulia G, Falendysh A, Orel Y and Pavliuchenkov M 2017 Structural improvements in a tank wagon with modern software packages *Procedia Engineering* **187** (Elsevier) pp 301–7
[18] Jeong D Y, Tang Y H and Perlman A B 2011 Semi-analytical approach to estimate railroad tank car shell puncture *Proc. of the ASME/ASCE/IEEE Joint Rail Conf. (Colorado: Cambridge)* 89457
[19] Kirkpatrick S W and Phan V 2012 Lading effects on tank car impact response *Proc. of the 2012 ASME Rail Transport Division Fall Technical Conf. (Omaha, NE)* 98557
[20] Barkan C P L, Rapik Saat M R, González III F and Treichel T T Cooperative research in tank car safety design *TR News* **286** (United States) pp 12–6
[21] Barkan C P L 2008 Improving the design of higher-capacity railway tank cars for hazardous materials transport: Optimizing the trade-off between weight and safety *J. of Hazardous Materials* **160**(1) (Elsevier) pp 122–34

Loading on the carrying structure of a flat wagon when transporting firing military equipment

O Fomin[1], G Vatulia[2] and A Lovska[3,4]

[1] Department of Cars and Carriage Facilities, State University of Infrastructure and Technologies, Kyrylivska str., 9, Kyiv, Ukraine, 04071
[2] Department of Structural Mechanics and Hydraulics, Ukrainian State University of Railway Transport, Feuerbach sq., 7, Kharkiv, Ukraine, 61050
[3] Department of Wagons, Ukrainian State University of Railway Transport, Feuerbach sq., 7, Kharkiv, Ukraine, 61050

[4] E-mail: alyonalovskaya.vagons@gmail.com

Abstract. The study deals with research through mathematic modeling of the dynamic loading on the carrying structure of a flat wagon in transportation of military equipment conducting fire. A 13-401 flat wagon loaded with a tank with combat load of 55 tons was taken as the prototype. The mathematical model was solved in MathCad software. It was established that the maximum acceleration on the carrying structure of a flat wagon in the horizontal plane was about 40 m/sec^2, and in the vertical plane – about 10 m/sec^2. The values obtained were considered in strength calculation for a flat wagon. The calculation was made using the finite element method in CosmosWorks software. It was established that the maximum equivalent stresses in the carrying structure of a flat wagon exceeded the admissible values twice. The results of the study will encourage engineers to design carrying structures of flat wagons intended for transporting military equipment firing in motion

1. Introduction

Prospects in the technical and economic development of European countries, as well as enhancing their national defense and security require higher efficiency of the rolling stock of railway transport, as a leading industry of the transportation system. Therefore, there is a need to implement highly efficient interoperable railway equipment in operation. In order to ensure their defense prospective European countries have to solve the problem of formation of special fleet of wagons to transport military equipment.

It is known that one of the most widely spread types of wagons for transportation of military equipment are flat wagons (figure 1 [1, 2]).

There are several ways to fix military equipment on a flat wagon:

- standard multiple-thread fixtures (1st method);
- metal lugs (2nd method);
- wooden support beams and bracing (standard) wires (3rd method);
- metal support brake shoes and wooden pads (4th method);
- wooden support beams and pads (5th method);
- tension cables and screw cramps (6th method).

All those methods, except the 5th one, are used for transporting equipment as part of military trains, including guarded ones [3]. Special railway cars for transportation of military equipment and research into the possibility to fire in motion are of considerable importance for ensuring national defense.

Figure 1. Flat wagons loaded with military equipment.

Calculation of capacity indexes for a flat wagon under transportation of containers and loading/unloading with the ACTS system was conducted in [4, 5].The strength calculation was conducted in the static setting in Nastran software. Numerical values of the designed loads on a flat wagon were taken according to the PNEN12663 and BN – 77/3532 – 40 Standards.

The structural features of a flat wagon for intermodal transportation are presented in [6]. The study outlines the general technological requirements in terms of organization of intermodal transportation and their advantages. Study [7] substantiates the efficiency of flat wagons for transportation of containers, including tank containers produced by Transmash (Russia).The loading capacity of a flat wagon is 73 tons and it can transport 1CC, 1C and 1CX containers. However, they do not study the possibility to transport military equipment firing in motion by flat wagons.

Research into the dynamics of a rail wagon with open charging floors is given in [8].The calculation is made in MSC Adams software. The study deals with research into the stability of a wagon against swinging-over along 250-m curves at different traffic speeds.

Dynamic loading on a flat wagon transporting military equipment is not considered.

Dynamic loading on a flat wagon in operational modes is defined in [9, 10].The studies present mathematical models which characterize the most unfavorable loading modes of carrying structures of flat wagons. However, loading on wagons, while transporting military equipment, is not determined in the studies.

The objective of the study is research into loading on the carrying structure of a flat wagon transporting firing military equipment conducting fire. To achieve the objective the following tasks were set:

- mathematical modeling of dynamic loads on the carrying structure of a flat wagon transporting firing military equipment;
- strength characteristics of the carrying structure of a flat wagon transporting firing military equipment.

2. Mathematical modeling of dynamic loads on the carrying structure of a flat wagon with a tank firing from it

A mathematical model was built to study the possibility to transport a military tank and conduct fire in motion. The study used the mathematical model given in [11].The model was improved with an additional degree of freedom – oscillations in the longitudinal plane. Therefore, the study deals with the movement of a flat wagon loaded with a military tank in longitudinal and vertical planes. The design diagram is presented in figure 2.The calculation was made in the plane coordinates. The firing process while the wagon was passing over a joint track irregularity has been considered.

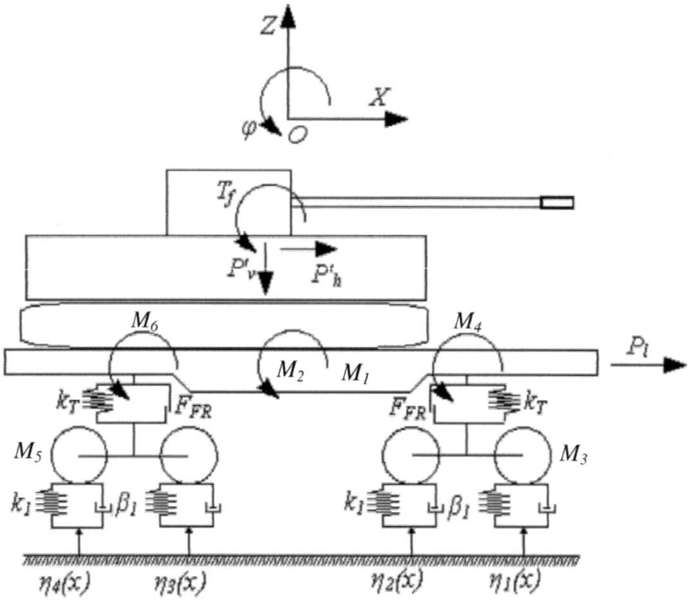

Figure 2. Design diagram of a flat wagon loaded with a military tank.

The equation of motion for the design model is as follows:

$$M_1 \frac{d^2}{dt^2} q_1 + M' \frac{d^2}{dt^2} q_3 = P_l + P'_h; \tag{1}$$

$$M_1 \frac{d^2}{dt^2} q_2 + C_{2,2} q_2 + C_{2,5} q_5 + C_{2,8} q_8 = -F_{FR}\left(sign\left(\frac{d}{dt}\delta_1\right) + sign\left(\frac{d}{dt}\delta_2\right)\right) + P'_v; \tag{2}$$

$$M_2 \frac{d^2}{dt^2} q_3 + C_{3,3} q_3 + C_{3,5} q_5 + C_{3,8} q_8 = F_{FR} l\left(sign\left(\frac{d}{dt}\delta_1\right) + sign\left(\frac{d}{dt}\delta_2\right)\right) + T_f; \tag{3}$$

$$M_3 \frac{d^2}{dt^2} q_4 = 0; \tag{4}$$

$$M_3 \frac{d^2}{dt^2} q_5 + C_{5,2} q_2 + C_{5,3} q_3 + C_{5,5} q_5 + B_{5,5} \frac{d}{dt} q_5 = F_{FR} sign\left(\frac{d}{dt}\delta_1\right) + k_1(\eta_1 + \eta_2) + \beta_1\left(\frac{d}{dt}\eta_1 + \frac{d}{dt}\eta_2\right); \tag{5}$$

$$M_4 \frac{d^2}{dt^2} q_6 + C_{6,6} q_6 + B_{6,6} \frac{d}{dt} q_6 = -k_1(\eta_1 - \eta_2) - \beta_1 a\left(\frac{d}{dt}\eta_1 - \frac{d}{dt}\eta_2\right); \tag{6}$$

$$M_5 \frac{d^2}{dt^2} q_7 = 0; \tag{7}$$

$$\begin{aligned} M_5 \frac{d^2}{dt^2} q_8 + C_{8,2} q_2 + C_{8,3} q_3 + C_{8,8} q_8 + B_{8,8} \frac{d}{dt} q_8 \\ = F_{FR} sign\left(\frac{d}{dt}\delta_2\right) + k_1(\eta_3 + \eta_4) + \beta_1\left(\frac{d}{dt}\eta_3 + \frac{d}{dt}\eta_4\right); \end{aligned} \tag{8}$$

$$M_6 \frac{d^2}{dt^2} q_9 + C_{9,9} q_9 + B_{9,9} \frac{d}{dt} q_9 = -k_1 a(\eta_3 - \eta_4) - \beta_1 a\left(\frac{d}{dt}\eta_3 - \frac{d}{dt}\eta_4\right), \tag{9}$$

where M_1, M_2 – weight and moment of inertia of the flat wagon supporting structure respectively; M_3, M_4 – weight and moment of inertia of the first bogie in the direction of travel respectively; M_5, M_6 – weight and moment of inertia of the second bogie in the direction of travel respectively; C_{ij} – elasticity characteristic of the vibration system elements; B_{ij} – pattern density function; a – half bogie wheelbase; q_i – specified coordinates corresponding to translational displacements relative to the longitudinal and vertical axes and angular displacements around the vertical axis; k_T – spring suspension rigidity; k_1 – rigidity of track; β_1 – damping factor; $\eta_i(x)$ – function that describes the track irregularity; δ_i – deformation of elastic elements of spring suspension; F_{FR} – absolute friction force in a spring complex; P_l – longitudinal loading on the automatic coupling supports; P_v', P_h' – loads on the carrying structure in vertical and horizontal planes in firing, respectively; T_f– moment on the carrying structure of a flat wagon with a tank firing from it.

Here the first equation characterizes a longitudinal displacement of the body, the second one – a vertical displacement of the body, the third – an angular, the forth, the fifth, and the sixth equations – displacements of the first body, respectively, and the seventh, eighth and ninth – displacements of the second body.

The matrix of elastic coefficients looks like:

$$C = \begin{vmatrix} 0 & 0 & 0 & 0 & 0 & 0 & 0 & 0 & 0 \\ 0 & 2k_T & 0 & 0 & -k_T & 0 & 0 & -k_T & 0 \\ 0 & 0 & 2k_T l^2 & 0 & k_T l & 0 & 0 & k_T l & 0 \\ 0 & 0 & 0 & 0 & 0 & 0 & 0 & 0 & 0 \\ 0 & -k_T & k_T l & 0 & k_T + 2k_1 & 0 & 0 & 0 & 0 \\ 0 & 0 & 0 & 0 & 0 & 2k_1 a^2 & 0 & 0 & 0 \\ 0 & 0 & 0 & 0 & 0 & 0 & 0 & 0 & 0 \\ 0 & -k_T & k_T l & 0 & 0 & 0 & 0 & k_T + 2k_1 & 0 \\ 0 & 0 & 0 & 0 & 0 & 0 & 0 & 0 & 2k_1 a^2 \end{vmatrix} \tag{10}$$

and the matrix of dissipative coefficients has the form:

$$B = \begin{vmatrix} 0 & 0 & 0 & 0 & 0 & 0 & 0 & 0 & 0 \\ 0 & 0 & 0 & 0 & 0 & 0 & 0 & 0 & 0 \\ 0 & 0 & 0 & 0 & 0 & 0 & 0 & 0 & 0 \\ 0 & 0 & 0 & 0 & 0 & 0 & 0 & 0 & 0 \\ 0 & 0 & 0 & 0 & 2\beta_1 & 0 & 0 & 0 & 0 \\ 0 & 0 & 0 & 0 & 0 & 2\beta_1 & 0 & 0 & 0 \\ 0 & 0 & 0 & 0 & 0 & 0 & 0 & 0 & 0 \\ 0 & 0 & 0 & 0 & 0 & 0 & 0 & 2\beta_1 & 0 \\ 0 & 0 & 0 & 0 & 0 & 0 & 0 & 0 & 2\beta_1 \end{vmatrix}. \tag{11}$$

The initial displacement and speeds were taken to be equal to zero. The input parameters of the model were technical characteristics of the carrying structure of a flat wagon, spring suspension, perturbation action, and engineering characteristics of a tank (table 1). The duration of the shock pulse action on a wagon from rail irregularity was taken to be equal to 0.1 sec at the speed 80 km/h. The calculation considered spring suspension parameters of an 18-100 bogie. There is a rule in artillery science that 3% out of the muzzle energy of a missile is consumed by the mechanical recoil energy [12].Therefore, on the basis of the work the authors obtained the loading on the carrying structure of a flat wagon transporting a tank firing from it. The rate of fire of a tank was taken to be 12 rds/min. The model considered one round in 5 sec.

Table 1. The input parameters of the model.

Name of the parameter	Value
CARRYING STRUCTURE OF A FLAT WAGON	
mass, t	11.5
moment of inertia, t·m^2	198.2
half base, m	4.86
BOGIE	
mass, t	4.3
moment of inertia, t·m^2	3.0
half base, m	0.925
rigidity of spring suspension of one bogie, kN/m	8000
coefficient of relative friction	0.1
TRACK CHARACTERISTICS	
damping factor, kN·sec/m	200
rigidity, kN/m	100000
amplitude of inequality, m	0.01
distance between the inequalities, m	25
TANK	
mass, t	55
length of the body with a gun, m	10.8
muzzle energy, mJ	24

Differential equations (1) – (9) were reduced to standard Cauchy problems. Then they were integrated by the Runge-Kutta method [13, 14]. On the basis of the calculation the authors defined the accelerations on a flat wagon during fire from the tank at various fire angles in horizontal and vertical planes. Besides, they considered the most unfavorable loading on the carrying structure of a flat wagon, particularly, movement over a joint track irregularity with simultaneous firing. The peak value of acceleration was about 40 m/sec^2. The maximum acceleration affected the carrying structure of a flat wagon at a fire angle of 30°. The pattern of change in accelerations on the carrying structure of a flat wagon at a fire angle of 30° is given in figures 3 and 4.

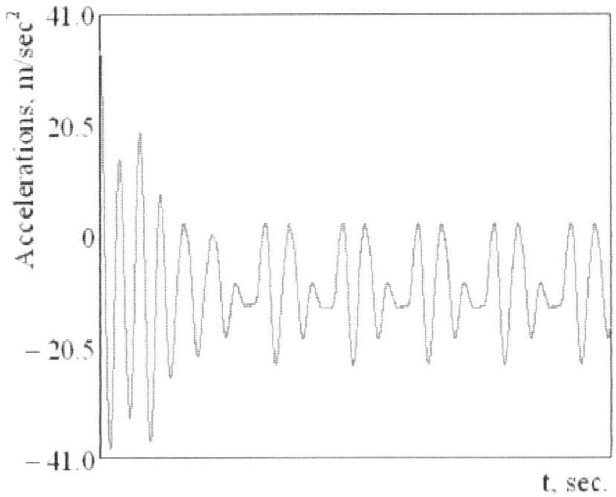

Figure 3. Accelerations in the mass center of the carrying structure of a flat wagon (the gun is at an angle of 30° relative to the longitudinal axle of a wagon) in the horizontal plane.

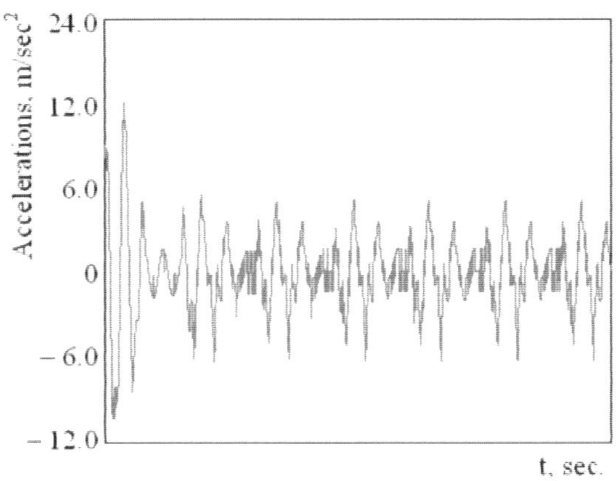

Figure 4. Accelerations in the mass center of the carrying structure of a flat wagon (the gun is at an angle of 30° relative to the longitudinal axle of a wagon) in the vertical plane.

3. Strength characteristics of the carrying structure of a flat wagon with a tank firing

The accelerations obtained were taken into account when determining strength values of the carrying structure of a flat wagon. The calculation was made for a 13-401 flat wagon. For modeling loads on the carrying structure of a flat wagon from a tank, the flat wagon was equipped with pads imitating the support part of caterpillar tracks (figure 5). The calculation was conducted with the finite element method in CosmosWorks software (figure 6). Spatial tetrahedrons were used as finite elements for the continuous model. The optimal number of elements was calculated with the graphical analytical method [15, 16]. The model consisted of 114,459 units and 353,604 elements. The maximum size of an element was 235.62 mm, and the minimum – 47.12 mm. The percentage of elements with a side ratio less than three was 24.7, and more than ten – 31. The minimum number of elements in the circle was 9, and a gain ratio of the size of an element was 1.8.

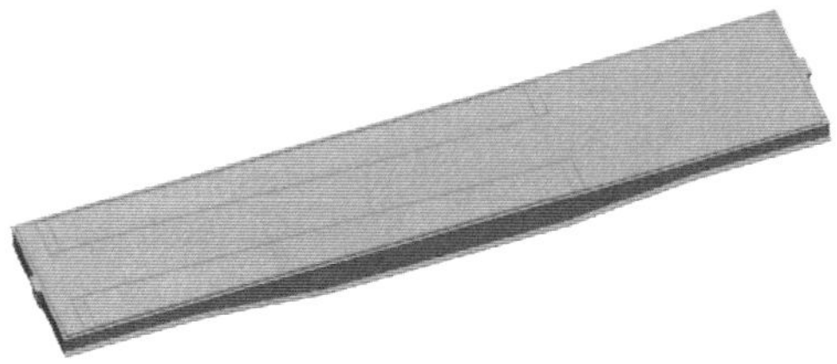

Figure 5. Spatial model of the carrying structure of a flat wagon.

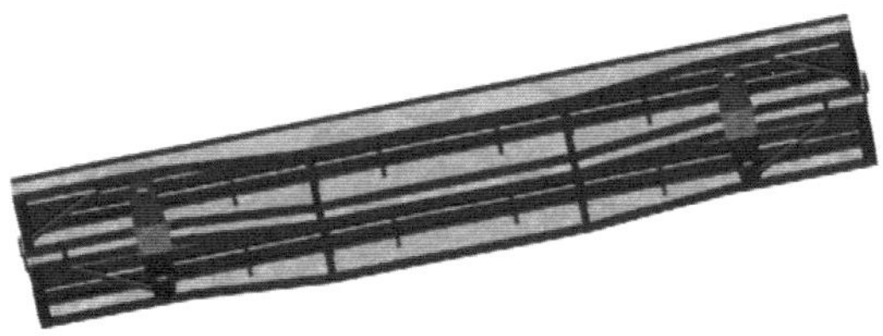

Figure 6. Finite element model of the carrying structure of a flat wagon.

The design diagram of the carrying structure of a flat wagon is given in figure 7. The design diagram takes into account that the longitudinal loading P_l 2.5 MN impacted the frontal supports of the automatic coupling device. The vertical load P_v from the gross weight of the tank and load P_f in the fixation areas impacted the carrying structure of a flat wagon. Due to the angular location of fixation elements, load P_f was decomposed. The model was fixed in the areas where it rested on the gear parts. Steel 09G2S was used as structural material for a flat wagon.

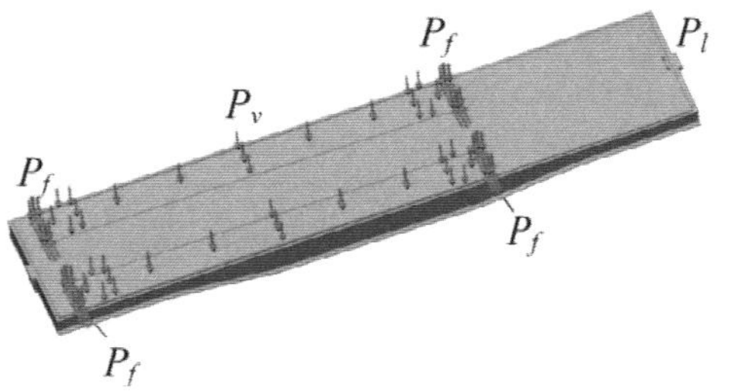

Figure 7. Design diagram of the carrying structure of a flat wagon.

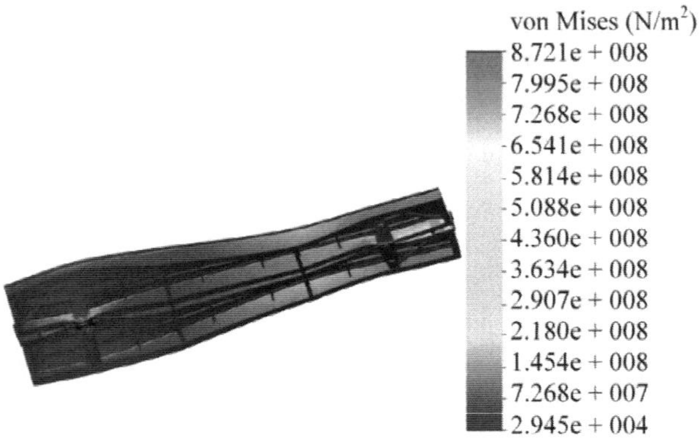

Figure 8. Stress strain behavior of the carrying structure of a flat wagon (deformation scale 50:1).

On the basis of the calculation the authors defined the basic strength parameters of the carrying structure of a flat wagon. The maximum equivalent stresses emerged in the areas where the carrying structure of the flat wagon rested on the bogies. Numerical values of the maximum equivalent stresses were about 870 MPa (figure 8). The maximum equivalent stresses exceeded the admissible values twice [17 – 19]. The maximum displacements were about 27.3 mm. The maximum deformations were $1.15 \cdot 10^{-2}$. Therefore, it is not possible for a flat wagon to carry a tank that is firing.

4. Conclusions

- The authors conducted the mathematical modeling of dynamic loads on the carrying structure of a flat wagon while conducting fire. Besides, they considered the most unfavorable loading on the carrying structure of a flat wagon, particularly, movement over a joint track irregularity with simultaneous firing. The peak acceleration on the carrying structure of a flat wagon in the horizontal plane was about 40 m/sec^2, and in the vertical plane – about 10 m/sec^2.
- The study defined the strength characteristics of the carrying structure of a flat wagon with a tank firing from it. The calculation was conducted with the finite element method in CosmosWorks software. The maximum equivalent stresses emerged in areas where the carrying structure of a flat wagon rested on the bogies and they were about 870 MPa; the maximum displacements were 27.3 mm, and the maximum deformations were $1.15 \cdot 10^{-2}$.

The research confirmed that the flat wagon does not have the needed strength to ensure safe transportation and simultaneous gun firing. The results of the study will encourage engineers to design carrying structures of flat wagons intended for transportation of military equipment firing in motion.

Acknowledgments

The research was conducted under a project for young researchers "Innovative approaches in designing resource-saving rail car structures based on refined dynamic loading and functional-adaptive flash-concepts", supported by the State Budget of Ukraine since 2020.

References

[1] 2017 New wagon certified at Uralvagonzavodhttps: //vsenovostint.ru/2017/04/11/novyiy-vagon-platformu-sertifitsirovali-na-uvz/ (In Russian)
[2] 2019 T-34 tanks in Irkutsk: echelon with 30 military weapons visited the capital of Eastern Siberia https://irkutsk.bezformata.com/listnews/tanki-t-34-v-irkutske-eshelon-s-30-boevimi/72165957/ (In Russian)
[3] 1987 Manual on the transport of troops by rail, sea, river and air transport (Cham: Moscow Military publication) p 124 (In Russian)

[4] Chlus K and Krason W 2011 Dynamic analysis of railway platform chassis model *Journal of KONES Powertrain and Transport* **182** pp 93–100
[5] Chlus K and Krason W 2012 Numerical standard tests of railway carriage platform *Journal of KONES Powertrain and Transport* **193** pp 59–64
[6] Nader M, Sala M, Korzeb J and Kostrzewski A 2014 Kolejowy wagon transportowy jako nowatorskie, innowacyjne rozwiązanie konstrukcyjne do przewozu naczep siodłowych i zestawów drogowych dla transportu intermodalnego *Logistyka* **4** pp 2272–9
[7] 2015 Switching over to the home platform *J. for partners Transmashholding* **3 09** pp 22–3
[8] Niezgoda T and Krason W and Stankiewicz M 2012 Simulations of motion of prototype railway wagon with rotatable loading floor carried out in MSC Adams software *Journal of KONES Powertrain and* Transport **194** pp 495–502
[9] Fomin O, Lovska A, Kulbovskyi I, Holub H, Kozarchuk I and Kharuta V 2019 Determining the dynamic loading on a semi-wagon when fixing it with a viscous coupling to a ferry deck *Eastern-European Journal of Enterprise Technologies* **8 2/7**(98) pp 6–12
[10] Lovskaya A and Rybin A 2016 The study of dynamic load on a wagon-platform at a shunting collision *Eastern-European Journal of Enterprise Technologies* **3** pp 4–8
[11] Domin Yu V and Chernyak G Yu 2003 Basics of carriage dynamics (Cham: Kiyiv KUETT) p 269 (In Ukrainian)
[12] Zelenskij E P, Smolyar B S and Belokur M N 1990 Textbook sergeant rocket troops and artillery (Cham: Moskow Military publication) p 227 (In Russian)
[13] Fomin O, Lovska A, Píštěk V and Kučera P 2020 Research of stability of containers in the combined trains during transportation by railroad ferry *MM Science Journal* **March** pp 3728–33
[14] Fomin O, Kulbovsky I, Sorochinska E, Sapronova S and Bambura O 2017 Experimental confirmation of the theory of implementation of the coupled design of center girder of the hopper wagons for iron ore pellets *Eastern-European Journal of Enterprise* Technologies **5 1**(89) pp 11–8
[15] Vatulia G, Falendysh A, Orel Y and Pavliuchenkov M 2017 Structural improvements in a tank wagon with modern software packages *Procedia Engineering* **187** pp 301–7
[16] Kitov Y, Verevicheva M, Vatulia G, Orel Y and Deryzemlia S 2017 Design solutions for structures with optimal internal stress distribution *MATEC Web of Conf.* **133**(1–3) 03001
[17] DSTU 7598:2014 2014 Freight wagons. General requirements for the calculation and design of new and upgraded rail wagons 1520 mm (non-self-propelled) p 162 (In Ukrainian)
[18] GOST 33211-2014 2014 Freight wagons. Requirements for strength and dynamic qualities p 54 (In Russian)
[19] DIN EN 12663-2-2010 2010 Railway applications – Structural requirements of railway vehicle bodies – Part 2: Freight wagons p 54

Research into the dynamic loading on the carrying structure of a flat wagon with lower center of gravity when transporting firing equipment

O Fomin[1], A Lovska[2,5], M Pavliuchenkov[3], S Sova[4]

[1] Department of Cars and Carriage Facilities, State University of Infrastructure and Technologies, Kyrylivska str., 9, Kyiv, Ukraine, 04071
[2] Department of Wagons, Ukrainian State University of Railway Transport, Feuerbach sq., 7, Kharkiv, Ukraine, 61050
[3] Department of Structural Mechanics and Hydraulics, Ukrainian State University of Railway Transport, Feuerbach sq., 7, Kharkiv, Ukraine, 61050
[4] Department of Railway, Road Transport and Handling Machines, East Ukrainian Volodymyr Dahl National University, Central av., 59-a, Sievierodonetsk, Ukraine, 93400

[5] E-mail: alyonalovskaya.vagons@gmail.com

Abstract. The article presents the study results of the dynamic loading of a flat wagon with a lower gravity center.The authors built a mathematical model that takes into account displacements of a flat wagon loaded with an anti-aircraft installation firing in motion. The solution of the differential equations of motion is carried out by using the Runge-Kutta method. The maximum vertical acceleration was about 5 m/sec^2 and almost independent of the fire angle. The horizontal acceleration was about 37 m/sec^2. The acceleration values obtained did not exceed the normative ones. The study defined the natural oscillation frequencies for the frame of a flat wagon with lower gravity center. It was established that the critical oscillation frequencies were within the admissible values. The results of the work will contribute to the creation of recommendations on the transportation of military equipment on wagons and the possibility of conducting military operations from them.

1. Introduction

The current development of transport infrastructure requires a higher operational efficiency for the railway transport being a leading industry. In terms of their structure, railway vehicles must not only ensure efficient freight transportation, but also be strategic reserve. One of the priority tasks for the railway transport is national defense. This can be achieved by introducing special-purpose wagons, the design of which allows them to carry equipment firing in motion.

Transportation of military equipment by rail is carried out mainly on flat wagons. In order for military equipment to be able to fire from a flat wagon in motion, appropriate research is required. Theoretically, the use of flat wagons with lower gravity center is the rational approach as such wagons have a better stability factor in comparison with that of standard wagons.

Therefore, it is important to do relevant research in this direction with the possibility of further implementation in the design of innovative carrying structures of wagons.

The structural features of the wagon for combined transport are described in [1]. The general requirements for ensuring the safety of combined transport are given. Analysis of a new generation flat wagon design is given in [2]. A special feature of a wagon is the possibility to adjust the effective length according to the dimensions of the freight transported.

It is important to note that in that research no attention was paid to the issues of wagon adaptation for use in military-strategic purposes.

A study of the freight wagon dynamics during their transportation by rail ferry is carried out in [3, 4]. The results of mathematic modelling are proven by computation. The issues of wagon dynamic loading and strength in using them for military-strategic purposes have not been considered in these papers.

Some requirements for carrying systems of railway wagons are presented in [5]. These requirements are suggested for the manufacture of new wagon structures, and also for ones under modernization. However, the requirements for dynamic loads on wagons transporting military equipment are not presented in this study.

The strength of wagon for transportation of timber freight (Model 17-494-01) is studied in [6]. The case of shock loading of a flat wagon was taken into account. However, the authors did not perform the mathematic modelling of dynamic loads on the flat wagon.

The strength of the frame for a rapid articulated flat wagon is estimated in [7]. Strength calculation was carried out by using modern software tools. The paper also presents the results of the fatigue strength calculation of a wagon frame.

The dynamic loads on wagons in train ferry transportation are defined in [8]. The dynamic loading values obtained were taken into account for calculation of the strength of wagon carrying structures. The dynamic loading of the wagons with firing military equipment is not studied.

The features of the strength simulation of a flat wagon for container transportation in the Nastran software are described in [9]. The main operating load conditions have been taken into account when doing this. The rationale for the operation of flat wagons for the removable rolling stock transport is carried out in [10]. This wagon is designed to carry heavy removable units.

However, these studies did not cover the specifics of the wagon loading during conditions above permitted standard, including military-strategic purposes.

The aim of the article is to describe the dynamic features of a flat wagon with a lower gravity center with an anti-aircraft installation firing from it in motion. To do this, the following tasks are set:

- to study the dynamics of a flat wagon with a lower gravity center with an anti-aircraft installation firing from it;
- to calculate the natural oscillation frequencies of the flat wagon frame with a lower gravity center with an anti-aircraft installation firing from it.

2. The main part of the research

To study the dynamics of a flat wagon with a lower gravity center with an anti-aircraft installation firing from it, mathematical simulation has been carried out. The studies were carried out in the XZ plane (figure 1). Three stages of freedom were taken into account: longitudinal plane oscillations, jumping oscillations and galloping oscillations.

It is taken into account that an anti-aircraft installation has muzzle energy of 89 kJ. It was assumed that the anti-aircraft installation did not move while conducting fire, and the carrying structure rested on two bogies of Model 18-100. The authors used the model created by Prof. Domin and Assoc. Prof. Chernyak, which is designed for an open wagon loaded with conditional cargo when using full carrying capacity [11]. Therefore, this model has been redone and takes into account the movement of a flat wagon with the anti-aircraft installation (1) – (6).

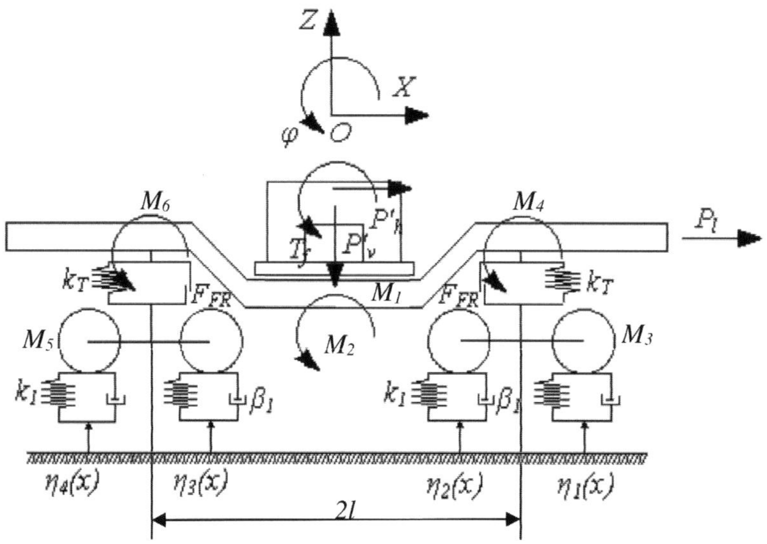

Figure 1. Design diagram of a flat wagon with a lower gravity center.

The movement of a flat wagon with the anti-aircraft installation is described by second-order differential equations:

$$M_1 \frac{d^2}{dt^2} q_1 + M' \frac{d^2}{dt^2} q_3 = P_l + P_h'; \tag{1}$$

$$M_1 \frac{d^2}{dt^2} q_2 + C_{2,2} q_2 + C_{2,5} q_5 + C_{2,8} q_8 = P_z; \tag{2}$$

$$M_2 \frac{d^2}{dt^2} q_3 + C_{3,3} q_3 + C_{3,5} q_5 + C_{3,8} q_8 = P_\varphi; \tag{3}$$

$$M_3 \frac{d^2}{dt^2} q_4 = 0; \tag{4}$$

$$M_3 \frac{d^2}{dt^2} q_5 + C_{5,2} q_2 + C_{5,3} q_3 + C_{5,5} q_5 + B_{5,5} \frac{d}{dt} q_5 = P_z^{T_1}; \tag{5}$$

$$M_4 \frac{d^2}{dt^2} q_6 + C_{6,6} q_6 + B_{6,6} \frac{d}{dt} q_6 = P_\varphi^{T_1}; \tag{6}$$

$$M_5 \frac{d^2}{dt^2} q_7 = 0; \tag{7}$$

$$M_5 \frac{d^2}{dt^2} q_8 + C_{8,2} q_2 + C_{8,3} q_3 + C_{8,8} q_8 + B_{8,8} \frac{d}{dt} q_8 = P_z^{T_2}; \tag{8}$$

$$M_6 \frac{d^2}{dt^2} q_9 + C_{9,9} q_9 + B_{9,9} \frac{d}{dt} q_9 = P_\varphi^{T_2}; \tag{9}$$

$$P_z = -F_{FR}\left(sign\left(\frac{d}{dt}\delta_1\right) + sign\left(\frac{d}{dt}\delta_2\right)\right) + P'_v; \quad (10)$$

$$P_\varphi = F_{FR}l\left(sign\left(\frac{d}{dt}\delta_1\right) + sign\left(\frac{d}{dt}\delta_2\right)\right) + T_f; \quad (11)$$

$$P_z^{T_1} = F_{FR}sign\left(\frac{d}{dt}\delta_1\right) + k_1(\eta_1 + \eta_2) + \beta_1\left(\frac{d}{dt}\eta_1 + \frac{d}{dt}\eta_2\right); \quad (12)$$

$$P_\varphi^{T_1} = -k_1(\eta_1 - \eta_2) - \beta_1 a\left(\frac{d}{dt}\eta_1 - \frac{d}{dt}\eta_2\right); \quad (13)$$

$$P_z^{T_2} = F_{FR}sign\left(\frac{d}{dt}\delta_2\right) + k_1(\eta_3 + \eta_4) + \beta_1\left(\frac{d}{dt}\eta_3 + \frac{d}{dt}\eta_4\right); \quad (14)$$

$$P_\varphi^{T_2} = -k_1 a(\eta_3 - \eta_4) - \beta_1 a\left(\frac{d}{dt}\eta_3 - \frac{d}{dt}\eta_4\right); \quad (15)$$

where M_i are the inertial coefficients of the oscillatory system elements (wagon frame and two bogies); C_i is the elasticity characteristic of the oscillatory system elements; B_i is the dispersion function; l is half of the distance between truck centres; a is a half of the bogie wheelbase; q_i are the generalized coordinates corresponding to the translational displacement relative to the longitudinal and vertical axes and the angular displacement around the vertical axis; k_T is a stiffness of spring set coils; β_i is the damping factor; F_{FR} is the absolute friction force in spring set coils of bogies; P_l is the longitudinal load acting on the front stops of the automatic coupling; P'_v, P'_h are the loads that are transmitted to the wagon frame when firing in the vertical and horizontal planes; T_f is the moment acting on the wagon frame when firing.

The input parameters to the mathematical model are given in table 1.

Table 1. The input parameters of the model.

Parameter	Dimension	Value
M_1	t	11.85
M_2	t·m^2	201.3
M_3, M_5	t	4.3
M_4, M_6	t·m^2	3.0
l	m	4.86
a	m	0.925
k_T	kN/m	8000
k_1	kN/m	100000
β_1	kN·sec/m	200
P_l	kN	2500

It is taken into account that the wagon moves along the track with irregularities. At the same time, an anti-aircraft installation is firing. The mathematical model was solved by using the Runge-Kutta method [12 – 14]. The initial displacement and speeds were taken to be equal to zero.

The accelerations acting on the flat wagon when firing from an anti-aircraft installation in vertical and horizontal planes are determined. The maximum vertical acceleration was about 5 m/sec^2; it was nearly independent on the fire angle (figure 2). And the horizontal acceleration was about 37 m/sec^2 (figure 3).

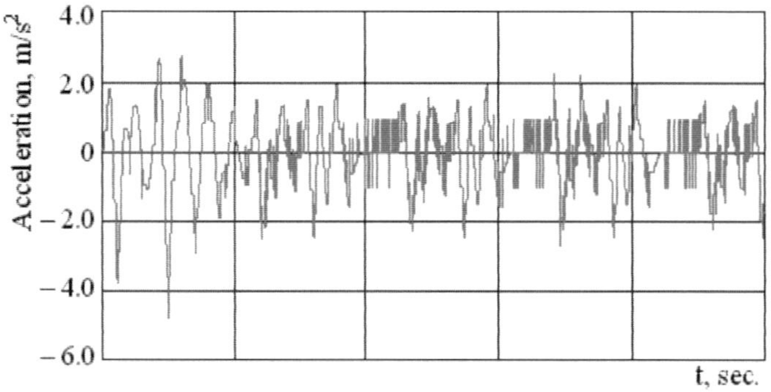

Figure 2. Vertical accelerations in the mass center of flat wagon.

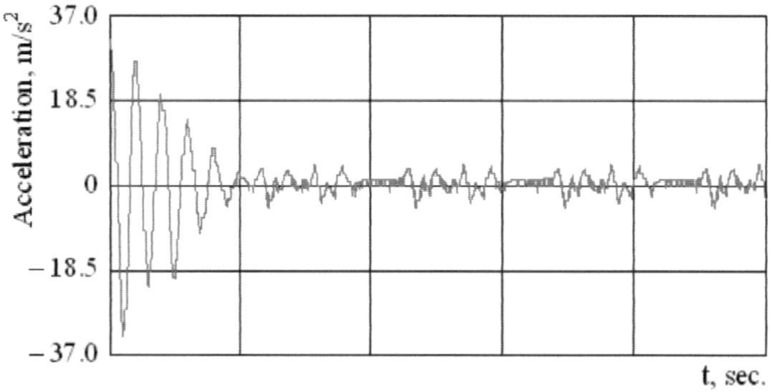

Figure 3. Horizontal accelerations in the mass center of flat wagon.

The acceleration values obtained did not exceed the normative values [15 – 17].

Also, within the framework of the study, a modal analysis of the supporting structure of a flat wagon has been carried out. The purpose of this analysis is to determine the natural frequencies of bending oscillations of the flat wagon supporting structure. In this case, as it is known, to ensure the safety of the flat wagon movement, the first natural frequency of bending oscillations of the supporting structure must be at least 8 Hz in the vertical plane at maximum load. The calculation was made with the finite element method [18 – 20] in CosmosWorks software [21, 22].

The finite element model of the flat wagon frame is shown in figure 4. Tetrahedrons were used to create it. The design diagram of the flat wagon frame is shown in figure 5.

Figure 4. The finite element model of the flat wagon frame.

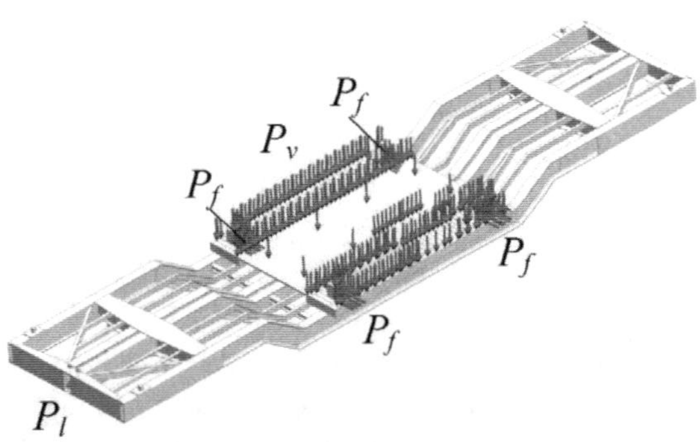

Figure 5. The design scheme of the flat wagon frame.

It is taken into account that the load P_l, equal to 2500 kN, acts on the front stops of the automatic coupling. The vertical load P_v initiated by the weight of the anti-aircraft installation and by the load P_f in the areas of fixing of the anti-aircraft installation to the wagon frame is also taken into account. The numerical values of the loads are shown in table 2. The model was fixed in the areas where it leaned on the gear parts. Steel 09G2S was used as structural material for a flat wagon. The calculation was carried out in quasi-statics, taking into account the application of dynamic loads obtained by mathematical simulation to the spatial model. The results of the calculation are given in table 3.

Table 2. The input parameters of the model.

The parameter	Dimension	Value
P_l	MN	2.5
P_v	kN	9.32
P_f	kN	
a horizontal component		10.2
a vertical component		1.18

Table 3 demonstrates that the natural oscillation frequencies are within the admissible values.

Table 3. The values of the natural oscillations frequencies of the carrying structure of a flat wagon of lower center of gravity.

Frequency number	Frequency, Hz	Frequency, rad/sec.
1	15.21	95.57
2	16.60	104.36
3	19.94	125.33
4	20.70	130.07
5	21.93	137.83
6	24.54	154.19
7	25.17	158.16
8	28.14	176.71
9	29.23	183.56
10	32.14	201.84

In subsequent studies in this direction, the authors plan to carry out a calculation of wagon dynamics taking into account the effect of track irregularities, as well as the stochastic behavior of shots from the anti-aircraft installation placed on it.

Based on the created computer model (figure 5), the dislocation fields of the acceleration acting on the flat wagon supporting structure with a tank firing from it are determined. It was found out that the maximum acceleration is concentrated in the middle part of the flat wagon supporting structure and is 4.3 m/s^2 (figure 6). The discrepancy between the values of acceleration obtained by mathematical and computer simulation in the vertical plane is 14%. In this case, the assessment of wagon motion can be characterized as “excellent” [15, 16].

Figure 6. Accelerations acting on the flat wagon supporting structure.

The work results will contribute to the creation of recommendations for the transportation of military equipment on wagons, and the possibility for the equipment to fire from it.

3. Conclusions

- The dynamics of a flat wagon with a lower gravity center with an anti-aircraft installation firing from it has been studied. The maximum vertical acceleration was about 5 m/sec^2; the maximum horizontal acceleration was about 37 m/sec^2. The accelerations obtained did not exceed the normative values.
- The natural oscillation frequencies of the flat wagon frame with a lower gravity center with an anti-aircraft installation firing from it has been calculated. When doing this, CosmosWorks software which implements the finite element method was used. The first natural frequency is 15.2 Hz. So, the critical oscillation frequencies are within acceptable limits. The dislocation fields of accelerations, which act on the flat wagon supporting structure with the equipment firing from it have been determined by computer simulation. The maximum acceleration was 4.3 m/s^2. The assessment of the wagon motion is "excellent".

The paper findings will contribute to the creation of recommendations for the transport of military equipment on wagons and the possibility for it to fire, as well as to the increase the efficiency of using railway transport for military strategic purposes.

Acknowledgments
The present study was conducted within the framework of the scientific project for young scientists "Innovative principles of creation of resource-saving structures of railroad cars by taking into account the refined dynamic loads and functional-adaptive flash concepts", which is funded by the state budget of Ukraine in 2020.

References

[1] Nader M, Sala M, Korzeb J and Kostrzewski A 2014 Kolejowy wagon transportowy jako nowatorskie, innowacyjne rozwiązanie konstrukcyjne do przewozu naczep siodłowych i zestawów drogowych dla transportu intermodalnego *Logistyka* **4** pp 2272–9
[2] 2016 WBN Waggon bau Niesky GmbH: Developing a flexible platform of freight wagons *Internationsl Edition* **1** p 46
[3] Fomin O, Lovska A, Kulbovskyi I, Holub H, Kozarchuk I and Kharuta V 2019 Determining the dynamic loading on a semi-wagon when fixing it with a viscous coupling to a ferry deck *Eastern-European Journal of Enterprise Technologies* **8 2/7 (98)** pp 6–12

[4] Fomin O, Lovska A, Píštěk V and Kučera P 2020 Research of stability of containers in the combined trains during transportation by railroad ferry *MM SCIENCE JOURNAL* **March** pp 3728–33
[5] Tkachenko V, Sapronova S, Kulbovskiy I, and Fomin O 2017 Research of resistance to the motion of vehicles related to the direction by railway *Eastern-European Journal of Enterprise Technologies* **5 7 (89)** pp 65–72
[6] Putyato A V, Pigunov A V and Belogub V V 2008 Strength calculation of the platform car body for the transport of timber cargo upon impact *Mechanics. Scientific Research and Educational Development* **2** pp 53–8 (In Russian)
[7] Bondarenko K V 2014 Strength assessment of the frame of a high-speed platform car *Transport of the Russian Federation* **3 (52)** pp 49–51 (In Russian)
[8] Lovskaya A and Rybin A 2016 The study of dynamic load on a wagon-platform at a shunting collision *Eastern-European Journal of Enterprise Technologies* **3** pp 4–8
[9] Krason W, Niezgoda T 2014 Fe numerical tests of railway wagon for intermodal transport according to PN-EU standards *Bulletin of the Polish Academy of Sciences Technical Sciences* **62**, **4** pp 843–51
[10] 2015 Switching over to the home platform Journal for partners *Transmashholding* **3 09** pp 22–3
[11] Domin Yu V and Chernyak G Yu 2003 Basics of carriage dynamics (Cham: Kiyiv KUETT) p 269 (In Ukrainian)
[12] Kapitsa M, Mikhailov E, Kliuiev S, Semenov S and Kovtanets M 2019 Study of rail vehicles movement characteristics improvement in curves using fuzzy logic mechatronic systems *MATEC Web of Conferences* **294**
[13] Fomin O, Gerlici J, Lovskaya A, Gorbunov M, Kravchenko K, Prokopenko P, and Lack T 2018 Dynamic loading of the tank container on a flat wagon considering fittings displacement relating to the stops *MATEC Web of Conferences* **234**
[14] Fomin O, Gerlici J, Lovskaya A, Kravchenko K, Prokopenko P, Fomina A, Hauser V 2018 Research of the strength of the bearing structure of the flat wagon body from round pipes during transportation on the railway ferry *MATEC Web of Conferences* **235**
[15] DSTU 7598:2014 2014 Freight wagons. General requirements for the calculation and design of new and upgraded rail wagons 1520 mm (non-self propelled) p 162 (In Ukrainian)
[16] GOST 33211-2014 2014 Freight wagons. Requirements for strength and dynamic qualities p 54 (In Russian)
[17] DIN EN 12663-2-2010 2010 Railway applications – Structural requirements of railway vehicle bodies – Part 2: Freight wagons p 54
[18] Vatulia G, Falendysh A, Orel Y and Pavliuchenkov M 2017 Structural improvements in a tank wagon with modern software packages Procedia engineering 187 pp 301–7
[19] Kitov Y, Verevicheva M, Vatulia G, Orel Y and Deryzemlia S 2017 Design solutions for structures with optimal internal stress distribution *MATEC Web of Conf.* **133(1–3)** 03001
[20] Kurowski P 2019 Engineering Analysis with SOLIDWORKS Simulation 2019 (Schroff Development Corporation, 2019) p 606
[21] Weber M, Verma G 2015 SolidWorks Simulation 2015 Black Book 2nd edition (CreateSpace Independent Publishing Platform) p 398
[22] Reyes A 2017 Beginners guide to SolidWorks 2018 (SDC Publications) p 232

Comparative analysis of the acoustic efficiency of classical and sonic crystal noise barriers

Elitsa Gieva[1], Ivelina Ruskova, Krasimir Nedelchev and Ivan Kralov

Technical University - Sofia, 8 Kliment Ohridski Blvd., 1797 Sofia, Bulgaria

[1]E-mail: gieva@ecad.tu-sofia.bg

Abstract. This study presents a comparative analysis of the acoustic efficiency of classical and sonic crystal noise barriers conducted by means of numerical simulation. One of the most common solutions for transport noise reduction is the use of expensive combined noise barriers of high degree of reflection and absorption. Sonic Crystals are a relatively new type of noise barrier under development and investigation. In this study both types of barriers are compared on the basis of their area of application, noise level attenuation behind them, frequency response, technology of manufacturing, maintenance and price. Some of the characteristics are obtained through numerical simulations and experimental data, while others are available in published research studies. On the basis of the results conclusions about the application of sonic crystal noise barriers are made.

1. Introduction

The comfort of the urban areas is one of the most important tasks to achieve when we talk about noise and environmental contamination. That is the reason why in recent years, many researchers have focused their attention on the construction and optimization of acoustic barriers. More than 55 dBA in the night hours and 65 dBA in the day hours should not be exceeded according to EU normative, however a large number of European citizens suffer higher noise levels. In fact, 20% of EU citizens during the day, and the 30% of citizens during the night suffer higher noise levels. The most commonly used method for sound insulation is the installation of acoustic screens between the noise source (transport infrastructure) and the receiver (the near urban areas). These classical Acoustic Barriers (AB) are made of continuous flat walls whose main noise control mechanism is reflection. However, some unwanted problems may arise due to this reflected noise energy and even cause a noise increase at protected areas, reducing the effectiveness of the installed ABs. To minimize these specular reflections, some solutions have been developed, such as the use of absorbent materials, the installation of inclined barriers, or scattering of the reflected noise [1-3]. However, the application of absorbent materials can be very expensive, and the use of inclined barriers could be technically complicated and impossible in some sites. Focusing on the solution of scattering the reflected noise, some proposals have been presented recently. One of the most widely accepted solutions is the use of new noise control devices based on technologically advanced materials. Sonic Crystals (SC) is one of these materials. SC are generally defined as heterogeneous materials formed by arrangements of acoustic scatterers embedded in air. Two of the many applications of these materials are their use as Acoustic Barriers (ABs) usually called Sonic Acoustic Screens (SCAS), and their use as sound diffusers in acoustic rooms to increase sound diffuseness [1-3].

1.1. Barrier types

All noise barriers should be designed in order to meet the requirements of the local environment; otherwise they can remain dull, only visual elements who damage the landscape quality. The visual quality of the barriers is also an important issue, which should be carefully considered, and this effect depends on the barrier height, the distance of the barrier from the receiver, etc.

Depending on the acoustic and non-acoustic requirements and the terrain configuration, the process of noise barrier selection should be carefully done. In figure 1 different types of noise barriers are presented.

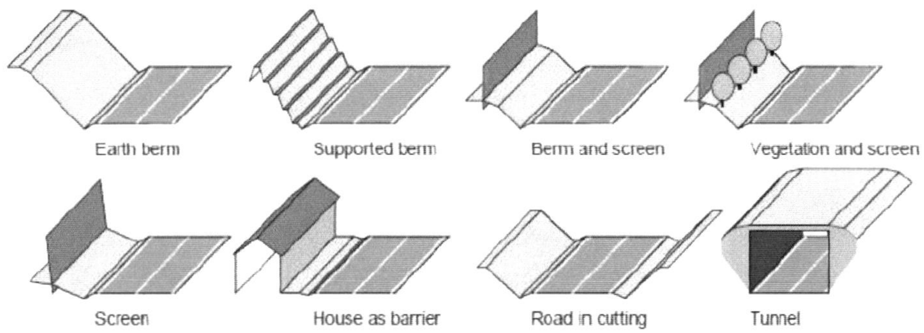

Figure 1. Different types of acoustic barriers [4].

Depending on the physical dimensions of the barrier and the location, an appropriate barrier material should be chosen. The materials can be divided into five groups, as follows:

- Timber acoustic materials: may be fully reflective or sound absorptive. Depending on the application the structure may be composed of absorbing materials covered with timber elements with a tongue and groove constructions (for sound absorptive elements, for example), or the timber elements can be jointed in order to avoid leaks (for reflective barrier, for example).
- Metal acoustic elements: the metal materials used here are mostly aluminium, steel or both and the construction is typically cartridge-type panels. Such barriers could be sound reflective or sound absorptive.
- Concrete acoustic elements: usually used for the structural part of the precast panels in concrete, and covered by porous concrete (cement or wood-fibre type).
- Transparent acoustic elements: they are commonly used for visual impact for the people, offer a view of the road for the drivers. They can be fully transparent or incorporated as part of a barrier constructed from other materials.
- Plastic or composite acoustic elements: usually the panels are made from plastic, which is reinforced with glass fibre. They can be sound reflective and absorptive. In order to be absorptive the front part of the panel should be perforated.

There are a lot of bad examples of barriers. The problems are various and can be manifested due to problems with materials, for example. Rust is the main problem for metallic barriers. Their structure-caissons are suitable for water from the rain to remain in it and this humidity may develop inside. Typical problems for timber barriers are warping, shrinkage and crocking. Over time the transparency of the transparent barriers diminishes and the main characteristic of the device is lost.

Innovative noise barriers solutions can be considered that use more innovative design or materials for construction of the acoustic elements, or perform some additional functions, such as power generation as addition to noise mitigation. Some examples for innovative sound isolation barriers with their advantages and disadvantages are described below.

- Combined noise and safety barriers: use both concrete and steel guardrails and serve as a safety barrier and an acoustic barrier, ensuring double advantage – to recover the mounted width and optimized bearing construction. One advantage of such barriers is that they can be placed closer to the noise source and thus protection increases significantly. Some of them are free-standing and do not need foundation for their construction. The main disadvantage is that the maximum tested height of such barriers is 5 meters with metal guardrail and 4 meters with concrete safety barriers.

- Photovoltaic noise barriers: they are used as combination of noise reduction and production of renewable energy, converting sound energy into electricity. The advantage here is that it limits the production of greenhouse gas emissions in the atmosphere. The photovoltaic panels can be directly integrated into the surface of the barrier or mounted onto the barrier. The advantages are environmental, economic and social. The disadvantages are related to the fact that they are not suitable for every situation because they are dependent upon the orientation of the barrier, the number of photovoltaic cells and the climate (the sunlight) and the energy use. The cost benefits should be carefully considered against the cost of the energy storage and transfer infrastructure.
- Noise barriers with TiO2 coating: such type barriers incorporate TiO2 coatings and thus employ photo catalysis, which is a chemical reaction where titanium oxide acts as a catalyst to eliminate nitrogen oxides. These type of barriers have limited effectiveness and success in practical trials.
- Inox/corten steel barriers: they are made of stainless steel or corten steel and have a lot of performance requirements. Their advantages include good environment insertion, positive perception by the population, mechanical strength and stability, ease of maintenance, easy removal of dirt and graffiti. The main disadvantage of these barriers is the cost.
- Sonic crystals: the sound attenuation characteristics could be tuned by appropriate selection of the lattice constant and cylinder radius. The height of these barriers can reach 6 m [4].

1.2. Sonic crystal basics
The first possibility, presented as an alternative to the classic acoustic screens, is the use of sound crystals (SCs). The basic unit of the ideal crystal structure can be defined as an identical group of atoms (in terms of composition, arrangement and orientation) in space (see figure 2).

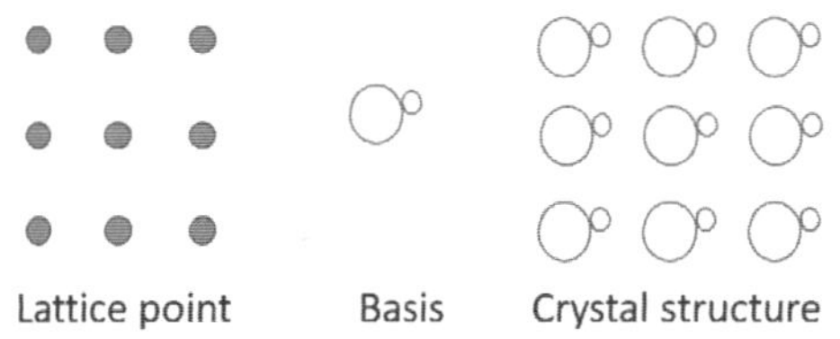

Figure 2. An ideal crystal structure [5].

SCs can be defined as a heterogeneous material consisting of a periodical array of acoustic diffusers embedded in an environment of different physical properties, where one of the materials is a liquid. Depending on the number of directions where the array periodicity exists, one dimensional (1D), two-dimensional (2D), and three-dimensional (3D) SC crystals (figure 3) can be defined.

1.3. Sound crystal acoustic screen (SCAS). Main characteristics of the sound crystals
The mechanism that prohibits the transmission of sound through the system is different from those prevailing in the classical acoustic screens (AS), in that case the main physical mechanism is the destructive Bregg Interference due to the Multiple Scattering (MS) process associated with the periodicity of system. These destructive disturbances lead to damping peaks. These peaks expand by increasing both the number and the diameter of the cylinders so that these periodic systems show ranges of frequencies associated with the periodicity of the structure where there is no wave propagation. These frequency ranges are called Band Gaps. Other physical mechanisms such as absorption or reflection can be added to the dividers to enhance damping capabilities.

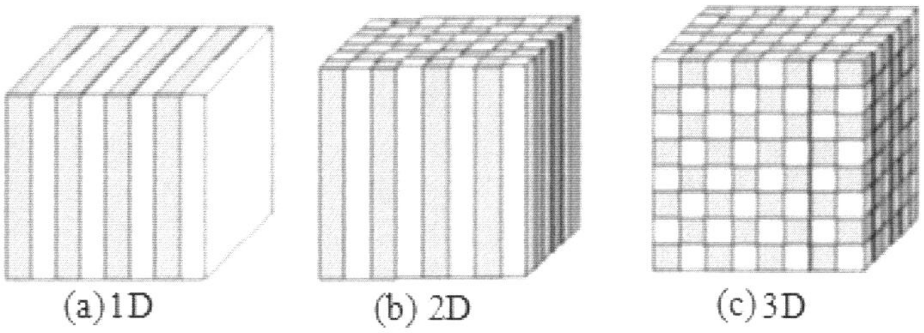

Figure 3. Schematic illustrations of crystal structure (a) one-dimensional, (b) two-dimensional and (c) three-dimensional [7-8].

By using group theory, it has been proven that there is only one unique 1D periodic system, five 2D and 14 different 3D grids. Most of the studies are focused on the sound crystal as a 2D arrangement for which the five different Bravais grids with their principal lattice vectors and angles are shown in figure 4.

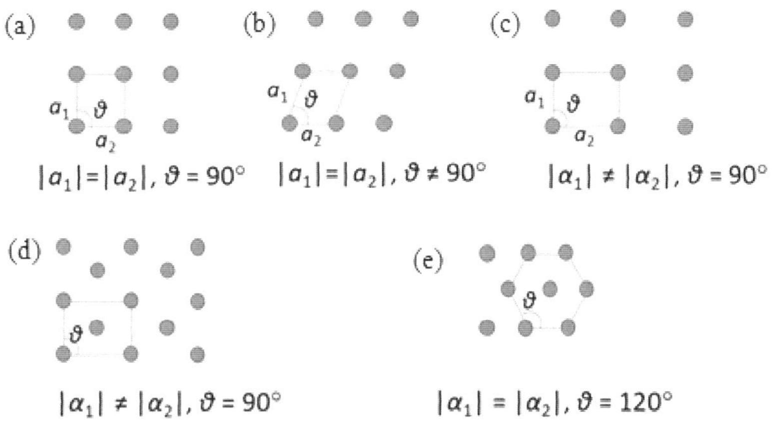

Figure 4. The five fundamental 2-Dimensionals Bravais lattices (a) square, (b) oblique, (c) rectangular, (d) centered rectangular and (e) hexagonal [6].

The sound crystals/Sonic crystals (SC) are non-homogeneous materials made from regular arrays of separators embedded in a liquid or air environment. An interesting feature of these materials is the occurrence of ranges of frequencies in which periodic structures are not supported. These ranges are known as band gaps (BG). The center frequency of the fBG bandwidth is determined by the grid constant α which represents the distance between adjacent diffusers:

$$f_{BG} = \frac{c}{2\alpha} \tag{1}$$

where c is the speed of the sound in the air.

Over the last two decades, an increasing number of studies and applications have been presented that analyze the distribution of acoustic waves inside sound crystals. Solid spacers, such as wood or aluminum, are used to study the physical behavior of sound crystals when the reflection of acoustic waves becomes a mechanism for creating bandwidth. In addition, sound crystals with mixed solid and absorbent materials are used to study the double reflection and absorption effect on the frequency response of the acoustic crystal. In fact, the use of soft cylinders is an alternative to solving the problem of angularity of acoustic attenuation. Resonance cylinders allow the sound to penetrate the elements of the periodic structure and as a result new acoustic properties may emerge. In fact, in urban areas, pedestrians and cyclists are in close proximity to road traffic or noise sources from urban transport. It is easy to protect with small devices not exceeding 1×1 m in vertical section. Engineering methods cannot produce accurate values for the acoustic damping of such barriers [7-8].

1.4. First and second generation sonic crystals

Acoustic screens of sonic crystals have developed various solutions such as the use of absorbent materials, the construction and installation of bending barriers or noise distraction that is reflected. The application of absorbent materials may be expensive and the installation of bending barriers is difficult and complicated. That is why most research aims to scatter the reflected noise and develop new materials such as sonic crystal.

The first experimental barrier for sound isolation by a periodic structure is in the form of a sculpture (made in 1995) in Madrid and consists of periodically spaced hollow cylinders. Their material is stainless steel and has a diameter of 0.029 meters. They are located in a constant grid and the distances between the cylinders are the same - 0.1 m (see figure 5). These cylinders are the sonic crystal, they can rotate around a vertical axis. The sound isolation is measured under different conditions. The acoustic barriers based on SC are usually called Sonic Crystals Acoustic Screens (SCAS) [9].

The first generation of SCAS are new noise reduction devices that use a new noise control mechanism based on the presence of a bandgap (see figure 6). These new barriers have some technological advantages over the classical acoustic barriers but also have a deficiency in noise coverage.

Figure 5. Minimalistic sculpture by Eusebio Sempere [9].

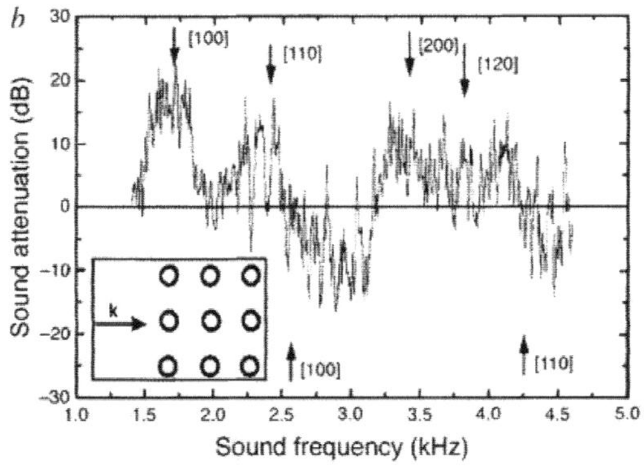

Figure 6. Sound attenuation as a function of frequency, also known as Insertion Loss [9].

In order to improve the SCAS characteristics and noise control capabilities, other solutions were proposed in 2011 by Romero-García. The design uses resonances or absorption mechanisms. A core of rigid cylinders with a slot along the entire length and a wrap of absorbent material are formed. The interior becomes an acoustic resonator cavity. The outer part with absorbent material contributes to noise suppression. Resonator cavities have damping peaks due to resonances. There are three mechanisms for noise control: band gap, absorption and resonance.

SCASs that use other noise control mechanisms, with the exception of band gap, are called second-generation SCASs [10].

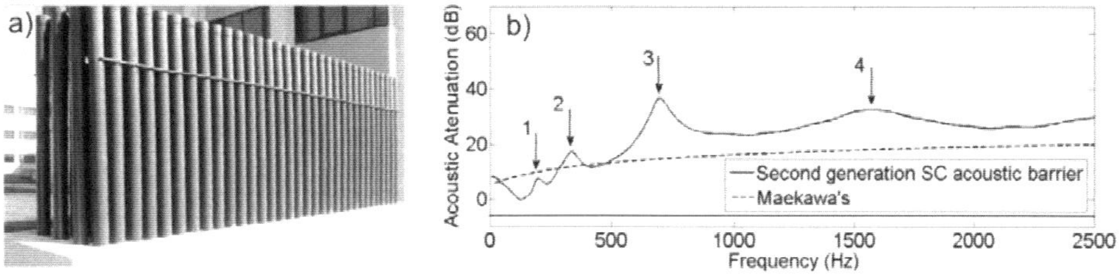

Figure 7. (a) Example of first generation SCAS; (b) Simulated attenuation spectrum of the SCAS [10].

Figure 7(a) shows an example of first generation SCAS and figure 7(b) shows the characteristics of the second generation SCAS. The peaks attenuation can be seen due to resonant cavity 1, whose position in the frequency range depends on the volume of the resonators. The damping peaks due to the periodicity of the array correspond to the band gap of the array 3, 4. The attenuation threshold due to the absorption used in each cylinder of the construction occurs from 500 Hz onwards [10].

This new generation of acoustic barriers allows for the design of specific acoustic screens to reduce noise. This design of the second-generation SCAS is protected by Spanish patents (Sánchez-Pérez et al., 2003). Comparative analysis of standard acoustic barriers and sonic crystals is made and detailed in the next section.

2. Comparison between classical acoustic absorber and sonic crystal barrier

2.1. Sonic Crystals and classical acoustic barrier

The modern way of life makes it easier for us in every respect, but at the same time there are negative sides - we are subject to pollution. A very common source of pollution is noise, which has reached alarming levels and leads to stress and a number of ailments in people who are exposed to it for a long time. Dealing with this problem is the subject of activity of many companies.

Noise in the physical sense is a sound with a complex inharmonious wave (chaotically collected tones), whose frequency and strength are not in definite and constant proportions in time; in the psycho-physiological sense - any acoustic irritation (impact), which is perceived as unpleasant, regardless of its acoustic qualities; in the patho-physiological sense - any acoustic signal that causes with its strength or duration temporary or permanent changes in the body (auditory analyzer and beyond).

The classical acoustic barriers (figure 8a) are strong and solid barriers between the highway and the houses that are located along this highway. They do not completely block the noise, but only reduce it by 5 to 10 decibels (dB), limiting noise traffic by almost half. Barriers can be made from earth embankments along the road, from high, vertical panels of different materials (wood, metal, acrylic glass, recycled plastic, etc.) or from a combination of earth embankments and panels. Earth embankments look most natural and are therefore attractive to contractors and are common. They reduce noise by about 3 dB more than vertical walls of the same height but making them requires a large amount of soil, especially if they have to be taller.

Barriers, on the other hand, take up much less space, but their height is usually limited (up to 8 m) due to structural and aesthetic reasons. The barrier reduces the noise that reaches the settlement from the busy highway by absorbing, transmitting and reflecting it back to the road or directing it along a long path above and around the barrier. The noise barrier must be high and long enough to "block" the road from the highway to the area to be protected - the "receiver". It should be noted, however, that it brings minimal benefits to houses on slopes that are higher than the highway or buildings that rise above the barrier.

The noise barrier can reduce noise by up to 5 dB when it is high enough to break the line of sight from the highway to the houses or the receiver. After refraction of the line of sight, an additional noise reduction of about 1.5 dB can be achieved for each meter of the height of the barrier.

The barrier must be at least 8 times longer than the distance between the houses or the receivers from the highway. This will reduce the noise coming from the ends of the barrier more effectively [11].

However, the use of acoustic barriers has some disadvantages. There are no such barriers which can distinguish the noise which has to be controlled. Second, the placement of continuous walls is also a problem. The heavy wind load can also produce some structural problems in viaducts with ABs installed in the case of high speed trains.

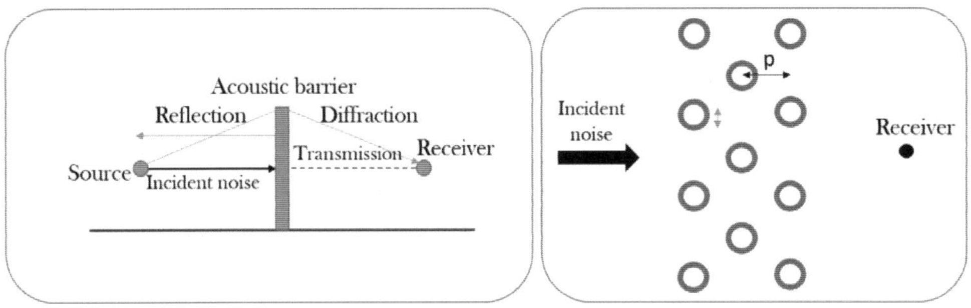

Figure 8. Sonic Crystals and classical acoustic barrier [12].

The placement of classical barriers can involve high costs from technical and economical points of view. For all these disadvantages, installation of classical acoustic barriers could be unsatisfactory in urban areas for transport noise control. In the last decade, the discovery of new materials has led to the development of new devices for noise control. Sonic Crystals (SC) can be defined as periodic arrays of cylindrical acoustic scatters with radius *r* separated by a predetermined lattice constant *p*, and embedded in air (figure 8b). SC performance is based on Bragg interferences due to a multiple scattering process [12], different from those previously known. As a consequence, there are frequency ranges, related to the periodicity of the medium, where the propagation of the waves is forbidden through the crystal. These frequency ranges are called band-gaps.

The first SCAS was theoretically proposed by [13], and the first prototype was designed and constructed by [14]. These devices are constructed using the Bragg interference as an unique damping mechanism and are formed by a set of solid reflectors embedded in the air. These screens are defined as the first generation of acoustic screens using sound crystal. However, the use of "bandgap zones" as unique mechanism for wave isolation is not enough to ensure a good performance of SCAS.

In recent years, the so called acoustic metamaterials have been developing, whose structure is characterized by periodic repetition of the shape of the structure and/or composites of building blocks or small elements of other materials intermittently located in one basic. The main advantages of the described metamaterials are the high degree of acoustic absorption by converting the sound energy into friction energy (and part of it in kinetic motion of built-in heavy elements) as well as the relatively easy design and synthesis of effective frequency acoustic sound absorption. In figure 9 a comparison between a second generation SC barrier and a classical acoustic barrier is presented. The acoustic attenuation in dB as a function of the frequency in Hz is clearly observed. A major disadvantage of the described metamaterials and structures is their high manufacturing cost, their relatively low mechanical strength and the need for a package to be incorporated in a number of cases.

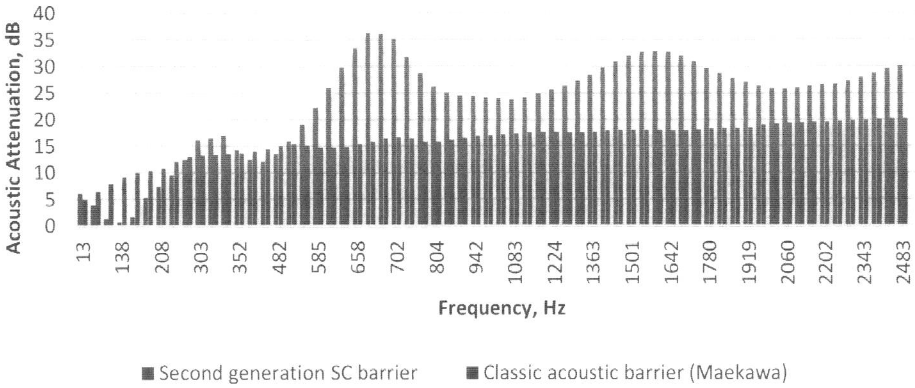

Figure 9. Acoustic attenuation as a function of the frequency compared for second generation SC barrier and classical acoustic barrier.

2.2. 3D Sonic Crystals acoustic barrier COMSOL model

In order to check and optimize the experiment, we have created a COMSOL model of 3D Sonic Crystals acoustic barrier. The parameters of the COMSOL model are given in table 1. We have 3 rows with 7 noise isolating tubes.

We have used Pressure Acoustic Frequency Domain and Solid Mechanics to study the designed models. The finite element method is used where the complex area can be divided into a series of small interconnected subregions in which PDEs are approximately solved. Each sub region of the area is called an element and the process of dividing the object/domain into a finite number of elements is called sampling.

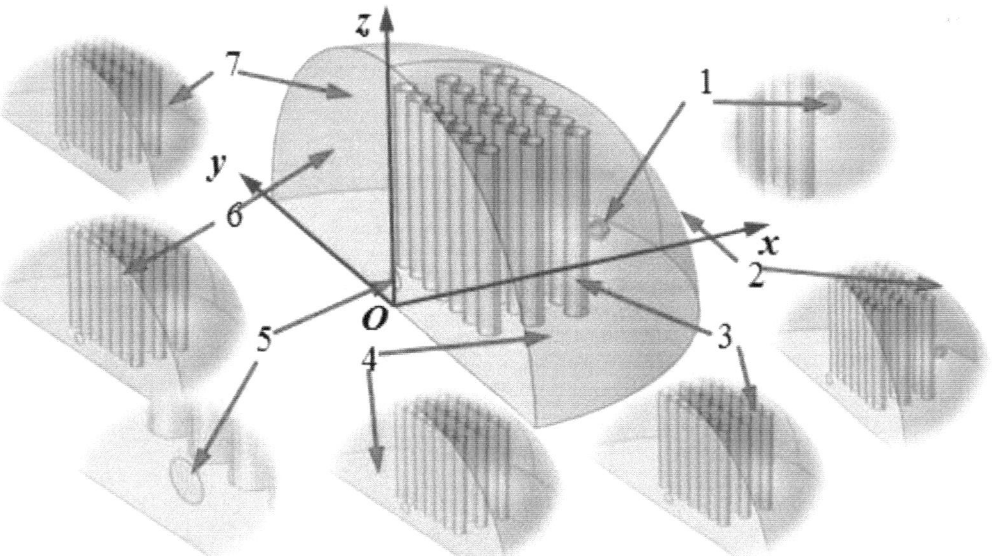

Figure 10. 3D model of a Sonic Crystal acoustic barrier:
1 – Acoustic receiver /microphone/; 2 – Surface around the sphere /Spherical Wave Radiation/; 3 – 3D Sonic Crystal acoustic barrier; 4 – Main surface of the Sonic Crystal acoustic barrier placement /Sound Hard Boundary/; 5 – Acoustic source; 6 – Front surface of the acoustic space /Symmetry/; 7 – Acoustic space.

Table 1. Design parameters.

	Parameter	Value
Ls	Distance between the barrier and the source	0.3 m
Lr	Distance between the barrier and the microphone	0.4 m
Dp	Tube diameter	0.16 m
dp	Thickness of the tube walls	0.004 m
Hp	Tube height	1.5 m
xp	Distance between the tubes in direction source-receiver (x)	0.16 m
yp	Distance between the tubes in direction perpendicular to the x-axis (y)	0.04 m
nr	Row number	3
nc	Column number	7
p_sourse	Source pressure	1 Pa
Rair	Radius of the acoustic area	2 m

Figure 11 shows the row and column arrangement and the highest attenuation at 500 Hz and 800 Hz. It is the same for the other frequencies, so not all of them are presented in the publication.

Figure 11 shows Noise attenuation in dB at a frequency of 500 Hz and 800 Hz from COMSOL. You can see the arrangement of the noise barrier in 3 rows and 7 columns, as described in the geometry of figure 10. After the modelling we have obtained results that were used to perform the experimental setup (shown in figure 12) and the comparison between 3D Sonic Crystals acoustic barrier COMSOL model and Experimental setup.

3D Sonic Crystals acoustic barrier COMSOL model shows good attenuation over a wide frequency range. This type of noise barrier is applicable at medium frequencies.

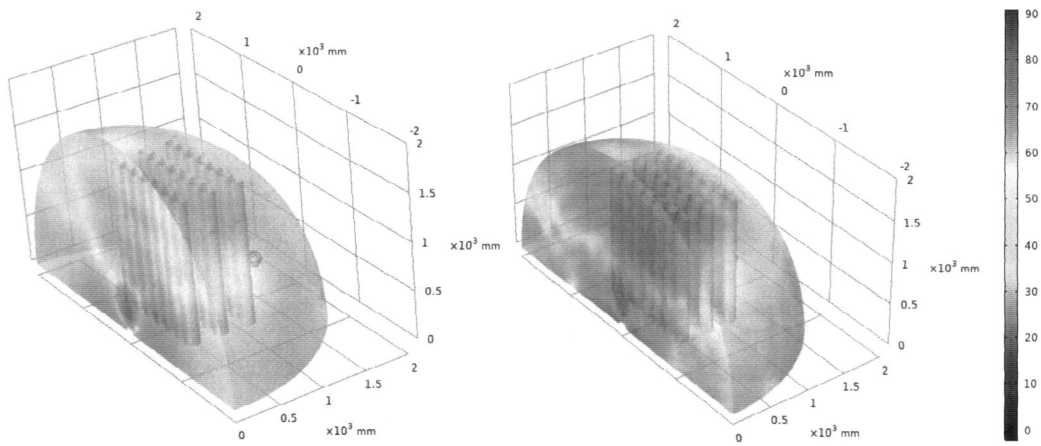

Figure 11. Noise attenuation in dB at frequency of 500 Hz and 800 Hz.

2.3. Sonic Crystals acoustic barrier measurement

The experimental setup repeats the optimal configuration based on the COMSOL model. Figure 12 shows the experimental setup.

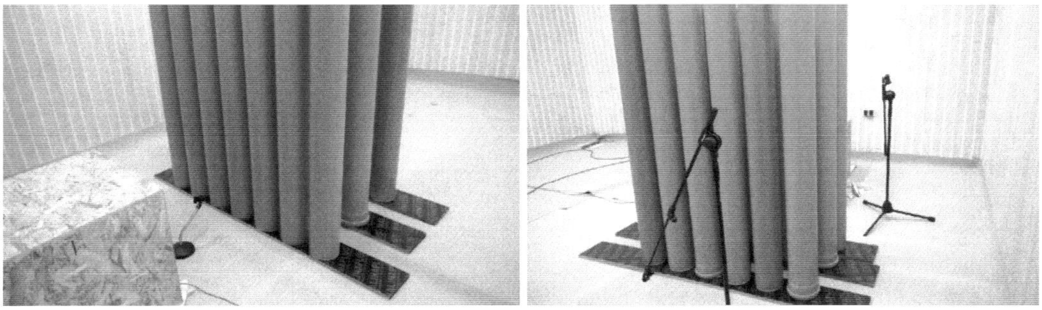

Figure 12. Experimental setup.

The parameters of the experimental setup that were used are those given in table 1. The data that are set in the model in COMSOL and those during the experiment are the same, as well as the set conditions. The distances to the source and the microphone are the same, as is the height when measuring. The material of the pipes, their dimensions and arrangement are the same as well.

A comparison of the results of the COMSOL model and the experimentally obtained result (figure 13) shows that at some frequencies there is a big difference in the obtained attenuation, such as at 50 Hz, 500 Hz and 1600 Hz. The difference is more than 30%.

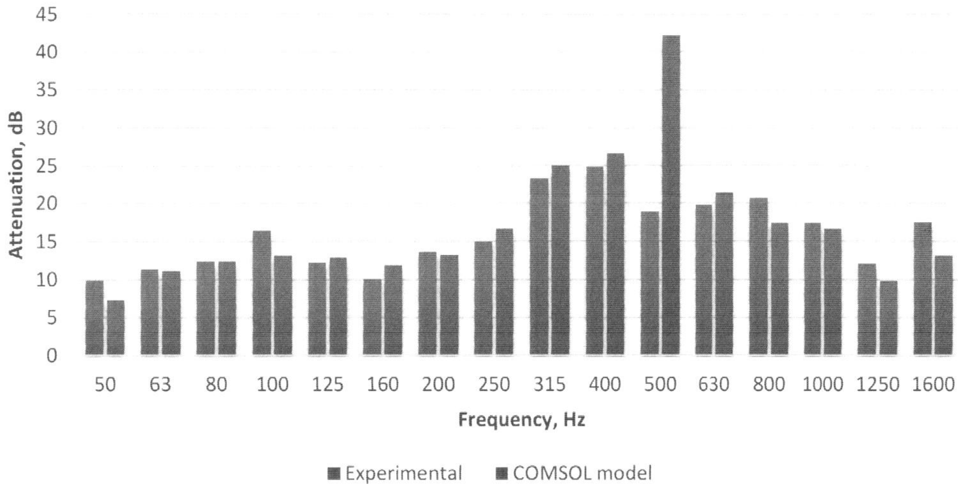

Figure 13. Noise attenuation as a function of the frequency from the experiment.

In table 2 the obtained results are compared and the difference in % of the COMSOL model and the real experiment are presented.

Table 2. Comparison of the attenuation between the COMSOL model and the experimental results at different frequencies.

Frequency, Hz	COMSOL model Attenuation, dB	Experimental results Attenuation, dB	Difference in %
50	7.244018	9.8242	35
63	11.10373	11.2729	1.5
80	12.36799	12.3915	0.19
100	13.14756	16.3877	24.6
125	12.90293	12.2052	5.4

(continued)

Table 2. *(continued)*

Frequency, Hz	COMSOL model Attenuation, dB	Experimental results Attenuation, dB	Difference in %
160	11.90689	10.0146	15.8
200	13.22491	13.6792	3.4
250	16.68946	14.9856	10
315	25.03462	23.3149	6.8
400	26.59061	24.8455	6.5
500	42.12851	18.8721	55.2
630	21.52336	19.7586	8.19
800	17.44579	20.7058	18.6
1000	16.70379	17.4253	4.3
1250	9.807703	12.0346	22.7
1600	13.15217	17.5173	33.1

The comparison between the two Sonic Crystal models, the designed and the experimental one, shows very promising results. The percentage difference between the two models at most of the frequencies is less than 10%.

3. Advantages and disadvantages

Sonic Crystal Acoustic Screens (SCAS) have several advantages over standard acoustic noise barriers:

- Use better and more advanced constructions of scatters. Various ways of controlling noise are used in these constructions. The presence of resonators allows for control of the resonance frequency and by changing the resonance geometry for control of the resonant frequency and even to have 2 or more resonance frequencies.
- The positions and distances between scatters in the sonic crystal array affect the bang gap, again by changing the geometry.
- Absorbent material is used, depending on the level of noise absorption.
- This type of barrier makes it possible to change the individual parameters and thus to correct and control only some noise control mechanisms. Depending on the type of noise, design the barrier at several levels and specify it. This will select a frequency range in which all noise control mechanisms will work and will be applicable to each type of noise. This advantage makes them highly competitive compared to classical acoustic barriers. SCAS may be designed on demand and used in various applications.
- With classic acoustic barriers mounted in the outdoor environment, the wind has a severely adverse effect, whereas in SCAS, the distribution and geometry allow the wind to pass without straining the barrier. Such a structure can solve structural problems at the classical acoustic barriers, which are further strengthened at their core against the impact of the wind.
- Another advantage of this type of acoustic barriers is their ability to position them both on the ground and in the air as their construction allows it.
- In SCAS, noise control methods are band gap, absorption and resonance frequencies. In this way a new technological procedure is introduced into the acoustic barriers. Their acoustic characteristics and reactions are pretty good compared to standard acoustic barriers. Therefore, these barriers can be used for protection in and outside cities to protect against traffic noise. Their open design and flexible structure designed for specific noises are the technological advantages of this type of barrier [10].

Acoustic noise barriers also have disadvantages:

- These types of barriers are more expensive than standard acoustic noise barriers.
- They are more difficult to design, several different features and parameters have to be calculated at the same time.

- The used materials must be studied and their technical characteristics as well as their aging and outdoor use must be known.
- They still do not have a common application like standard acoustic barriers.
- Their dimensions and size are much larger than those of standard barriers.
- Transporting them is difficult.

4. Conclusions

In this article we have presented a comparative analysis of the noise protection efficiency of standard noise barriers and type of sound crystals. Numerical simulations are presented in this study. One of the most commonly used solutions to reduce transport noise is the use of expensive combined noise barriers with a high degree of reflection and absorption. Sound crystals are a relatively new type of noise barrier under development and research. This article compares both types of barriers on the basis of their field of application, noise attenuation, their frequency range of application, production technology, maintenance, transport and price. Some of the results and characteristics are obtained by numerical simulations and experimental data, while others are obtained from published studies. Based on the obtained results, analyses and conclusions are made about the advantages and disadvantages of SC noise barriers and their application in comparison with the classic noise barriers.

The obtained results show that sound crystals can be used to improve the noise protection efficiency of standard barriers. There are a large number of materials on the market for the construction of noise barriers, including natural materials such as wood. Based on the results, one can think about designing new types of noise barriers, such as those that use sound crystals. In addition to new materials, we have investigated various shapes and arrangements of barriers in order to improve the attenuation of these noise barriers.

Acknowledgments

This work is supported by Project "Identification and investigation of the acoustic parameters of the noise-insulating barriers by advanced engineering", Contract No. КП-06-Н37/10 dated 06.12.2019.

References

[1] Mohsen S, Henry A S and Anton S R 2019 A review of energy harvesting using piezoelectric materials: state-of-the-art a decade later (2008–2018) *Smart Mater. Struct.* **28** 113001 p 62
[2] Mechano Transformer Corporation 2006 A thin type piezo actuator: Macro Fiber Composite (MFC) http://www.mechano-transformer.com/en/products/mfc.html Accessed on: 15.06.2020
[3] Noh H M 2018 Acoustic energy harvesting using piezoelectric generator for railway environmental noise *Advances in Mechanical Engineering* **10**(7) pp 1–9
[4] De Leo A, Marcocci S and Vanhooreweder B 2016 State of the art noise barriers in CEDR experience *INTER-NOISE, Hamburg* pp 4160–70
[5] Thompson D 2009 *Railway noise and vibration* (Elsevier) p 518
[6] More H 2020 *Bravais lattices* https://thefactfactor.com/facts/pure_science/chemistry/physical-chemistry/bravais-lattices/7443/ Accessed on: 15.06.2020
[7] Banov S and Kralov I 2003 *Noise in transportation technics* (TU-Sofia) (In Bulgarian) p 135
[8] Belnikolovski B and Abuid B 1994 On the dynamic characteristics of a mechanism with two-stage spur gears *American Society of Mechanical Eng.* (Petroleum Division PD) **64**(7) 261
[9] Chong YB 2012 *Sonic crystal noise barriers* PhD thesis (The Open University)
[10] Peiró-Torres M D P, Redondo J, Bravo J M and Sánchez Pérez J V 2016 Open noise barriers based on sonic crystals. Advances in noise control in transport infrastructures *Transportation Research Procedia 12th Conf. on Transport Engineering (CIT) (Valencia, Spain)* pp 392–8
[11] Yankova A 2008 Noise barriers on highways *Stroiteli* **V**(3) April (In Bulgarian)
[12] Chen Y Y and Ye Z 2001 Theoretical analysis of acoustic stop bands in two-dimensional periodic scattering arrays *Phys. Rev.* E **64** 036616

IOP Conf. Series: Materials Science and Engineering 1002 (2020) 012014 doi:10.1088/1757-899X/1002/1/012014

[13] Kushwaha M and Rahnejat H 1997 Virtual prototype testing applied for non-linear dynamic analysis of a vehicle tailgate *J. Multi-body Dynamics: Monitoring and Simulation Techniques (Wiley-Blackwell)* p 315
[14] Sánchez-Pérez J, Rubio C, Martinez-Sala R, Sánchez-Grandía R and Gomez V 2002 Acoustic barriers based on periodic arrays of scatterers *Appl. Phys. Lett.* **81** pp 5240–42

Numerical investigation of the mechanical behaviour of a structural element containing a self-healing agent

T Zhelyazov[1,3] and R Ivanov[2]

[1] Department of Mechanics, Faculty of Transport, Technical University of Sofia, 8 Kl. Ohridski Blvd., Sofia 1000, Bulgaria
[2] Section "Seismic Engineering", Department "Seismology and Seismic Engineering", National Institute of Geophysics, Geodesy and Geography – Bulgarian Academy of Sciences, Acad. G. Bonchev St., bl. 3, Sofia 1113, Bulgaria

[3] E-mail: todor.zhelyazov@tu-sofia.bg

Abstract. This paper presents a virtual experiment on the behaviour of a self-healing material from the family of cement-based composites, that contains a healing agent. A numerical model of a specimen made of a cement-based material, and containing a healing agent is subjected to the loading configuration of the four-point bending test, whereby the forces are applied in a quasi-static way. The response of the healed specimen is compared to the response of a specimen that doesn't contain a healing agent. For the specimen that contains a healing agent, homogenization techniques are used to determine the characteristics of the equivalent material (cement-based composite / healing agent) in zones where damage and macro-cracking have occurred, i.e., zones in which the healing agent has been activated. The main result of this contribution is the formulation and validation of a numerical simulation approach suitable for modelling the mechanical behaviour of self-healing cement-based composites.

1. Introduction

Increasing demands in exploitation along with the desire to extend the service life of structures, machines, and vehicles boost the research in many fields of industry and science. Material science provides new materials and advanced techniques for refurbishment, such as self-healing. For example, ceramic–matrix composites gain positions as structural components of hypersonic and rocket engines [1], cement-based composites for strengthening or repair of structures [2]. Significant research has also been done on the use of nanomaterials for enhancement of the performance of composites [3]. Self-healing is a technique for refurbishment applicable to a wide range of materials, such as metals [4], asphalt [5], etc.

Concrete has been known to possess self-healing properties ever since it became widely used as a construction material more than a hundred years ago. The concept of self-healing is a very appealing one, since it offers a way of overcoming the inherent flaw of concrete structure to develop cracks due to various reasons followed by gradual structural deterioration. Following the first formal report on self-healing concrete, [6] the research on the subject has been very active, [7] and references therein. The two mechanisms by which self-healing occurs – autogenous and autonomous have been identified and research has been conducted in both directions encompassing all physical phenomena and technological challenges involved, e.g. hydration and crystallization and the water supply needed for it, healing by epoxy resins, the delivery of healing agent by capsules, the role of moisture air and heat in the healing process, the reaction of the healing agent with the cementitious matrix or with a hardener, vascular systems for healing agent delivery etc., and their efficiency and feasibility, [7, 8]. Research on numerical simulation methods to complement the experimental investigations has followed suit and has been equally diverse, [9]. Most notably, a coupled hydration – elastic damage model has been developed and used to simulate a three-point bending test of a concrete specimen, [10], a model incorporating all relevant material behaviour such as mechanical damage, creep, shrinkage, thermal effects, self-healing and the interactions among them is presented in [11]. The fracture behaviour of encapsulation-based self-healing concrete is studied in [12], whereby the influence of parameters such as the volume fraction of capsules, and their shell thickness is incorporated. In [13], the Solidification-Microprestress-Microplane model for concrete has been extended to apply to self-healing and used to simulate the recovery of mechanical properties in cement-based composites.

The aim of this contribution is to demonstrate how a Damage Mechanics based computation incorporated into finite element code, can be extended and used to assess the effect of self-healing on the behaviour of a structural member made of cement-based composite, subjected to mechanical loading.

2. Finite element modelling

A four-point bending test is simulated. The model is 0.40 m long, 0.10 m deep and 0.10 m wide, the span is 0.30 m, and the two loading lines are at 0.10 m from the model mid-point, figure 1.

The finite element mesh and the support conditions are shown in figure 2.

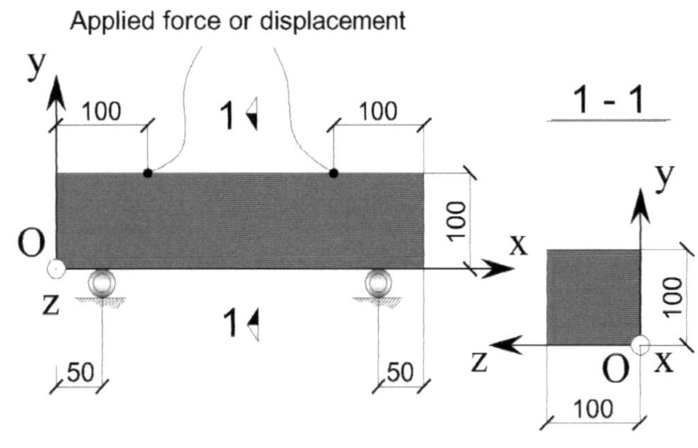

Figure 1. Geometry of the test specimen (all dimensions are in millimetres).

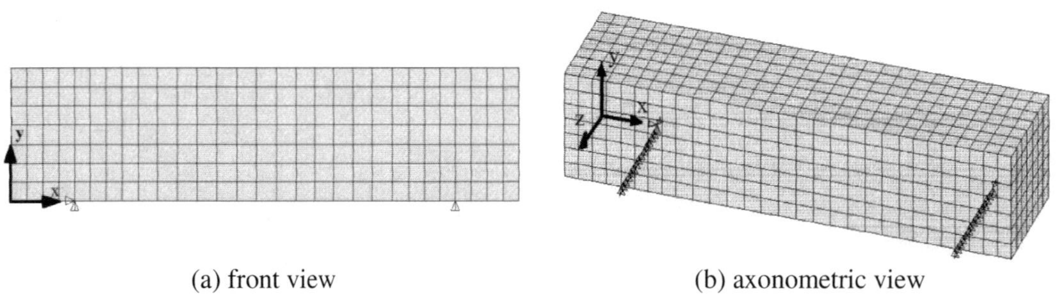

(a) front view (b) axonometric view

Figure 2. Finite element model of four-point bending test.

The loading is applied stepwise as prescribed displacements to the relevant nodes on the top surface of the model, thus simulating a displacement-controlled test.

All nodes of the two supports are restrained in the global Y direction. In order to prevent rigid body motion, one node of the left-hand support is restrained in the global X and Z directions, while one node of the right-hand support is restrained in the global Z direction.

The cement-based composite has an initial modulus of elasticity E (of the undamaged material) of 29,000 MPa and a Poisson's ratio ν of 0.22. The mechanical characteristics of the healing agent are as follows: $E_h = 7{,}050$ MPa and $\nu_h = 0.39$.

The cement-based composite is assumed to have fragile, strain-softening behaviour which is modelled by coupling linear isotropic elasticity with damage [14]:

$$\sigma_{ij} = \frac{\nu}{(1+\nu)(1-2\nu)} E(1-D)\varepsilon_{kk}\delta_{ij} + \frac{1}{(1+\nu)} E(1-D)\varepsilon_{ij} \tag{1}$$

In equation (1), ε_{kk} is the trace of the strain tensor ε_{ij}, and δ_{ij} are the components of the unit tensor. The scalar damage variable is split into a tensile, and a compressive component [15]:

$$D_t = 1 - \frac{\varepsilon_{inf}(1-A_t)}{\varepsilon_{eqv}} - \frac{A_t}{\exp[B_t(\varepsilon_{eqv}-\varepsilon_{inf})]} \tag{2}$$

$$D_c = 1 - \frac{\varepsilon_{inf}(1-A_c)}{\varepsilon_{eqv}} - \frac{A_c}{\exp[B_c(\varepsilon_{eqv}-\varepsilon_{inf})]} \tag{3}$$

The material constants A_t, B_t, A_c, B_c and ε_{inf} should be identified via curve fitting. The equivalent strain ε_{eqv} is defined as follows:

$$\varepsilon_{eqv} = \sqrt{\langle\varepsilon_1\rangle^2 + \langle\varepsilon_2\rangle^2 + \langle\varepsilon_3\rangle^2} \tag{4}$$

In equation (4) $\varepsilon_i, (i = 1..3)$ are the principal strains, and the McAuley brackets applied to a given quantity q are typically defined as follows:

$$\langle q\rangle = \frac{1}{2}(q + |q|) \tag{5}$$

Damage accumulates provided the condition

$$\varepsilon_{eqv} \geq \varepsilon_{inf} \tag{6}$$

is met.

A subroutine is developed to implement the nonlinear constitutive relation defined by equations (1) – (4) into the general-purpose finite element code ANSYS Mechanical APDL (®, ™). The strain-softening response of the cement-based composite material is obtained by incremental finite element analysis. For each increment of the analysis, the stress and strain distributions are determined. The tension and compression components of the damage variable are calculated and for the finite elements where damage accumulation is detected (i.e., where the criterion provided by equation (6) is met), the material constants are modified, according to equation (1). Furthermore, finite elements in which the critical value of the damage variable is reached are deactivated and do not contribute to the overall rigidity of the structural member in the subsequent increments.

It is assumed that a healable crack propagates in zones where the critical value of the damage variable is reached ($D = D_{max}$). As to the relation between the loading, cracking and healing, two scenarios are considered: a) the loading rate is very slow, so a newly-formed crack is filled with healing agent which has cured completely before the next loading step, and b) the loading rate is very fast, so healing cannot take place during the loading time. Instead, healing of all cracks occurs simultaneously, and the structure is again subjected to loading only after healing has completed.

For a specified finite element, the volume of cracks is predicted by subtracting from the volume of the deformed finite element $V_{d,FE}$ (prior to deactivation) the volume of the same finite element $V_{i,FE}$ in the initially generated finite element mesh,

$$V_{crack} = V_{d,FE} - V_{i,FE} \tag{7}$$

Finite elements containing cracks filled with healing agent are homogenized (see, for example [16]):

$$\frac{\mu_{eff}-\mu_1}{\mu_{eff}+\mu_1} = f\frac{\mu_2-\mu_1}{\mu_2+\mu_1} \tag{8}$$

$$\frac{(\lambda_{eff}-\mu_{eff})-(\lambda_1+\mu_1)}{\lambda_{eff}+\mu_{eff}+\mu_1} = f\frac{(\lambda_2+\mu_2)-(\lambda_1+\mu_1)}{\lambda_2+\mu_2+\mu_1} \tag{9}$$

In equations (8) and (9) μ_1, λ_1 and μ_2, λ_2 are the Lamé coefficients for the cement-based composite and for the healing agent, respectively; f stands for the volume fraction of the healing agent,

$$f = \frac{V_{crack}}{V_{i,FE}} \tag{10}$$

For the considered case, for practical purposes, after the micro-crack initiation and the activation of the healing agent, the homogenization may be avoided by simply restoring the initial mechanical characteristics of the cement-based composite. In the subsequent loading, the restored mechanical properties will be maintained until the tensile strength of the healing agent is reached. This argument will hold as long as the volume of the crack is small compared to the volume of the finite element, i.e., the mesh is relatively coarse.

3. Results and discussion

A comparison between the response of a test specimen without a healing agent and a test specimen subjected to the action of a healing agent is shown in figure 3. Both specimens have the same geometry, identical mechanical properties, and are subjected to a displacement-controlled quasi-static loading.

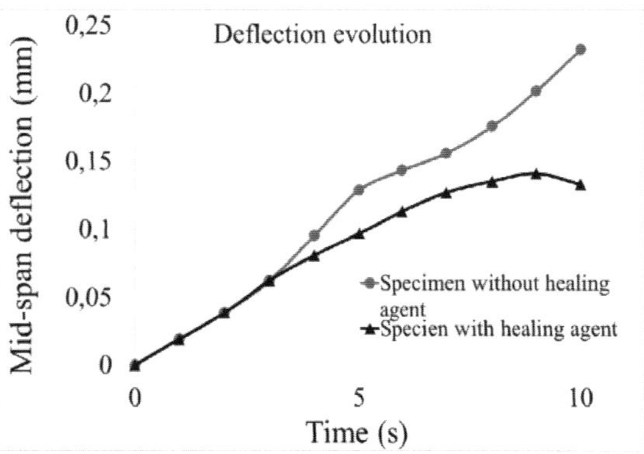

Figure 3. Evolution of the mid-span deflection of a specimen without healing agent and a specimen with healing agent. Absolute values of the mid-span deflection are plotted. According to the convention for the coordinate system used (see figure 1), the mid-span deflection should be negative.

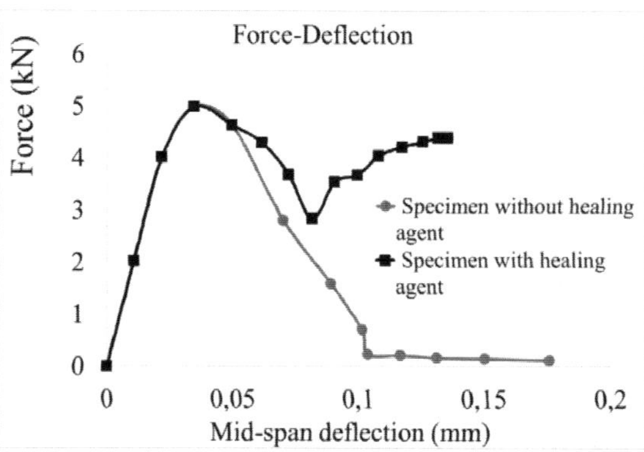

Figure 4. The predicted effect of the healing agent: comparison between the mechanical responses of a specimen containing a healing agent and a specimen without a healing agent. Absolute values of the mid-span deflection are plotted.

It can be seen that with the damage accumulation and cracking, the specimen which is not subjected to the action of a healing agent loses rigidity, which results in a more rapid increase of the mid-span deflection. The same trend is confirmed by figure 4 from a different perspective; with the loading and the healing action being concurrent, the specimen preserves its load-bearing capacity much better than its non-healing counterpart.

Both figure 3 and figure 4 illustrate the continuous action of the healing agent. The healing agent starts to act as soon as the cracks' width reaches a specific, predefined value (equivalent to a specified value of the damage variable in the model). The healing agent fills the macro-crack as soon as it forms ($D = D_{max}$). It should be noted that this scenario can be considered realistic only for a loading rate smaller than the curing rate of the healing agent.

Figure 5 illustrates another loading scheme. The test specimen is loaded until a specified amount of damage is reached. The specimen is then unloaded and healed. Afterwards, the specimen is loaded again. The mechanical response shown in figure 5 is obtained for a different set of model constants (A_t, B_t, A_c, B_c).

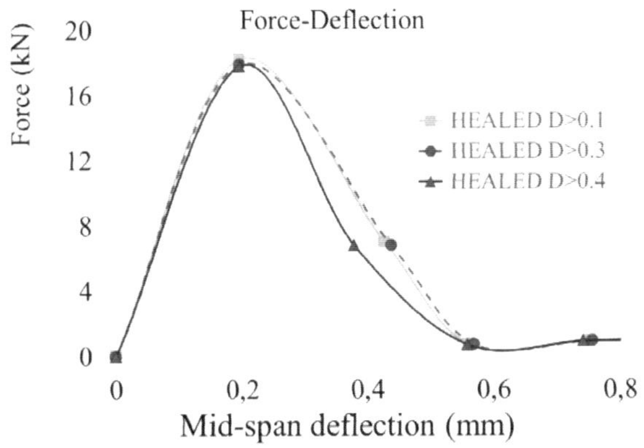

Figure 5. Mechanical response of specimens, subjected to partial healing after a preloading phase. Absolute values of the mid-span deflection are plotted.

The healing process is illustrated in figure 6 and figure 7. During loading, mechanical damage is initiated and accumulated in the specimen, which results in a degradation in the initial mechanical properties of the material. This can be seen in figure 6, where all finite elements affected by damage, i.e., all finite elements in which Young's modulus has become less than the initial one, because of damage accumulation are plotted in light blue. Due to the action of the healing agent, the initial mechanical properties are partially restored. Figure 7 shows the damaged specimen from figure 6, after having been subjected to partial healing. Again, finite elements of degraded material properties are plotted in light blue, whereas finite elements in which the initial material properties are maintained (or restored due to the action of the healing agent) – in blue.

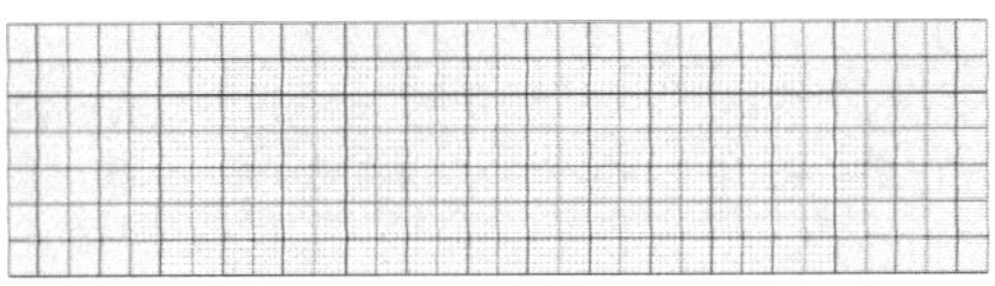

Figure 6. Damaged specimen. All finite elements affected by damage are displayed in light blue.

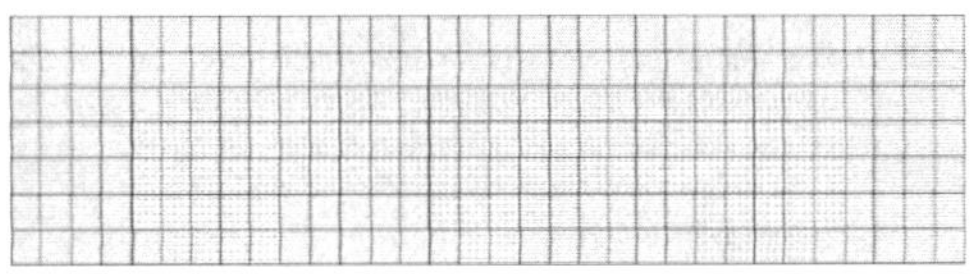

Figure 7. Partially healed specimen: the initial mechanical properties of some of the damaged finite elements are restored.

It should be noted that the model chosen for the cement-based composite is not designed to account for the damage-induced anisotropy; a scalar damage variable is used in the constitutive relation.

It has been pointed out above that after the micro-crack initiation, the healing agent is activated and fills immediately the newly-formed crack when the limit value of the damage variable is reached. At this point, the medium within the finite element is relieved from stress due to the formation of a macro crack, which is healed once the agent cures, so a portion or all of the initial rigidity of the medium should be restored. How much should be restored is a question that can only be clarified by experimental evidence.

Typically, it is assumed that damage localization takes place. However, if damage localization is considered as a process, any formation of a micro-crack is equivalent to formation of a damaged zone within the finite element. The healing agent acts only in the finite elements in which the value of the damage variable is larger than a specific value, i.e., it is tacitly assumed that the healing agent cannot circulate in cracks of width less than a specified value. Thus, the restored rigidity within the finite element should be inferior to the initial characteristics of the material at this location.

As pointed out, the input for the strain-softening model is provided by the strain distribution obtained for each increment of the nonlinear finite element analysis. In the context of this study, the stress distribution is relevant only for the investigation of possible failure in the joint formed by the healing agent after the partial healing. The first principal stress determined in the healed finite element is compared with the tensile strength of the healing agent. The stress distributions for two consecutive increments of the numerical simulation are plotted in figure 8.

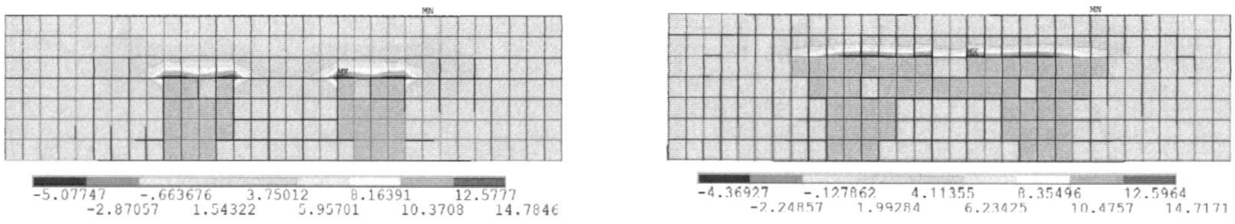

(a) increment "i" (b) increment "i+1"

Figure 8. First principal stress (N/mm^2) in the specimen for two consecutive increments of the finite element simulation of the partially healed specimen.

In figure 8, the zones containing deactivated finite elements are shown in grey. It can be seen that these zones propagate, as the simulation progresses.

4. Conclusions

Various scenarios of the action of the healing agent on a specimen made of a cement-based composite and subjected to a quasi-static four-point-bending test, have been investigated numerically: a) a continuous action of the healing agent throughout the loading history and b) partial healing after a preloading phase and subsequent loading. The effect of the healing agent has been demonstrated in the finite element simulations. The mechanical behaviour of the specimens is characterized by the reaction force-deflection relationships.

The novelty of this paper is in the implementation of the Damage Mechanics based approach for modelling the mechanical response of structural members made of cement-based composite and subjected to self-healing.

Clearly, experimental validation of the numerical predictions is required. Moreover, an experimentally obtained input is needed for the better calibration of the material constants. In this context, the numerical investigation in the contribution focuses on a case study that can be, relatively easily, reproduced experimentally. The results of such characterization tests are required for the calibration of the damage-based model in its current form.

Further research work includes: (i) a mesh-sensitivity study, as a finer mesh will possibly provide more accurate results in terms of crack patterns obtained by finite element analysis; (ii) finite element modelling and simulation of various benchmark examples to validate the model for a broader range of geometries and load cases; (iii) investigation of the relationship between the crack width and the damage variable; study of the influence of these parameters and the effect of the healing agent on the mechanical response of a damaged structural member subjected to partial healing for practical purposes.

References

[1] Padture N P 2016 Advanced structural ceramics in aerospace propulsion *Natural Materials* **15** pp 804–9
[2] Mechtcherine V 2013 Novel cement-based composites for the strengthening and repair of concrete structures *Construction and Building Materials* **41** pp 365–73
[3] Yang H, Cui H, Tang W, Li Z, Han N, Xing F 2017 A critical review on research progress of graphene/cement based composites *Composites Part A: Applied Science and Manufacturing* **102** pp 273–96
[4] Ramasamy P, Stoica M, Ababei G, Lupu N and Eckert J 2020 Soft Ferromagnetic Bulk Metallic Glass with Potential Self-Healing Ability *Materials* **13** 1319
[5] Norambuena-Contreras J, Arteaga-Perez L-E, Guadarrama-Lezama A Y, Briones R, Vivanco J F and Gonzalez-Torre I 2020 Microencapsulated Bio-Based Rejuvenators for the Self-Healing of Bituminous Materials *Materials* **13** 1446
[6] Dry C M 1994 Matrix cracking repair and filling using active and passive modes for smart timed release of chemicals from fibers into cement matrices *Smart Materials and Structures* **3** pp 118–23
[7] Tittelboom K and Belie N 2013 Self-healing in cementitious materials – a review *Materials* **6** pp 2182–2217
[8] Ferrara L, Mullem T, Alonso M C, Antonaci P, Borg R P, Cuenca E, Jefferson A, Ng P L, Peled A, Roig-Flores M et al 2018 Experimental characterization of the self-healing capacity of cement based materials and its effects on the material performance: a state of the art report by COST Action SARCOS WG2 *Construction and Building Materials* **167** pp 115–42
[9] Mauludin L M and Oucif C 2019 Modeling of self-healing Concrete: a review *J. Applied Computational Mechanics* **5**(3) pp 526–39
[10] Granger S, Pijaudier-Cabot G and Loukili A 2007 Mechanical behavior of self-healed ultra-high performance concrete: from experimental evidence to modelling *Proc. 6th international conference on fracture mechanics of concrete and concrete structures, FraMCos-6*, ed Carpinteri A, Gambarova P G et al (London: Taylor & Francis)
[11] Hazelwood T, Jefferson A D, Lark R J and Gardner D R 2015 Numerical simulation of the long-term behaviour of a self-healing concrete beam vs standard reinforced concrete *Engineering Structures* **102** pp 176–88
[12] Mauludin L M, Zhuang X and Rabczuk T 2018 Computational modeling of fracture in encapsulation-based self-healing concrete using cohesive elements *Composite Structures* **196** pp 63–75
[13] Luzio G, Ferrara L and Krelani V 2018 Numerical modeling of mechanical regain due to self-healing in cement based composites *Cement and Concrete Composites* **86** pp 190–205
[14] Lemaître J and Dasmorat R 2004 *Engineering Damage Mechanics: Ductile, Creep, Fatigue and Brittle Failures* (Berlin/Heidelberg: Springer)
[15] Mazars J J 1984 *Damage mechanics application to nonlinear response and failure behavior of structural concrete* Ph. D. Thesis (Paris: Paris 6 University)
[16] Mei J, Liu Z, Wen W and Sheng P 2007 Effective dynamic mass density of composites *Physical Review B* **76** 134205

Development of a method for studying the impact of the time reserve value on the reliability of the train schedule based on the epidemiological SIR model

S Panchenko[1], A Prokhorchenko[2,5], O Dekarchuk[1], D Gurin[1], D Mkrtychian[3] and V Matsiuk[4]

[1] Department of automation and computer remote control train traffic, Ukrainian State University of Railway Transport Feierbakh sq., 7, Kharkiv, 61050, Ukraine 4
[2] Department of Management of Operational Work, Ukrainian State University of Railway Transport Feierbakh sq., 7, Kharkiv, 61050, Ukraine
[3] Department of manage freight and commercial work, Ukrainian State University of Railway Transport Feierbakh sq., 7, Kharkiv, 61050, Ukraine
[4] Department of Transport Technology and Process Control Traffic, State University of Infrastructure and Technologies, Ivana Ohienka str., 19, Kyiv, 03049, Ukraine

[5] E-mail: prokhorchenko@kart.edu.ua

Abstract. It is important to find rational values of time reserves to compensate for delays during train movement to ensure train schedule reliability. This study proposes to investigate the impact of different categories of trains and their time reserve on schedule reliability. The main purpose of the study is to develop a method for studying the extent of the impact of time reserve on train schedule reliability based on the epidemiological model. Mathematical train delay propagation at the section model has been developed on the basis of the Susceptible – Infected – Removed (SIR) epidemiological model of adaptation type (taking into account the effect of using different train categories. It is suggested to use a binary genetic algorithm to search for parameters of delay propagation rate in the SIR model. According to a specially developed data acquisition algorithm, empirical data on the propagation of train delays at the station were obtained. Experimental studies have been carried out to search for train delay propagation velocity coefficients in the SIR model. The train movement restoration results during a delay at the station are given to study the impact of the time reserve value on train schedule reliability, taking into account the speed recovery delay set in the SIR model. The simulation results confirm the developed mathematical model adequacy.

1. Introduction

One of the important ways to increase the efficiency of rail transportation of goods and passengers is to ensure high quality indicators of basic transport services as a result of both constructive [1] and organizational measures [2, 3]. However, in railway systems with mixed traffic of passenger and freight trains with partial adherence to the timetable, there is a problem of ensuring the reliability of the transportation process. Improving the reliability of work schedules of train traffic is one of the important indicators of organizational decisions to ensure the competitiveness of railway transport. To achieve these indicators in the process of using the developed train schedule, it is necessary to conduct studies on the effect of train delays that may occur on the stability of movement of all other categories of train flows. This will allow, at the stage of developing a working train schedule, to provide for temporary compensation zones with the possibility of restoring train movements, which are delayed with the least impact on network traffic. It should be noted that there are almost no effective methods for numerically determining the impact of trains of various categories and the size of their reserve time on the reliability of the train schedule without observing the freight train departure timetable. Taking into account that the mutual impact of train flows, in conditions of delays, on a railway section is nonlinear, characterized by a significant level of uncertainty, complicates the task of formalizing this process. Under such conditions, the goal is to develop a research method that will formalize the process of propagation of train delays on a railway line, taking into account the mutual impact of trains of various categories and the size of their reserve time. This will make it possible to assess the impact of delays that may occur on the section under the conditions of existing values of time reserves for finding rational parameters of the train schedule. As a result, based on the research results, it is possible to develop a method that will reduce subjectivity in making decisions on establishing the values of time reserves for various categories of trains with a mixed traffic system without observing the freight train departure schedule, and will increase the accuracy and speed of assessing the reliability of the train schedule before its implementation. This will ensure the stability of movement, high speed of trains, increase the capacity of the section and reduce the impact of delays on the stability of the line. Thus, studies aimed at improving the reliability of working train schedules for railway systems without adhering to the freight train departure timetable are relevant and have practical value.

2. Literature review and problem statement

A lot of work has been devoted to research in search of new effective methods for analyzing the effect of time reserves on the reliability of train schedules. In particular, in his studies author of [4] compares the effectiveness of various strategies for applying time reserves in Danish rail timetables to the recommendations of the UIC. In [5], based on an analysis of train delays on the railways of Holland, Switzerland, and Germany, threshold delays were determined for various categories of trains in these countries. There is research [6] focused on review of the state of practice and existing schedule planning procedures based on the rules of railway administrations from several European countries. A number of shortcomings were identified in the development of a motion schedule, one of which is the lack of effective modeling tools, optimization and data analysis methods for improving schedules. As part of the development of research methods for studying the impact of the time reserve value on the reliability of the train schedule, the accuracy of predicting the propagation of train delays in the network is important. Based on the analysis of various approaches for predicting the spread of train delays in the network, the following are distinguished: deterministic, stochastic, statistical.

Using the deterministic approach, it is possible to study the network response to delays, but its application does not allow for a realistic assessment of the dynamics of the distribution of delays of railway networks in which the planned train schedule is completely or partially missing [7]. The stochastic approach allows for more accurate and adequate prediction of the propagation of the delay [8], but requires verified input data regarding the parameters of the propagation of delays, which are not always possible to obtain. Many studies focus on statistical analysis of completed train schedules to find statistical patterns in the propagation of delays [9]. In addition, all approaches can be divided according to the level of detail into micro- and macro-modeling. Simulation models based on the principles of micromodeling provide more accurate results, but knowledge of the normative train schedule is important for their work. In the railway system of Ukraine and the like, freight trains depart from forming stations "on readiness" without observing the timetable, which complicates the analysis of the fulfillment of the train schedule for deviations from the standard transit time standards. It is for this type of movement that it is important to increase the reliability of the transportation process.

According to the analysis, which made it possible to identify the advantages and disadvantages of various approaches to modeling, the proposed macroscopic approach to modeling the propagation of train delays is proposed for a railway system with partial observance of train schedules. One of the promising directions for modeling the propagation of train delays at the macro level is the use of modified epidemiological mathematical models used to simulate epidemics, the spread of viruses, and the like. A successful example of the application of the epidemiological model is [10], which proposed a mathematical SIR model for formalizing the congestion propagation on urban transit networks by a congestion susceptibility recovery process. However, the use of such a model for determining the time reserves in the schedule of movement of railway networks of mixed movement of passenger and freight trains without observing the latest timetables is impossible. The study [11] proposed an SIS model of epidemic spreading, which is able to reproduce the empirical distribution of train delays. This approach suggests the possibility of several delays for one train. The simulation results in terms of scheduled train traffic in the Italian railway network confirmed the adequacy of the mathematical model in the real processes of delay propagation. This proves the effectiveness of the chosen research area. Given that the macroscopic approach to modeling the propagation of train delays can more adequately describe the uncertainty of the parameters of the train flow system without observing the network traffic schedule, the main goal of the study is to develop a method for studying the impact of the time reserve on the reliability of the train schedule based on the epidemiological model.

3. Materials and methods

For railway systems not observing the freight train schedule, to which the Ukrainian railway belongs, the rules for establishing time reserves in the threads of trains of various categories are quite general, which requires studies of the impact of trains of various categories and the size of their time reserve on the reliability of the schedule on the section, taking into account that the most unexplored indicator is the time for recovery in the thread, distributed by the method of compensation for movement [12]. Compensation for movement consists of using the difference between the minimum train time for the section and the fact that the planned schedule contains a component – the regular addition time, which is attached to the train time for each section in accordance with the requirements of the infrastructure operator. The amount of addition time (time reserve) in different railway systems varies depending on the distance the train follows, the time it takes and the station to park. This reserve of time, when the train is late, allows for surge operation to accelerate the train in comparison with the planned schedule in order to reduce delay and avoid scenarios of cascading delays with significant costs. Figure 1 shows one of the possible practical examples of the application of surge operation in the event of an initial delay in the train schedule on a conditional double-track section A-B.

Upon departure from station B, freight train No. 2002 experienced an initial delay of t_{delay} = 14 minutes with two scenarios. Delay involves the development of a second scenario in which the stop of train No. 2002 at the station is inevitable in order to perform the overtaking operation (with 24 minutes of downtime) of passenger trains following a pack according to the normative schedule with the highest priority. If there is a reserve of time – compensation during movement, the course of events may change into scenario 1, which provides for the transfer of an order from the dispatcher to the freight train operator No. 2002 to accelerate the latter on the first stretch B-I of the second train by 2 minutes, on the second j-f – by 4 min; on the third f-e – by 5 minutes; on the fourth e-d – by 5 minutes; on the d-c – by 4 minutes, which will allow freight train No. 2002 to arrive at station A in accordance with the regulatory train schedule. One example of the development of scenarios in the practical conditions of the implementation of the motion graph proves the effectiveness of the existence of a reserve in the form of compensation time in the graph thread. However, if too long standby time is selected, the duration of the train occupying the infrastructure of the section increases, which leads to a decrease in the district speed and throughput of the section. Under such conditions, it is important to solve the problem of finding rational values of the compensation time in the threads of the train schedule of various categories. This will allow us to find a balance between the reliability of the traffic schedule and the practical throughput of the railway section.

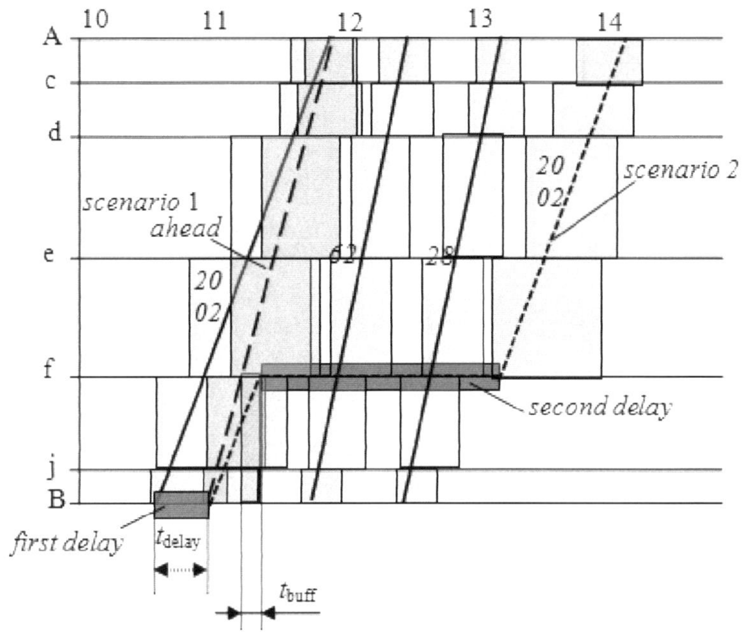

Figure 1. Scheme of train movement scenarios when a delay occurs on a double-track section: scenario 1 – implementation of the ahead operation; scenario 2 – overtaking with passenger trains.

To solve the problem of finding rational values of the compensation time in threads, it is proposed to apply the SIR-model (Susceptible-Infected-Removed model), which is used to model the spread of infectious diseases in a closed society, the spread of viruses in a computer network and cases of planned exposure on an agent in social networks [13]. Adapting this model to the problem of the impact of train delays on the railway network, it is possible to imagine the spread of infectious diseases as the process of spreading train delays and the associated effect on the movement of trains on the railway network. Within the framework of the adopted statement of the problem, to describe the effect of the time reserve in the train schedule threads of various categories, it is proposed to present the total number of trains in the calculation section in three states, each of which represents an outstanding stage in the development of delay propagation.

Thus, the total number of train threads (freight and passenger) should be divided into groups: $S(t)$ – delay-sensitive trains, which at time t follow the schedule but can be delayed when moving on the section (Susceptible) $I(t)$ – train, following with a delay infecting the movement of other trains, passing on a delay (Infected) $R(t)$ – train, during the passage through the railway section with a delay, compensated for the delay, followed the schedule, and did not cause the occurrence of delays in further movement (Removed). Figure 2 shows a graph of the process of transition of trains between states on the site.

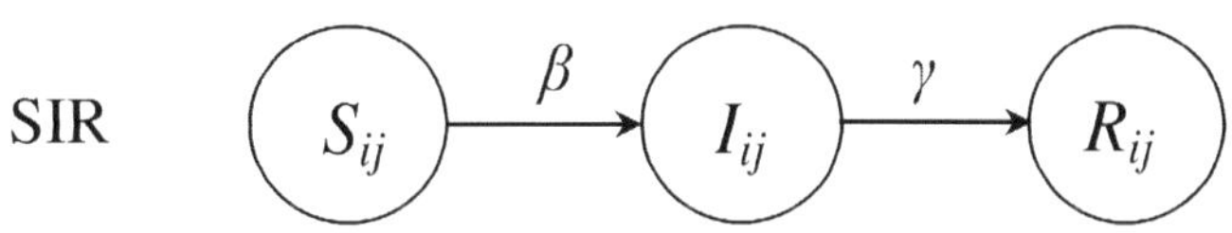

Figure 2. Graph of the transition process of trains between states in the section.

Given the importance of taking into account the different priority of traffic on a section of trains of various categories, it is proposed to expand the state groups into subgroups-classes in accordance with their category r. For a mixed model of site operation, in this study it is proposed to take into account the following categories of trains: *r*=1 – passenger trains; *r*=2 – community trains, *r*=3 – freight trains. According to current rules [14], the priority hierarchy corresponds to the indicated sequence of categories. It has been suggested that trains of each class have the same characteristics of delay transmission and motion recovery processes.

According to [15] a mathematical model can be written as a system of differential equations of the form

$$\begin{cases} \dfrac{dS^r(t)}{dt} = -\sum_l \beta^{r,l} S^r(t) I^l(t); \\ \dfrac{dI^r(t)}{dt} = \sum_l \beta^{r,l} S^r(t) I^l(t) - \gamma^r I^r; \\ \dfrac{dR^r(t)}{dt} = -\gamma^r I^r, \end{cases} \quad (1)$$

where S^r – the number of opportune trains in class r on the section, trains; $\beta^{r,l}$ – propagation speed of the delay from class r trains to class l trains in the area where $r = l, r, l \in R$; I^l – the number of infected class l trains in the section, trains; R^r – the number of class r trains that followed with a delay, but removed it, trains; γ^r– delay recovery speed in class r trains on a section; t – time or simulation step, hour. All system parameters have positive values.

The system of equations has a standardization condition of the form $N^r = S^r(t) + I^r(t) + R^r(t)$, where N^r – the total number of class r trains in the section $N = \sum N^r$, N – the total number of trains of all classes in the section. The initial conditions for the system (1) are the level $(S_0^r, I_0^r, 0)$, where S_0^r, I_0^r – the number of favorable and delayed class r trains in the section at $t = 0$, respectively.

Important parameters of the system (1) are the propagation rate of delay from trains $\beta^{r,l}$ and the recovery rate of delay in class r trains in section γ^r. In this study, it is proposed to conduct studies in the following sequence: at the first stage, develop a method for searching for parameters $\beta^{r,l}$ on the delay data of the real section, and then model the propagation of the delay with different delay recovery rates.

It is proposed to use the method of evolutionary calculations, in particular, the binary genetic algorithm, BGA [16], to search for parameters of the delay propagation velocity in the system of differential equations (1). In general, the parameter search problem $\beta^{r,l}$ can be represented as an optimization problem of the form

$$\begin{aligned} &F(\beta) \to \min \\ &g(\beta) = 0, \\ &h(\beta) \le 0,\ \beta \in \Re^n, \end{aligned} \quad (2)$$

where F – the objective function estimates between empirical values of the number of infected (delayed) trains I^r and model I_m^r obtained by solving the system of differential equations (1) taking into account the given initial conditions corresponding to real ones; $g(\beta) = 0$; $h(\beta) \le 0$ – the restriction of equality and inequality, respectively, corresponding to the problem statement by solving the system (1). Within the framework of solving the optimization problem (2) in the system (1), the parameters $\gamma^r = 0$.

To implement the optimization problem (2) within the framework of the genetic algorithm, the model variables are presented in the form of a fixed-length chromosome, taking into account the restrictions on their ranges and putting them into one numerical vector

$$C^h = \left(\beta^{1,1}, \beta^{1,2}, \beta^{1,3}, \beta^{r,l}, \dots, \beta^{3,3}\right), \quad (3)$$

where $h = \overline{1,K}$ – the chromosome number C, $\beta^{r,l} \in \{3 \times 3\}$ or $C^h = [1 \times 9]$, and the variable has limitations $0 \leq \beta^{r,l} < 1$. The solution of the system of differential equations (1) within the fitness function of the BGA is proposed to be performed using the fourth-order Runge-Kutta method [17]. The criterion for evaluating the selection of variables in the optimization problem (2) applied the average absolute error (MAPE) between the empirical and the results [18]

$$F(\beta) = \frac{1}{M} \sum_{M,r} \frac{|I^r(t) - I^r_m(t)|}{I^r_m(t)} 100\% \rightarrow \min \tag{4}$$

where the indicator $F(\beta)$ is not defined for zero values $I^r_m(t)$, M – the number of rows of vectors of experimental data of the sample $\langle \bar{X}_q, \bar{y}_q \rangle$, $q = \overline{1,M}$; $I^r(t)$ – output sample vector $\langle \bar{X}_q, \bar{y}_q \rangle$, $I^r(t) \in \bar{y}_q$; $I^r_m(t)$ – the result of solving the system of differential equations (1) under initial conditions $(S^r_0, I^r_0, 0)$, corresponding to the q-th line of the sample $\langle \bar{X}_q, \bar{y}_q \rangle$, $(S^r_0, I^r_0, 0) \in \bar{X}_q$.

4. Results

To be able to obtain accurate data on the distribution of the number of delayed trains over time on the calculated railway section, taking into account the various categories of trains, a special data acquisition algorithm was developed. To generate train delays of various categories and their distribution in the traffic schedule, an optimization mathematical model [19] was used in this work. According to the developed procedure, the simulation of the distribution of train delays in the work schedule for 2017-2018 years was performed on the Liubotyn - Sovnarkomivska regional branch of the Southern Railway branch of Ukrzaliznytsya JSC. This section is a bottleneck in the movement of trains to stations of the Kharkov railway junction.

In the experimental studies, three scenarios were constructed according to which the delay of one train from the adopted class r was simulated with constant parameters of the infrastructure and the normative schedule. The delay was modulated in the odd direction of movement for the departure from the Lyubotyn station with a value of 25.17 minutes, according to the research of [19]. Initial conditions at $t = 0$ for modeling $N^{r=1} = 9$ trains, $N^{r=2} = 4$ trains, $N^{r=3} = 32$ trains. Visualization of the graphs of the dependences of the number of delayed trains of different classes with a cumulative total from the moment the model delay of the passenger train occurs is shown in figure 3.

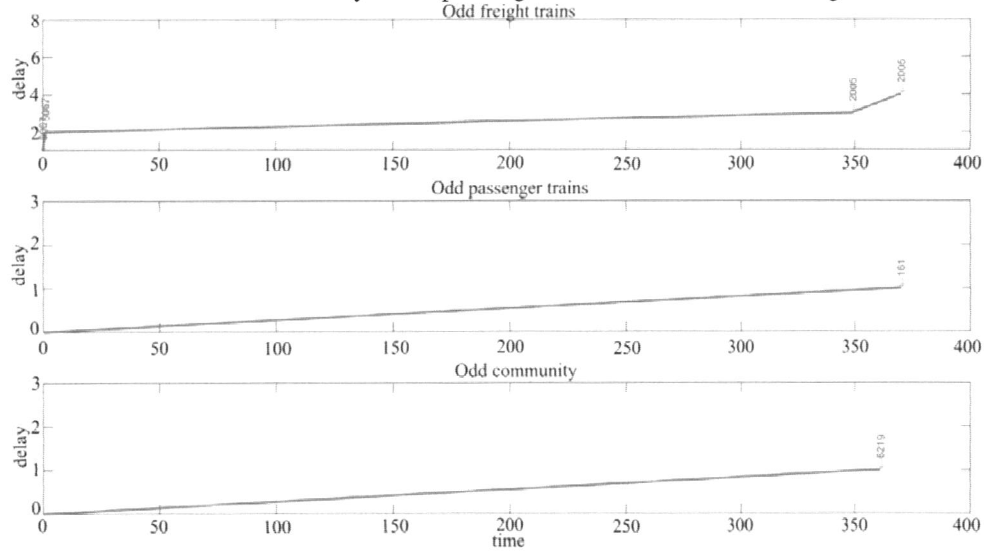

Figure 3. The dependence of the number of delayed trains of different classes with a cumulative total from the moment of model delay of passenger train No. 161 in the odd direction of movement on the Liubotyn - Sovnarkomivska section.

With the delay of one community train, the consequences are much less compared to the delay of the passenger train, which has a higher priority (figure 4).

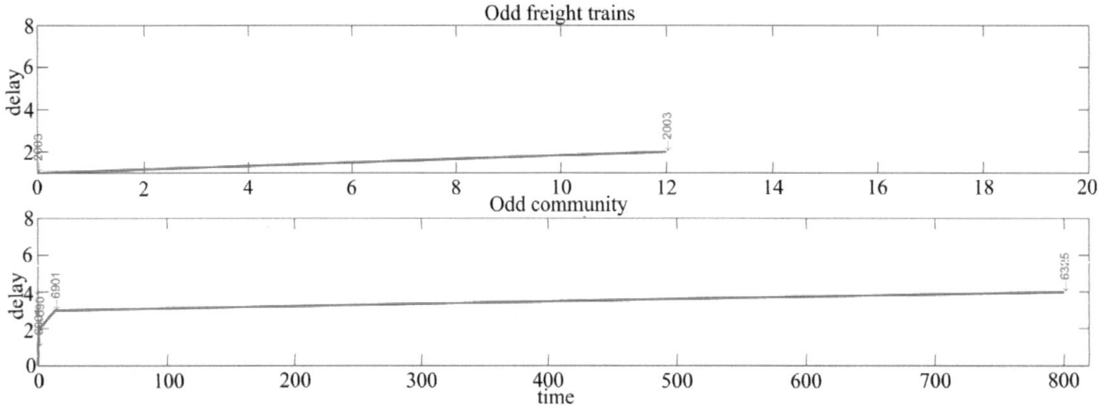

Figure 4. The dependence of the number of delayed trains of different classes with a cumulative total from the moment of model delay of the passenger train No. 161 in the odd direction of movement.

Delay of a freight train caused a cascade of delays of other freight trains in the section with the same high priority. Having the highest priority, passenger trains avoided delays, while one community train fell into the cargo delay zone, which was a consequence of its delay (figure 5).

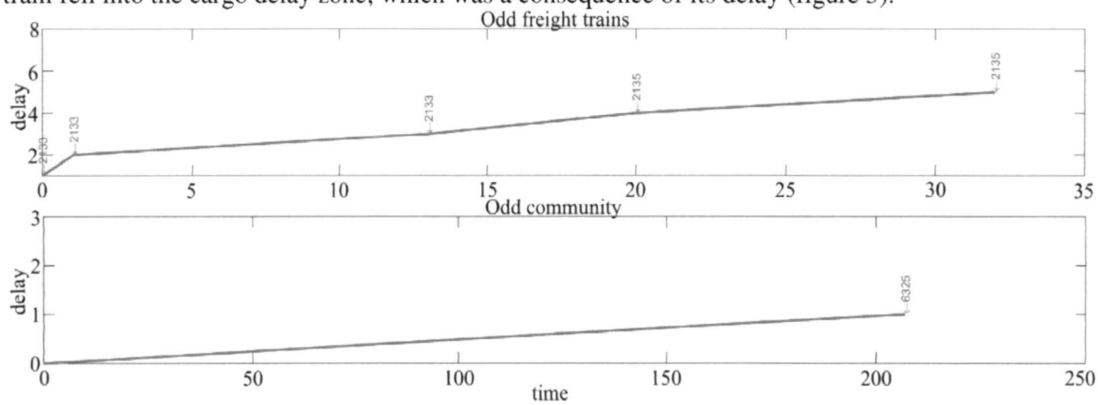

Figure 5. The dependence of the number of delayed trains of different classes with a cumulative total from the moment of model delay of the freight train No. 161 in the odd direction of movement.

The obtained data are presented as a set of vectors $\langle \bar{X}_q, \bar{y}_q \rangle$ for converting time from minutes to hours. The data are used to search for delay propagation velocity parameters in a system of differential equations by decoupling the optimization model (2) using the BGA method in the MATLAB environment. Visualization of the dynamics of changes in the values of the fitness function of BGA is shown in figure 6.

On the basis of the results of the decoupling of the optimization model (2), the following coefficients are found; $\beta^{1,1} = 0.0004$; $\beta^{1,2} = 0.0011$; $\beta^{1,3} = 0.0001$; $\beta^{2,1} = 0.0012$; $\beta^{2,2} = 0.1784$; $\beta^{2,3} = 0.0015$; $\beta^{3,1} = 0.0001$; $\beta^{3,2} = 0.0001$; $\beta^{3,3} = 0.0023$. The magnitude of the coefficients characterizes the impact of some categories of trains on others, and allows for numerical evaluation of the mutual impact of trains of different priorities in the train schedule.

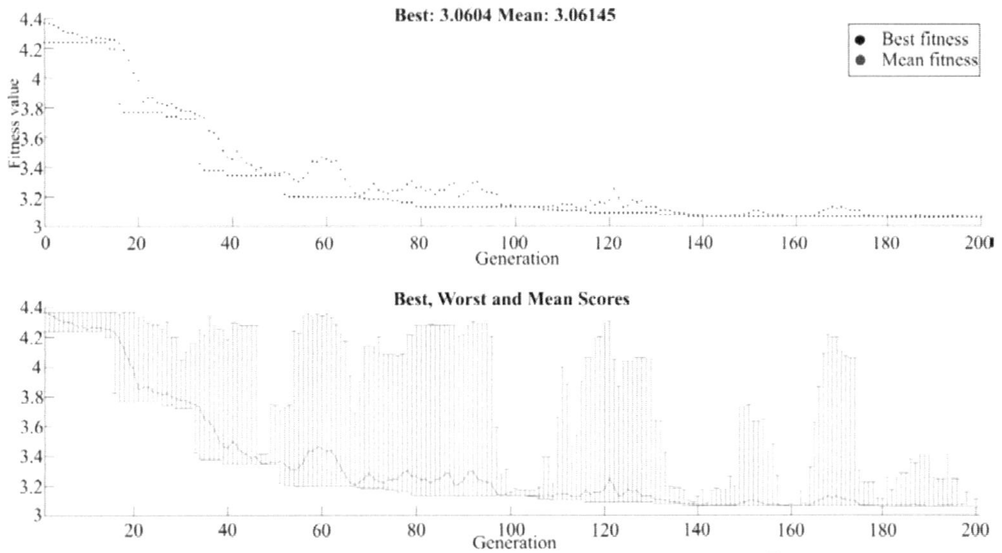

Figure 6. A graph of the best and average values of the objective function F versus the number of BGA iterations in the process of selecting delay propagation velocity coefficients.

According to the given input conditions $\langle \bar{X}_q, \bar{y}_q \rangle$, three delay scenarios ($q = 3$), the empirical results of which are shown in figures 3-5 and model results are obtained, which are shown in figure 7. The average absolute error (MAPE) between empirical and with the results is 3%, which is a high result.

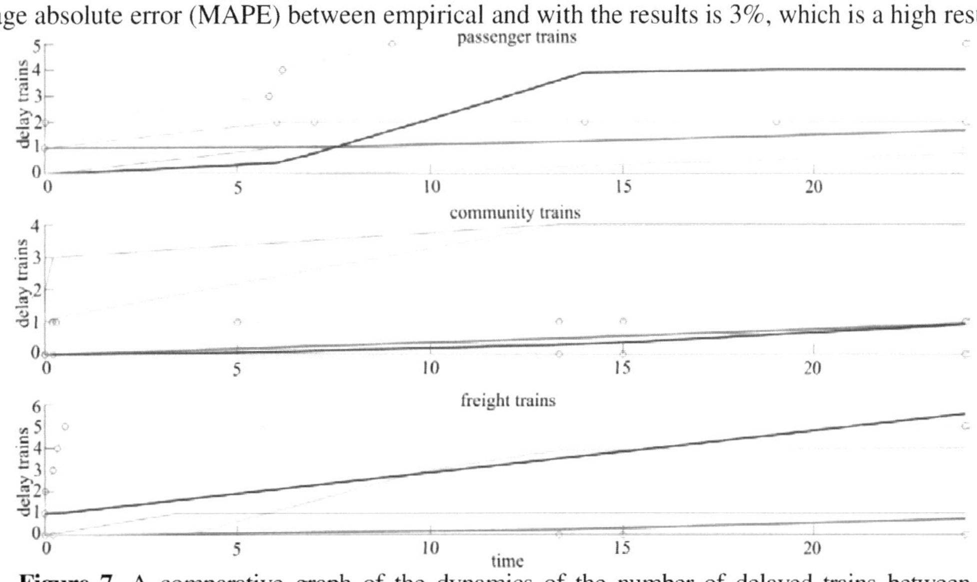

Figure 7. A comparative graph of the dynamics of the number of delayed trains between empirical values I^r and model I^r_m according to the three delay scenarios in accordance with $\langle \bar{X}_q, \bar{y}_q \rangle$ the calculated section: thin line with circles – empirical values I^r; bold line – model values I^r_m; red colour – passenger trains r=1; green colour – community trains r=2; blue colour – freight trains r=3.

To study the effect of the time reserve on the reliability of the train schedule, the experimental results of modeling the restoration of train movement in the area under study were carried out. It is proposed to change the initial conditions – to delay 5 passenger trains out of the 9 trains available, $I^{r=1}$ (t=0) = 5 trains. Based on current standards, use the following parameters of the delay recovery speed for trains of class r in the section $\gamma^{r=1} = 0.05$; $\gamma^{r=2} = 0.0667$; $\gamma^{r=3} = 0.10$. Initial conditions at t =

0 for modeling $N^{r=1} = 9$ trains, $N^{r=2} = 4$ trains, $N^{r=3} = 32$ trains. The coefficients of the propagation velocity of the delay $\beta^{r,l} \in \{3 \times 3\}$ are taken in accordance with the given values of the decoupling of the optimization problem in Section 6 of this article.

The solution of the developed system of equations of the SIR model makes it possible to obtain the results of the restoration of train movement in the event of a delay of five passenger trains on the Liubotyn - Sovnarkomivska line (figure 8).

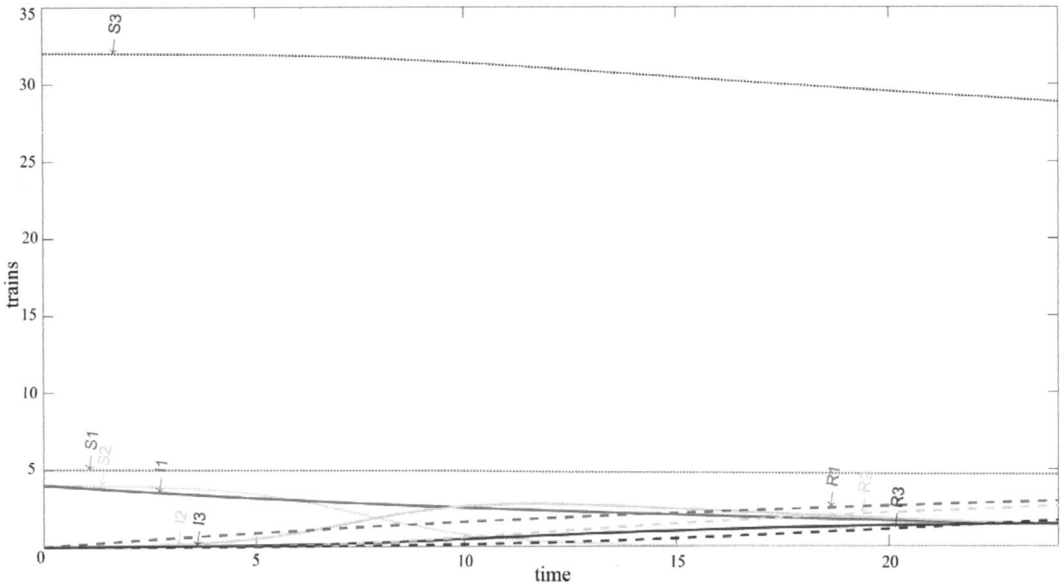

Figure 8. Graph of the dynamics of changes in the state of train flows under the conditions of restoration of train movement during the day (t = 24 hours) when there is a delay of five passenger trains on the Liubotyn – Sovnarkomivska line: red – passenger trains *r*=1; green – community trains *r*=2; blue – freight trains *r*=3.

The initial conditions and the value of the states of trains of different classes *r* on the line at *t* =24 hours are shown in table 1.

Table 1. The results of modeling the restoration of train movement during the day (t = 24 h) in the event of a delay of five passenger trains on the Liubotyn – Sovnarkomivska line.

States	Passenger trains, *r*=1	Community trains, *r*=2	Freight trains, *r*=3
Susceptible, S^r(t=0)	4	4	32
Infected, I^r(t=0)	5	0	0
Removed, R^r(t=0)	0	0	0
Susceptible, S^r(t=24)	3.699	0.003	28.735
Infected, I^r(t=24)	1.683	1.356	1.483
Removed, R^r(t=24)	3.618	2.641	1.783

The numerically determined effect of trains of various categories and the value of their time reserve on the reliability of the traffic schedule on the line was investigated for the first time. Of the five trains that were delayed, with a compensated recovery time of 20 minutes, it was possible to restore movement of 47% of the total number of trains detained. At the same time, the delay of five passenger trains (the highest priority of traffic) led to delays of other trains. The set recovery time for suburban and important trains made it possible to restore movement of 51% and 83% of trains, respectively, of the total number of delayed trains.

5. Discussion and conclusion

According to the results of this research, the process of propagating train delays on the railway line was formalized. This allowed to investigate how trains of different categories and the size of their time reserves were affected the reliability of the traffic schedule. The obtained results of modeling the recovery of train movement on the section indicate that the proposed method for studying the impact of the amount of the reserve time on the reliability of the train schedule based on the modified epidemiological SIR model permits to obtain numerical results with acceptable accuracy based on the macroparameters of the transportation process. The advantage of the proposed research method based on macromodeling is its versatility and the ability to gain knowledge about complex nonlinear processes of interaction between train flows in the event of delays from the historical data of previous periods of operation of the section. The proposed method for finding the coefficients of the propagation speed of the delay in the system of differential equations of the SIR-model allowed for the first time to numerically determine the mutual interaction of trains of different priority in the train schedule. It has been found out that the transfer rate of delay from passenger trains with the highest priority freight trains is 0.0001, while the transfer rate between freight trains is much higher and is 0.0023. This can be attributed to the better planned timetable for passenger trains with large intervals from freight trains. Whereas freight trains are quantitatively significantly predominant, and their movement is poorly planned, which leads to significant mutual interaction. Based on the simulation results, for the first time, the numerically determined impact of the availability of trains of various categories and the size of their reserve time on the reliability of the traffic schedule on the line. With a compensation time set in the schedule of 20 minutes for passenger trains and a delay of five, it is possible to restore the movement of 47% of trains of the total number of detained ones. The established recovery time for suburban and freight trains allowed 51% and 83% of restored trains to be, respectively, of the total number of detained trains. The simulation results confirmed the adequacy of the developed method.

The disadvantages of the developed method include the need to adjust the parameters of the mathematical SIR model for each area of the investigated. To improve this approach, it is necessary to conduct additional research on the application of machine learning methods to obtain knowledge from the historical data of the operation of lines. It will increase the versatility of the approach to conducting research on grouping in terms of the operational parameters of various sections on the network. This will make it possible to reduce the dimension of the problem and speed up the calculations by using the found parameters of the SIR model for lines that are similar in terms of performance to each other.

From a practical point of view, the application of the proposed mathematical SIR-model, taking into account the different priorities of train traffic, will allow to automate the complex process of searching for rational values of compensation time on the lines of trains of various categories for train schedules of railway lines within the software applications. This will allow to increase the punctuality and reliability of regulatory train schedules.

References

[1] Fomin O, Lovska A, Pistek V and Kucera P 2020 Research of stability of containers in the combined trains during transportation by railroad ferry *MM Science Journal* **03** 3728

[2] Matsiuk V et al 2019 Improvement of efficiency in the organization of transfer trains at developed railway nodes by implementing a "flexible model" *Eastern-European Journal of Enterprise Technologies* **2/3**(98) pp 32–9

[3] Moroz M M, Korol S O and Boiko Y O 2016 Social traffic monitoring in the city of kremenchuk *Actual Problems of Economics* **175** pp 385–98

[4] Schittenhelm B 2011 *Planning with timetable supplements in railway timetables* (Trafikdagepå Aalborg Universitet) p 18

[5] Kroon L G, Dekker R and Vromans M J C 2007 Cyclic railway timetabling: a stochastic optimization approach, railway optimization *Springer Verlag* **4359** pp 41–66

[6] Olsson N O E and Hiselius L W 2018 The Planners' perspective on train timetable errors in Sweden *Journal of Advanced Transportation* **8502819** 17

[7] Kariyazaki K, Hibino N and Morichi S 2013 Simulation analysis of train operation to recover knock-on delay under high-frequency *13th WCTR (Rio de Janeiro: Brazil)* pp 1–14

[8] Yuan J, Goverde R M P and Hansen I A 2006 Evaluating stochastic train process time distribution models on the basis of empirical detection data *WIT Transactions on State of the Art in Science and Engineering* **40** pp 95–104

[9] Rößler D, Reisch J and Kliewer N 2019 Modeling delay propagation and transmission in railway networks *14th Int. Conf. on Wirtschaftsinformatik (Siegen, Germany)* pp 98–112

[10] Zeng Z and Li T 2018 Analyzing congestion propagation on urban rail transit oversaturated conditions: a framework based on SIR epidemic model *Urban Rail Transit* **4** pp 130–40

[11] Monechi B, Gravino P, Di Clemente R and Servedio V D P 2018 Complex delay dynamics on railway networks from universal laws to realistic modeling *EPJ Data Science* **7** 35

[12] *UIC leaflet 406 R Capacity UIC International Union of Railways* 2013 (France 2e édition) p 60

[13] Diekmann O and Heesterbeek J A P 2000 *Mathematical Epidemiology of Infectious Diseases* (Chisteter: UK/John Wiley&Sons)

[14] *Rules of technical operation of railways of Ukraine* 1996 (Kyiv: Ministry of Transport of Ukraine) p 132

[15] Watson R K 1972 On an epidemic in a stratified population *Journal of Applied Probability* **9** pp 659–66

[16] Sivanandam S and Deepa S 2008 *Genetic Algorithm Implementation Using Matlab. In: Introduction to Genetic Algorithms* (Berlin, Springer)

[17] Hunt B R, Lipsman R L, Osborn J E and Rosenberg J M 2019 *Differential Equations with Matlab* (3rd Edition) 304

[18] Spanos A 1999 *Probability Theory and Statistical Inference: Econometric Modeling with Observational Data* (University of Cyprus) p 844

[19] Butko T, Prokhorchenko A, Golovko T and Prokhorchenko G 2018 Development of the method for modeling the propagation of delays in noncyclic train scheduling on the railroads with mixed traffic *Eastern-European Journal of Enterprise Technologies* **1/3(91)** pp 30–9

Study of the operation of taxi cars in an emergency situation

Dimitar Grozev[1]**, Ivan Georgiev and Dimitar Georgiev**

University of Ruse „Angel Kanchev", 8 Studentska Str., 7017 Ruse, Bulgaria

[1] E-mail: dgrozev@uni-ruse.bg

Abstract. This paper examines the transportation work of taxi companies operating in the city of Ruse in an emergency situation caused by the COVID-19 epidemic when the number of requests for taxis decreased by 60%, about 30% of drivers stopped working due to reduced workload. The aim of the present study is to identify changes in their work intensity. The taxi operating system is modelled as a queuing system with failures with non-stationary incoming flow of requests and a variable number of service channels. After collecting and processing statistical information about the received requests, the incoming flow is modelled as a periodic nonlinear function. The main parameters of the study are: density of failures, average number of taxi cars occupied per day and intensity of serviced requests. The object of the study is a full working day of 24 working hours, taking into account the reduced intensity of transportation work. Even at the highest peaks the load coefficient is below 0.1, which means that in the intervals with the highest load, the service occupancy is below 10% of those working at the moment. In the interval from 0:00 h to 7:00 h a load factor slightly above 0.04 is observed. An application has been created in the environment of MATLAB.

1. Introduction

In the late 2019, the first case of a coronavirus called SARS-CoV-2 was identified in Wuhan, central China. This happened after many people developed pneumonia in December for no apparent reason and the treatment of which is not affected by traditional methods. Almost all countries in Europe, North America, Asia-Pacific reported new cases of 2019-nCoV [1]. The incubation period of the virus is between 2 and 14 days, but it is transmitted before the onset of symptoms, as well as a few days after cure. This has led the governments of almost all countries to take various measures to reduce the infection by reducing social contacts and increasing the distance between people. Such measures to distance and limit social contacts in order to limit and slow down the spread of the COVID-19 epidemic have major economic and financial implications and should be the subject of research.

The COVID-19 coronavirus pandemic and the measures taken to curb it have reduced both private and urban passenger journeys, including taxis.

Analysis of data in the city of Ruse shows about six times a decrease in travel by urban passenger transport during the state of emergency.

The number of requests for taxis decreased by 60%. This affects the organization of work in dispatching points and taxi companies. The number of employees in the control points decreases by 50%. About 30% of drivers stop working due to reduced workload. Passengers are mainly transported under contracts with corporate clients. All operating vehicles are equipped with protective Plexiglas transparent partitions between the front and rear seats, with the aim of separating passengers from the driver as the best precaution against contamination with COVID-19. All preventive measures against infection have been taken for the drivers - equipment with masks and gloves, and treatment of taxis with disinfectants [2, 3].

The purpose of the present study is to establish the change in work intensity in the emergency imposed by COVID-19, to determine the main parameters of taxi activity in emergency situations and to determine the load factor of a car for a day. The created model makes it possible to simulate different variants of work organization. This will allow you to calculate the load factor of a server per day for these options. Thus, it will be possible to choose the option that will satisfy the management of the company to the maximum.

2. Literature review

Restrictions imposed in some cities on evening and night travel have further limited the work of taxi companies.

In recent years, the development of road transport is happening at a very fast pace. Due to its good technical and operational qualities – being faster and more flexible - it does not require in certain cases special road conditions. It turns out to be more convenient for a number of shipments. This predetermines its rapid development, which in turn leads to an increase in traffic motor vehicles. In a state of emergency, the pace of this process is greatly reduced.

An important characteristic of the transport flow is the intensity of traffic. It represents the number of cars that have crossed a section of road per unit time [4]. The intensity of movement changes during the hours of the day, week, month and year. The traffic on the street network of a city depends on the number of vehicles in motion, the territory of the city, the density of the street network and how the connections between the traffic and pedestrian flows on one or several levels are made [5].

Due to the sharply reduced intensity of traffic on the transport network, many problems arise related to the organization, management and safety of traffic. These problems are of different nature and complexity [6].

There are a total of about 2,800,000 vehicles registered in Bulgaria, which are roughly divided into the following groups [7]:

- light passenger – 2,200,000;
- minivans and vans incl. minibuses – 400,000;
- light goods up to 3.5 tons – 50,000;
- public transport buses – 6,000.

Passenger taxi transport is an integral part of the public transport system in the modern city. It has a number of advantages over the mass urban transportation of passengers between different areas of the city - non-connection with a pre-established route network (i.e. the ability to perform transport on routes freely chosen by passengers), the ability to perform transport "from door to the door", 24 hours a day and others.

The need for taxis arises when, as a result of the growth of cities, the distance between the center and its outskirts exceeds the so-called "pedestrian zone". It is usually estimated as the time for walking from the outskirts to the city center. This necessitates the study and research of passenger flows served by taxi.

Renting a taxi can be done at:

- the street,
- taxi ranks, through a control room and
- with mobile application.

Renting a taxi directly from the street is the least used case by passengers.

Using a taxi rank to rent a taxi is more reliable for the passenger than the first method. The taxi ranks are a special area with places for approaching and waiting for taxis and passengers. They are in places close to the places of occurrence of requests for taxi services (stations, ports, shopping malls, theaters). The taxi rank is equipped with a distinctive sign and is remotely connected to the central dispatching point or taxi station for submission and management of the execution of requests for taxi transport. Taxi ranks are also located on the streets, squares of cities, where there is a large flow of passengers and in places with a low degree of passenger flow. In terms of operation and technology, the taxi ranks meet the expectations that the time for renting a taxi is less than the time for renting a taxi on the street. Observations show that the use of a taxi service from the taxi rank is more than 60% of all possibilities.

The main technological problem of the organization in the two considered forms of renting a taxi is that the passenger and the free taxi have different locations. As a result, time and resources are wasted. This is the reason for the introduction of the technology for pre-rental of a taxi on request [4]. Crucial to the introduction of this technology on the market is the availability of an affordable connection. The accessibility of the connection for passengers is based on modern communication connections. For taxi transport, the accessibility of the connection is based on the possibility to use modern communication and information systems.

The purpose of the study is to determine the main parameters of taxi activity in emergency situations and to determine the load factor of a car for a day.

In the European Union, over 60% of the population lives in cities. Just under 85% of the European Union's gross domestic product is generated in cities. Taxis are an important part of the urban transport system in modern cities. As of January 2020, 151 companies are registered in the Municipality of Rousse with a permit for taxi activity. The total number of taxis operating on the territory of the Municipality of Rousse is 597. The average number of taxis in a company is 3.95 cars, and the number of taxis per 1000 people is 4.18.

For the city of Ruse the number of companies which use autonomous radiofrequency communication systems is 4, indicated in table 1.

Table 1. Taxi companies for the city of Ruse with autonomous radiofrequency communication systems to 30.03.2020.

№	Company
1	Taxi 2222 – to4nite
2	8806 - taxi 6
3	Orion Taxi 8800
4	Milanov Taxi 8111

At the same time, the number of trips with clients in emergency situations in most companies has decreased by nearly 60%. Table 2 shows the number of calls received by 4 taxi companies in the city of Ruse for the period 13.04.2020 - 17.04.2020.

Table 2. Number of applications received by dates in 4 taxi companies in the city of Ruse for the period 13.04.2020 - 17.04.2020.

Date	Taxi 2222 – to4nite		8806 - taxi 6		Orion Taxi		Milanov Taxi 8111	
	Day shift	Night shift	Day shift	Night shift	Day shift	Night shift	Day shift	Night shift
13.04.2020	368	162	72	113	59	75	58	51
14.04.2020	344	118	75	63	57	41	44	43
15.04.2020	369	96	90	51	71	32	42	28
16.04.2020	341	109	72	74	60	47	43	31
17.04.2020	353	106	57	69	57	43	36	33
№ of cars	112		84		56		49	

The table shows that on average one car for "Taxi 2222 - to4nite" has about 3.2 calls per taxi during the day shift and 1.05 calls during the night shift. For "8806 - taxi 6" this number is respectively - about 0.9 calls for day shift and about 0.8 calls for night shift. For Orion Taxi - 1.1 calls per day and 0.85 for the night shift and for Milanov Taxi 8111 - about 0.9 calls for the day shift and about 0.8 calls at night. This shows that with an 8-hour work shift, most of the time the cars stay in taxi ranks waiting to be hired by a customer or move through the streets of the city in waiting for an order. The table shows that in the conditions of implementation of emergency measures, the calls have decreased by almost 60% and the number of cars by almost 30%.

To describe the mode of operation of a taxi company, considered as a queuing system, it is necessary to know the characteristics of the incoming flow of orders, considered as a stochastic process, the intensity of service, the maximum allowable queue length and the number of service units [8-10].

For the incoming flow of taxis we can make the following prerequisites:

- ordinary flow – the probability of entering two or more taxi cars for an elementary time interval is infinitely small compared to the probability of entering only one taxi car, this means that cars arrive in singles, not in pairs, triples, at the same time.

- flow without consequences – the number of taxis entering the system for a time interval Δt does not depend on how many cars have already arrived, i.e. does not depend on the prehistory of the studied phenomenon (the flow without subsequent action (Poisson flow).

- stationary / non-stationary flow – for sufficiently long periods of time - 1 month, 6 months, 1 year, etc. with certain conditions, stationarity of the incoming flow can be assumed, i.e. the probability of a certain number of taxis appearing in a given, sufficiently large time interval depends only on the length of this interval. In the general case at arbitrary periods, the flow is non-stationary. This stands out well for a period of 1 working day (24 hours).

3. Materials and methods

For the intensity of service, it was found out that the service time of an order is a relatively constant value and is 8.5 min. in time intervals $7^{00}h - 10^{00}h$ *and* $16^{00}h - 19^{30}h$, and during the rest of the day it is 5.5 min. The number of service channels (servers) during the day is variable. It is also known that at a time when all servers are busy, the received order is rejected. To study the operation of the system, it is necessary to find the probability that there are currently a number of vehicles in the system with the server running

$$P_k(t)=?,\ k=\overline{0,n},\ t\in[0, T],\ 26\leq n(t)\leq 120, \tag{1}$$

where one full working day of 24 hours is taken for a unit period. The beginning of the working day coincides with the astronomical beginning of the day (0 hours and 0 minutes), the end of the working day, with the end of the astronomical day (24 hours and 0 minutes).

For the model such as a queuing system, the following can be summarized: a non-stationary flow of requests with density arrives at a queuing system with serving channels. The service time of the request is a random variable with an indicative distribution and a parameter that is a partially constant function [11, 12]. The number of service channels also changes around the clock, i.e. An order that arrives at the time of busy channels leaves the system as unserved, i.e. failure system.

The specified defining system is of type (M / M / n) in non-stationary mode, with a variable number of chànnels. To describe a system of this type, the following system of Kolmogorov differential equations (Erlang-Kolmogorov) is valid [13,14]:

$$\frac{dP_0}{dt}=-\lambda(t)P_0(t)+\mu(t)P_1(t)$$
$$\frac{dP_k}{dt}=P_{k-1}(t)\lambda(t)-\big(\lambda(t)+k\mu(t)\big)P_k(t)+\mu(t)(k+1)P_{k+1}(t) \quad (2)$$
$$\frac{dP_n}{dt}=P_{n-1}(t)-n(t)\mu(t)P_n$$

where $n(t)$ – number of running servers currently t, $P_k(t)$ – the probability that t currently has a k client in the system (k servers are occupied), $\lambda(t)$ – the current flow rate at the moment t, $\mu(t)$ – the service speed of a server at the moment t.

4. Results and Discussion

The data on the received requests (by hours) for 24 hours during one week are given in table 3.

Table 3. Number of requests in the interval from 31.03 to 06.04.2020.

hour	31.03. 2020	01.04. 2020	02.04. 2020	03.04. 2020	04.04. 2020	05.04. 2020	06.04. 2020	Average value
00-01	16	14	16	16	30	37	16	20.71
01-02	12	14	12	12	21	30	10	15.85
02-03	7	8	8	8	18	23	7	11.28
03-04	10	8	8	8	16	18	8	10.85
04-05	9	8	10	9	13	15	6	10.00
05-06	10	11	12	12	17	20	14	13.71
06-07	24	24	24	25	18	20	29	23.42
07-08	41	45	47	48	29	20	52	40.28
08-09	55	57	54	45	31	23	63	46.85
09-10	51	53	55	54	40	29	58	48.57
10-11	51	48	51	47	42	31	58	46.85
11-12	48	50	42	43	38	31	46	42.57
12-13	43	38	42	43	37	25	40	38.28
13-14	49	43	46	45	38	37	51	44.14
14-15	43	44	37	40	37	32	43	39.42
15-16	43	38	46	51	31	32	46	41.00
16-17	47	41	36	49	34	32	44	40.42
17-18	47	47	43	47	40	33	44	43.00
18-19	40	38	38	47	40	34	38	39.28
19-20	29	36	40	48	45	30	32	37.14
20-21	30	32	32	41	43	29	26	33.28
21-22	26	27	29	39	42	24	24	30.14
22-23	22	26	24	36	38	25	24	27.85
23-24	20	20	20	34	40	27	27	26.85

For modeling it is expedient to choose a relatively elementary function having periodicity. This is appropriate given the daily fluctuations. To approximate the average values of the received applications for a period of 24 hours, the method of least squares was used. The model should be as simple as possible, but should reflect the most characteristic behavior of the real flow. The following trigonometric order (Fourier series with 7 terms) is chosen as a model, linear with respect to the required coefficients:

$$\lambda(t)=a_0+a_1\cos\left(\frac{2\pi t}{24}\right)+b_1\sin\left(\frac{2\pi t}{24}\right)+a_2\cos\left(\frac{4\pi t}{24}\right)+b_2\sin\left(\frac{4\pi t}{24}\right)+a_3\cos\left(\frac{6\pi t}{24}\right)+b_3\sin\left(\frac{6\pi t}{24}\right) \quad (3)$$

where $\lambda(t)$ – the approximate current flow rate t, a_i, $i=\overline{0,3}$, b_j $j=\overline{1,3}$ – coefficients of the approximating function.

Figure 1 shows the average data of the inflow for one day of 24 hours for one week and the density of the inflow approximated using the method of least squares.

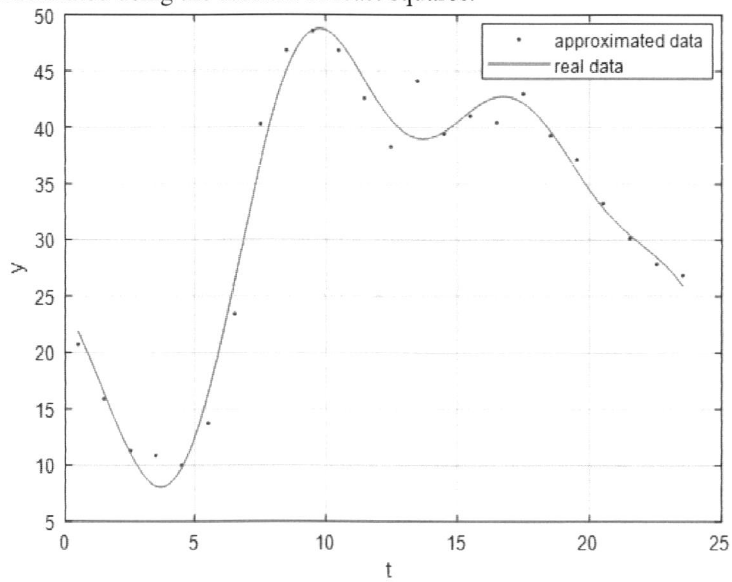

Figure 1. Graph of the inflow density approximated using LSM (Least squares method) and observed average inflow data for one day of 24 hours during one week.

The coefficients $a_0, a_1, b_1, a_2, b_2, a_3, b_3$ calculated according to LSM, as well as their confidence intervals are given in table 4. All coefficients are statistically significant (confidence intervals do not contain 0). The coefficient of determination is also statistically significant, which means that the chosen model is adequate.

Table 4. Values of the coefficients of determination and their confidence intervals.

Value of the coefficients in the model	Confidence interval of the coefficients, guaranteed with probability $\gamma = 0.95$
a_0=32.16	(31.21;33.11)
a_1=-12.9	(-14.24;-11.56)
b_1=-7.73	(-9.07;-6.39)
a_2=1.07	(0.12;2.01)
b_2=-7.34	(-8.64;-6.002)
a_3=3.79	(2.45;5.13)
b_3=2.29	(0.95;3.64)

It is known about the speed of service that during the busiest part of the day, it takes about 8.5 minutes between the service of one request, i.e. for one hour the average speed of service is 120/17 = 7.0588 requests per hour, and during the rest of the day, the requests are served for 5.5 minutes, i.e. the speed is 120/11 = 10,909 requests per hour.

$$\mu(t)=\begin{cases} \frac{120}{11}, & 0\leq t<7 \\ \frac{120}{17}, & 7\leq t<11 \\ \frac{120}{11}, & 10\leq t<16 \\ \frac{120}{17}, & 16\leq t<19.5 \\ \frac{120}{11}, & 19.5\leq t<24 \end{cases} \quad (4)$$

Statistics for one week on the number of servers running during the day are given in table 5.

Table 5. Number of servers running during the day for the interval 31.03 to 06.04.2020.

Hour	31.03. 2020	01.04. 2020	02.04. 2020	03.04. 2020	04.04. 2020	05.04. 2020	06.04. 2020	Average value
00-01	22	28	27	24	53	63	27	34.85
01-02	21	31	21	27	31	52	21	29.14
02-03	12	14	10	9	36	37	16	19.14
03-04	19	7	9	17	29	27	9	16.71
04-05	17	11	19	16	21	37	16	19.57
05-06	13	14	25	26	29	41	27	25.00
06-07	41	52	52	36	31	36	51	42.71
07-08	71	76	80	82	43	30	90	67.42
08-09	99	103	96	88	53	36	109	83.42
09-10	93	92	96	106	66	50	101	86.28
10-11	94	88	81	86	70	58	102	82.71
11-12	80	85	72	81	61	49	81	72.71
12-13	76	61	66	72	62	34	73	63.42
13-14	80	70	83	77	59	58	95	74.57
14-15	71	79	61	69	68	57	77	68.85
15-16	75	65	81	94	56	58	80	72.71
16-17	89	71	61	85	60	57	79	71.71
17-18	79	89	77	86	69	57	78	76.42
18-19	72	68	63	73	67	60	62	66.42
19-20	49	63	73	83	83	51	60	66.00
20-21	58	63	60	68	74	55	47	60.71
21-22	41	44	58	62	70	36	43	50.57
22-23	39	49	40	64	72	46	45	50.71
23-24	37	39	27	60	69	44	48	46.28

Given the statistics, the number of running servers is approximated integer. The length of the time intervals in which the change in the number of servers is more dynamic are shorter (a shorter time interval of 1 hour cannot be taken due to the specifics of statistical measurements), those in which the dynamics is more small are longer.

$$n(t)=\begin{cases} 35, & 0\le t<1 \\ 29, & 1\le t<2 \\ 18, & 2\le t<5 \\ 25, & 5\le t<6 \\ 43, & 6\le t<7 \\ 80, & 7\le t<11 \\ 71, & 11\le t<18 \\ 64, & 18\le t<21 \\ 49, & 21\le t<24 \end{cases} \quad (5)$$

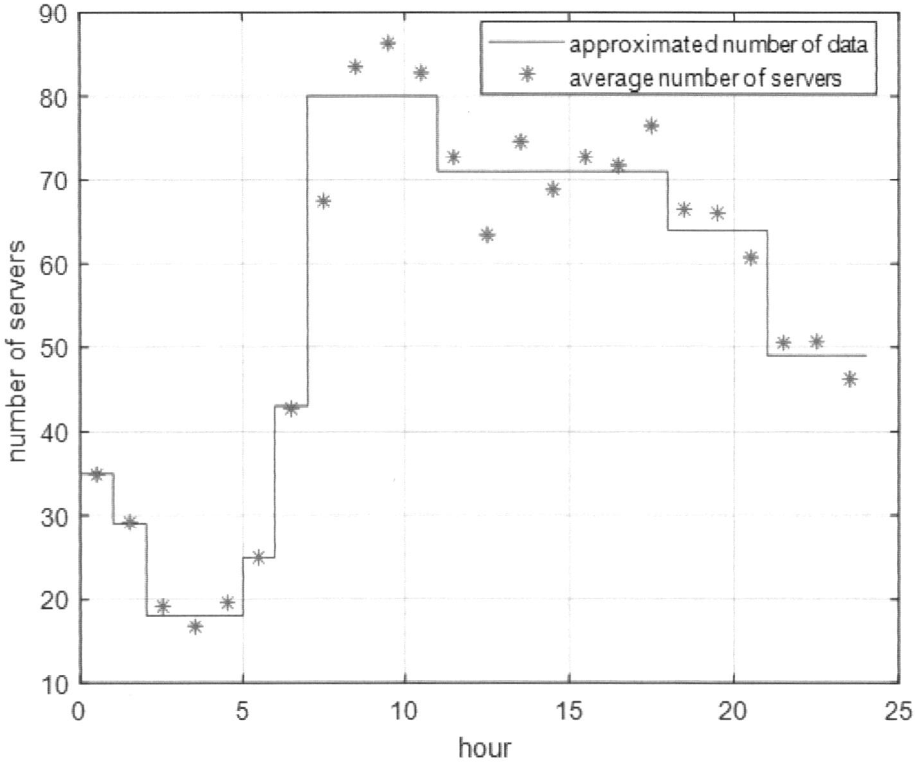

Figure 2. Graph of the average number of measured working taxis (servers) during the day and graph of the function $n(t)$, integer approximating this number.

The computational features related to system (2) are the following:
- large dimension system (in general);
- the system is of the "stiff system" type (rigid system of equations);
- the number of equations is variable, in different intervals of integration.

This can be summarized as: a system with a variable number of solid, differential equations with a large dimension [15]. Special numerical methods have been developed to overcome these difficulties. A Matlab program has been developed to solve system (2), using the built-in "solver" ode15s, implementing the Geer method [16]. When entering $\lambda(t), \mu(t), n,$ the application returns a numerical solution of $P_k(t)$. The accuracy is artificially increased compared to the default for the solver – from 10^{-6} to 10^{-9} for absolute error and from 10^{-3} to 10^{-5} for relative error [20]. The integration interval is divided into sub-intervals $t_0, t_1, \dots t_{i-1}, t_i, t_{i+1}, \dots t_N$, so that $[t_i, t_{i+1}]$, $\mu(t)$ and $n(t)$ are constants in each sub-interval.

The initial state of the system $P_k(t_0)$ is unknown. It is known that this type of processes are stable and after a long time enter a regular mode of operation. Therefore, can be take any initial state. The values of $P_k(t)$, at the end of the time interval $[t_{i-1}, t_i]$, become initial values for the subinterval $[t_i, t_{i+1}]$. If for a given time interval there are more equations than the previous interval, then for the initial conditions of these equations are take the zero value. The integration of the system in the interval needs to be done not once, but many times, and after each integration the values at the end of the period become the initial values for the next integration. Thus, the probability functions $P_k(t)$ begin to trend to their regular values. After several integrations, it was found out that only after 5-6 periods functions $P_k(t)$ enter a regular mode (for two adjacent periods, they remain the same). The accuracy is also increased here as the integration is done for 20 periods, as the difference of all $P_k(t)$ in the last and penultimate period is less than 10-8 for each t. Table 6 shows the time intervals $[t_{i-1}, t_i]$, the values of the function μ, as well as the number of service channels n in the given interval.

Table 6. Time intervals $[t_{i-1}, t_i]$, values of function μ, number of service channels n in the given interval.

Interval of time in the day $[t_{i-1}, t_i]$	Average value of service speed μ	Average number of service servers n
[0,1]	120/11	35
[1,2]	120/11	29
[2,5]	120/11	18
[5,6]	120/11	25
[6,7]	120/11	43
[7,11]	120/17	80
[11,16]	120/11	71
[16,18]	120/17	71
[18,19.5]	120/17	64
[19.5,21]	120/11	64
[21,24]	120/11	49

Figure 3 shows the probabilities of having a certain number of cars in the system for one day. It can be seen that the probability of having less than five requests is the highest during most of the day. In the intervals from 9:00 h to 11:00 h and from 16:00 h to 20:00 h the probability of having between 6 and 15 requests is the highest. Throughout the day, the probability of having more than 16 requests at a given time is practically equal to 0.

The graph in figure 4 shows the average number of taxis occupied by customers for a period of 24 hours. There are two peaks, the larger one is between 8 and 11 o'clock, and the smaller one is between 16-19 o'clock. The smallest peaks are in the interval 3-4 o'clock in the morning.

Figure 5 shows the load factor of a server (taxi) during the day. It can be seen that even at the largest peaks the coefficient is below 0.1, which means that in the intervals with the highest load, the server occupancy is below 10% of those working at the moment. In the interval from 0:00 h to 7:00 h a load factor slightly above 0.04 is observed.

Even at the highest load peaks, it can be seen that the state of emergency has a serious impact on the operation of taxis.

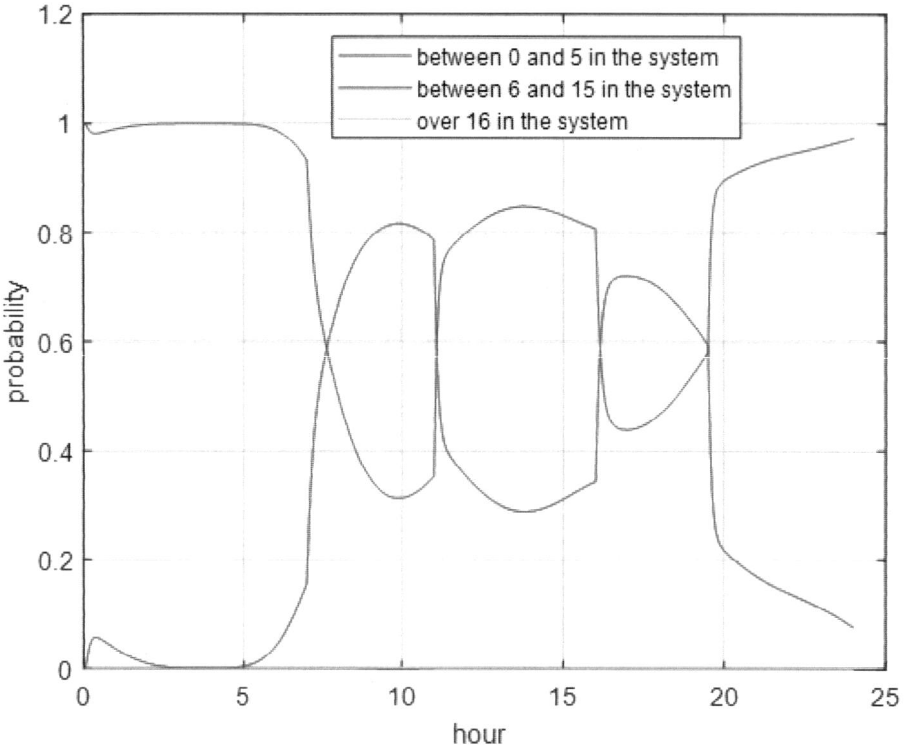

Figure 3. Probabilities to have between 0 and 10, 11-21 and over 22 service requests for a period of 24 hours, respectively.

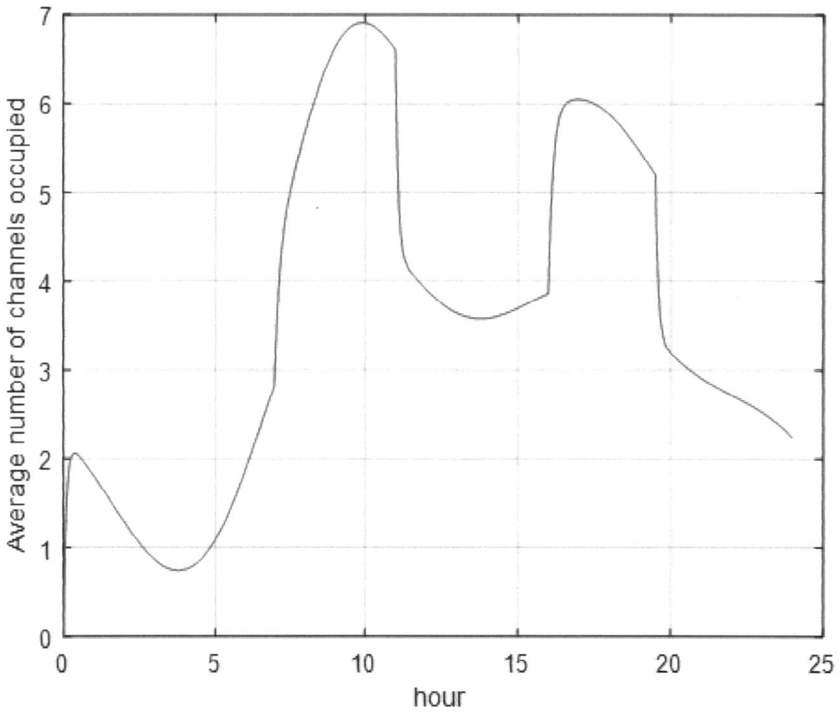

Figure 4. Average number of occupied servers (taxis) for a period of 24 hours.

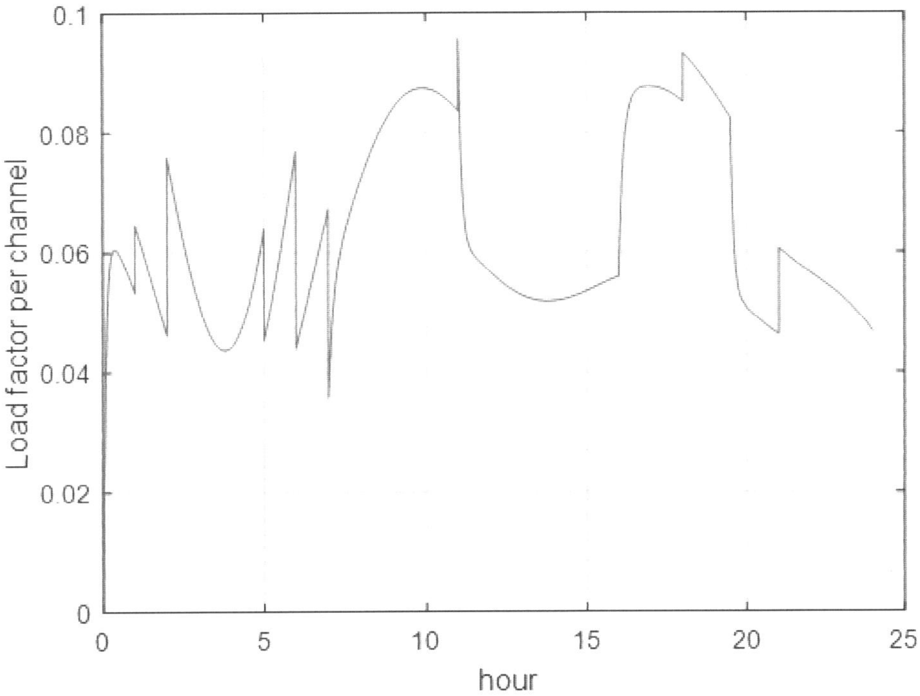

Figure 5. Load factor of one server (taxi) for a period of 24 hours.

5. Conclusion

1. On average, one car for "Taxi 2222 - to4nite" company has about 3.2 calls per taxi during the day shift and 1.05 calls during the night shift. For "8806 - taxi 6" this number is respectively - about 0.9 calls for day shift and about 0.8 calls for night shift. For Orion Taxi - 1.1 calls for day and 0.85 for night shift and for Milanov Taxi 8111 - about 0.9 calls for day shift and about 0.8 calls for night.
2. With an 8-hour work shift, most of the time the cars stay in taxi ranks waiting to be hired by a customer or move on the streets of the city in search of an order.
3. In a state of emergency, applications have dropped by almost 60% and the number of used taxis by nearly 30%.
4. There are two peaks in the number of occupied cars, the largest is between 8 and 11 o'clock (around 7 o'clock), and the smaller is between 16-19 o'clock (around 18 o'clock). The smallest peak is in the interval 3-4 o'clock in the morning (average below one).
5. The load factor of a taxi during the day at the highest peaks is below 0.1, which means that in the intervals with the highest load, the occupancy of the servers is below 10% of those currently working. In the interval from 0:00 h to 7:00 h a load factor of slightly above 0.04 is observed.
6. The present study makes it possible to establish the change in the intensity of work in the emergency situation imposed by COVID-19 and to determine the basic parameters of taxi activity in emergency situations and to determine the load factor of a car for one day.
7. The created model makes it possible to simulate different variants of work organization. This will allow for calculation of the server load factor per day for these options. Thus, it will be possible to choose the option that will satisfy the company's management to the maximum.

Acknowledgments
The paper reflects the results of the work on the project No 2020-FT-02, funded by the National Science Fund of The University of Ruse.

References

[1] Confirmed 2019-nCoV Cases Globally. // 30 January 2020, https://www.who.int/docs/default-source/coronaviruse/situation-reports/20200130-sitrep-10-ncov.pdf?sfvrsn=d0b2e480_2
[2] Eichenbaum M S, Rebelo S and Trabandt M 2020 The macroeconomics of epidemics *Technical report, National Bureau of Economic Research*
[3] Correia S, Luck S and Verner E 2020 Pandemics depress the economy, public health interventions do not: Evidence from the 1918 flu. Available at SSRN: https://ssrn.com/abstract=3561560
[4] BDS 16578: 1987 *Road Roads. Determination of traffic intensity* (Sofia, Bulgaria: Bulgarian Institute for Standardization) (In Bulgarian)
[5] Pencheva V, Asenov A, Grozev D, Angelova R and Georgiev I 2018 Analysis of the traffic intensity of cargo vehicles in the border points *Transport Problems* **13**(4) pp 23-36
[6] Stoilova S 2012 Network models for traffic management of vehicles *J. Machines. Technologies. Materials* **1** pp 47–56
[7] *Statistical Reference Book* 2018 (Republic of Bulgaria: National statistical institute) Available at: hhttp://www.nsi.bg/sites/default/files/files/publications/StatBook2018_en.pdf
[8] Sztrik J *Basic Queueing Theory* 2016 (GlobeEdit)
[9] Zukerman M 2016 *Introduction to Queueing Theory and Stochastic Teletraffic Models* (EE Department, City University of Hong Kong)
[10] Grozev D, Milchev M and Georgiev I 2019 Analysis of the load on the taxi system in a medium-sized city *IOP Conf. Ser.: Mater. Sci. Eng.* **664** 012035
[11] Mitrofanova A 2007 *Continuous Times Markov Chains. Poisson Process. Birth and Death Process* (New York University)
[12] Pencheva V, Asenov A, Grozev D, Stoyanov P, Georgiev I 2019 Study of the daily irregularity on specific routes, servicing the passenger stops in Ruse Bulgaria *Transport Problems* **14**(4) pp 5–19
[13] Haghighi A and Mishev D 2013 *Difference and Differential Equations with Applications in Queueing Theory* (Wiley) p 424
[14] Grozev D, Pencheva V, Georgiev I and Beloev I 2018 Investigation of the operation mode at Ruse-Danube Bridge border checkpoint considered to be a mass service system with incoming flow of automobiles at a non-stationary mode of operation *MATEC Web of Conferences* **234** 06003
[15] Gautam N 2012 *Analysis of Queues: Methods and Applications* (CRC Press) p 802
[16] Shampine L F and Reichelt M W 1997 The MATLAB ODE Suite *SIAM J. Sci. Comput.* **18**(1) pp 1–22

Driving strategy for minimal energy consumption of an ultra-energy-efficient vehicle in Shell Eco-marathon competition

Tsvetomir Gechev[1] and Plamen Punov

Department of combustion engines, automobiles and transport, Technical University of Sofia, 8, Kl. Ohridski blvd., 1756, Sofia, Bulgaria

[1] E-mail: tsv.gechev@gmail.com

Abstract. The paper focuses on the different driving strategies (driving cycles) for minimisation of the energy consumption of an ultra-energy-efficient electric vehicle developed by students for Shell Eco-marathon competition. The vehicle runs on hydrogen fuel and completes a set track with a 1420 m length and zero vertical deviation which is used for the 2019 European edition of the contest. A dynamic simulation model of the vehicle is developed taking into account vehicle resistance forces. A simplified model of the propulsion system is also described and used in the simulation. The propulsion system consists of a hydrogen fuel cell with a 1 kW rated power output, electric motors, electric converters, motor controllers and transmission. Furthermore, a strategy overview is proposed by which the vehicle complies with all necessary strategy limitations deriving from the competition rules. The various driving strategies, that comply with the competition average speed and differ from one another by the number of motors and transmission ratio, are simulated. The optimisation is done by changing the maximal and the minimal no-load speeds of the vehicle. A comparison of the energy usage is therefore conducted based on these strategies and the optimal one is chosen. In conclusion, additional driving strategies are proposed, in case more vehicle manoeuvrability or simpler driving style are desired during the race.

1. Introduction

The road transport sector produces around a fifth of the annual greenhouse gases (GHG) in Europe and is primarily responsible for the bad air quality in cities [1]. Unlike other sectors of the economy, transport has not experienced a gradual decline of the emissions and in the overall energy consumption in recent years [2]. Europe is not only a big consumer of vehicles, but also a big producer, with about 6.1% of the working people in EU being employed in the sector [3]. Furthermore, G. Fontaras revealed in his profound study [4], that the official laboratory emission values in Europe, more notably the ones of CO_2, are significantly lower than the values obtained from the actual on-road vehicle performance. Having all this in mind, the European Commission has developed a new low-emission mobility strategy, aiming at making transportation less polluting and safer, not only by bettering the efficiency of the newly produced vehicles from the contemporary vehicle fleet (the 2020-21 reference CO_2 emission target for new cars is 95 g per km [5]), but also by deploying vehicles with zero on-board GHG emissions [1, 6].

Hydrogen represents a serious pretender to replace the fossil fuels, as the hydrogen-based renewable energy production in Europe has gradually increased and is directly dependant on economic growth [7]. Particularly in the automotive applications, hydrogen is used in fuel cell electric vehicles (FCEVs) where it is transformed into electricity by the fuel cell (FC), whose only other products are heat and water [8].

Among the various types of FCs, the Proton Exchange Membrane (PEM) FC has proven to be the most suitable choice for such applications, being highly efficient [9, 10]. The frequent starts and stops, the sudden changes and the widely varying power demand require from the fuel cell systems to fast adapt to the operating conditions. Such conditions could easily be coped up with if the fuel cell system is hybridised using an additional energy storage such as batteries and/or supercapacitors (SCs) [9, 10]. Moreover, this allows for the storage of regenerative energy, since the fuel cell alone could not handle the regenerative braking energy flux. Such systems are more robust as they also benefit the fuel economy and further improve the dynamic performance of the FC. In [11] Fathabadi proves their advantage by experimentally studying the performance of a FCHEV powered by 90 kW FC and 600 F SCs, which provided a 435 km range with 5.4 kg hydrogen consumption as the maximum vehicle speed was 158 kmh^{-1} and the vehicle weight – 1800 kg.

The various possible approaches to building a fuel cell system with an auxiliary energy storage were of an interest to many scientific researches. The architecture of the hybrid systems with the choice of energy management strategies and a road model cycle are of a particulate significance for the efficiency of the vehicle. In [12] for example, Marzougui et al. proposes an energy management algorithm for a FC hybrid vehicle with a battery and SCs as energy storage system, which aims at regulating the energy provided from all three electric sources. The algorithm is evaluated on New European Driving Cycle (NEDC) and the results confirm its effectiveness in terms of power demand and energy efficiency. In [13] Carignano proposes an energy management strategy for a light commercial vehicle, tested on the standard Manhattan Bus Cycle (MBC) and the Buenos Aires Bus Cycle (BABC), while [14] investigates the effectiveness of SCs hybrid electric propulsion system for the same vehicle type on NEDC and ECE-15 cycles. [15] proposes a methodology for sizing of the energy storage system validated for the Worldwide Harmonized Light Vehicle Test Cycle (WLTC), while Kaya in his work [16] examines two new control strategies of FCHEV on standard "stop-go" and "uphill" road models. In [17] a multiphysics model for driving strategy optimisation of an Urban Concept vehicle is proposed, which led to maximal performance of 500 Wh/100km with 25 kmh^{-1} average speed.

Most of the studies were conducted on standardised driving cycles with set acceleration curves at any time of the cycle. They also mostly estimated the energy management of light duty vehicles, where the advantages of a hybrid propulsion systems are obvious due to the large power fluctuation and the availability of regenerative braking. Very few, though, are the researches focusing on real the driving strategy (driving cycle), notably for ultra-energy-efficient vehicles, where there is a limited speed range with and a very low efficiency of regenerative braking.

Firstly, this study is focused on the driving strategy architecture for an ultra-energy-efficient vehicle, powered by a FC and SCs as an additional energy source, which operates on a set track and with a set average speed. Then it proposes the most suitable driving strategies for energy efficiency, vehicle manoeuvrability and driving feasibility on the same track, by optimising the maximal and minimal no-load speed of the vehicle, and the transmission ratio, while maintaining the constant average speed. The energy efficiency optimisation is calculated for the electric power output, without taking into account the necessary power output of the FC, its efficiency and the efficiency of the SCs.

2. Competition and vehicle overview

In Shell Eco-marathon school and university students establish teams in which they develop a vehicle from its initial stages of design up to the final stages of production and assembly. These vehicles are divided into two categories: Prototype and Urban Concept, as each of them aims at reaching the lowest possible energy consumption, yet fulfilling different design requirements – Prototypes are made as slick and drop-like as possible, minimising resisting forces and maximising energy efficiency, while Urban Concept vehicles are created more similar to the conventional vehicles. The vehicles in each category could be electric – using a battery or a fuel cell as an energy source, or with an internal combustion engine (ICE) – running on gasoline, diesel or CNG fuel.

The main target for the teams is to cover a certain distance for a set amount of time by using minimal energy/fuel. As of 2019, the main European competition has been held at Mercedes-Benz World track in London, Great Britain. The track is 1420 m long and has no change in the elevation profile. An official track completion is recorded only when the vehicle covers 11 complete laps (15 620 m) in less than 39 minutes, hence the average speed is about 25 kmh^{-1}. Each team is given four attempts at completing the track. In order to recreate operation in urban conditions, Urban Concept vehicles must stop at the finish line once per lap.

Technical University of Sofia has a twelve-year-old tradition of taking part in the Shell Eco-marathon competition. The latest car was built in 2019 to compete in the Urban Concept – hydrogen fuel cell category in the same year in London (see figure 1). The car is significantly lighter than its predecessors due to the utilisation of a fully carbon-fibre body. According to the competition rules, the maximum vehicle weight without the driver is 225 kg, while the driver should be at least 70 kg. Additionally, hydrogen powered vehicles could use only supercapacitors as an energy storage system. The energy which is stored in them at the finish line of each lap has to be at least as much as that at the starting line. This is controlled by altering the SCs voltage. Moreover, regenerative braking is not allowed, as it overcomplicates the propulsion system and is also less efficient when integrated in vehicles of such size.

Figure 1. Vehicle of TUS Team.

In addition, the realisation of a hybrid propulsion system with a fuel cell and SCs, as well as the possibility of overloading of the motors, is possible by utilisation of buck/boost DC/DC convertors. Figure 2 represents a simple schematic of the vehicle propulsion.

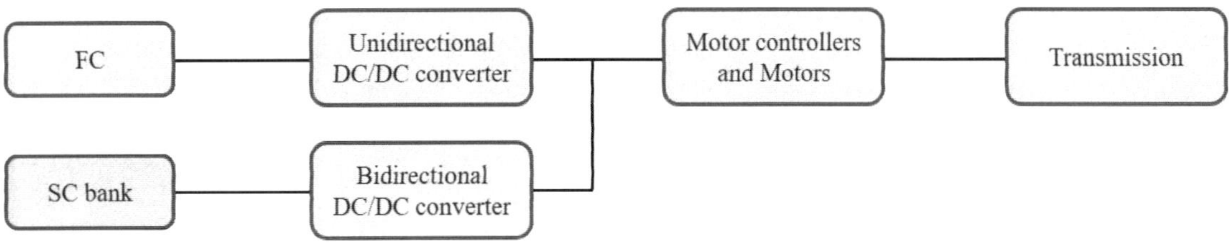

Figure 2. Car propulsion schematic.

3. Simulation model

3.1. Vehicle dynamic model

The power demand and energy consumption of the vehicle are calculated by means of a dynamic vehicle model. The force balance in longitudinal direction is given in (1), taking into account the traction force, rolling resistance, air resistance and inertia forces [8]. Due to the constant track profile, grade resistance is not included in the equation.

$$F_{tr} = F_i + F_{roll} + F_{aero}. \tag{1}$$

Therefore, the following equation (2) is used for estimating the instantaneous vehicle acceleration, where m is the vehicle mass and $\ddot{x}$ – its acceleration:

$$m\ddot{x} = F_{tr} - F_{roll} - F_{aero}. \tag{2}$$

The traction force is calculated in (3) by using the electric torque of the motor/motors M_{mot}, the transmission ratio (TR) i_{tr}, the transmission efficiency η_{tr}, and the dynamic radius of the wheel r_{roll}. The transmission efficiency is assumed constant as a single stage spur gear is used. The dynamic wheel radius is assumed equal to the static one, because of the low vehicle weight and the high tyre pressure.

$$F_{tr} = \frac{M_{mot} i_{tr} \eta_{tr}}{r_{roll}}. \tag{3}$$

The rolling and air resistances are calculated according to equations (4) and (5), where f_0 is rolling resistance coefficient at low speeds, k is an additional coefficient, ρ_a is air density, S_v is frontal area of the vehicle, c_x is drag flow coefficient and $\dot{x}$ – vehicle speed. In [11] the aerodynamic characteristics of this designed vehicle body were studied and the CFD results were then validated for a scaled model in a wind tunnel.

$$F_{roll} = mg(f_0 + k\dot{x}), \tag{4}$$

$$F_{aero} = \frac{1}{2}\rho_a S_v c_x \dot{x}^2. \tag{5}$$

The total mass of the vehicle and the inertial moments of all moving parts are calculated as a sum of the gross mass of the vehicle and the mass equivalent to the inertia moments, also taking into account the transmission ratio. The main parameters of the car are displayed in table 1, where Δm represents the reduced rotating masses.

Table 1. Main parameters of the vehicle.

Vehicle type	m_v, kg	Δm, kg	r_{roll}	f_0	k	S_v, m^2	C_x	η_{tr}	i_{tr}
Urban Concept	170	10.05	0.275	0.004	6.10^{-6}	0.795	0.136	0.95	200/12 or 200/13

3.2. Traction motor and propulsion

The motors used in our simulation are Maxon RE65, 36 V nominal voltage, 250 W nominal power, brushed DC with permanent magnets, that utilise coreless technology so that the hysteresis losses are minimised. The efficiency is about 90-91 % and the motors mostly function in overload mode due to their limited operation time during the race. The motor controllers have current limitation I_{mot} set at 9.5 A and the motors are overloaded by altering the input voltage U_{mot} up to 44-45 V. The maximum permitted motor voltage according to the competition rules is 48 V. By overloading the motors by voltage, the torque M_{mot} and angular velocity ω_{mot} are also increased above the nominal. Their correlation with the motor power P_{mot} is given in (6).

$$P_{mot} = \frac{M_{mot} \pi n_{mot}}{30}. \tag{6}$$

The motor torque M_{mot} and friction losses M_f are calculated according to (7) and (8), as k_m is motor torque constant and I_0 is no load current constant:

$$M_{mot} = k_m I_{mot} - M_f, \tag{7}$$

$$M_f = k_m I_0. \tag{8}$$

Motor efficiency $\boldsymbol{\eta_{mot}}$ and electric power $\boldsymbol{P_{el}}$ are calculated as follows:

$$\eta_{mot} = \frac{\pi n_{mot}}{30} \frac{(M_{mot}+M_f)}{U_{mot} I_{mot}}, \tag{9}$$

$$P_{el} = \frac{P_{mot}}{\eta_{mot}}. \tag{10}$$

The angular velocity of the motor n_{mot} is estimated by the following equation, where k_n is speed motor constant and $\frac{\Delta n}{\Delta M}$ is speed/torque gradient.

$$n_{mot} = k_n U_{mot} - \frac{\Delta n}{\Delta M} M_{mot}. \tag{11}$$

The motor controller ESCON 70/10 is used for motor management, due its high flexibility of control, most notably the possibility of closed loop current control, angular position control and speed control. In our simulation the torque of the motor is controlled by changing of its current I_{mot}, which is dependant on the level to which the acceleration pedal of the vehicle is engaged. A 100 % pushed acceleration pedal represents $I_{mot} = 9.3$ A. The main parameters and constants of the motor are shown in table 2, where P_N, U_N, I_N, M_N and n_N, are nominal values. The motor controller parameters are given in table 3, where U_{CC} is nominal operating voltage and I_{cont} is continuous output current.

Table 2. Main parameters of the electric motor.

Type	P_N, W	U_N, V	I_N, A	M_N, Nm	n_N, rpm	k_n, rpmV^{-1}	k_m, NmA^{-1}	I_0, A	η_{max}
DC RE 65	250	36	9.32	751.10^{-3}	3700	113	84.4.10^{-3}	0.6	0.87

Table 3. Main parameters of the motor controller.

Type	P_{max}, W	U_{cc}, V	I_{cont}, A	I_{max}, A	η_{max}
ESCON 70/10	700	10-70	10	30 (< 20 s)	98 %

3.3. Simulation limitations and strategy overview

3.3.1. Simulation limitations

In the process of generating different driving strategies and comparing their energy efficiency it is important to set proper simulation limitations so that the results are realistic and the strategies are not only feasible, but also fulfilling the rules of the competition. The main conditions defining the simulation are:

- Completion of 11 laps, S=1420 m each (total 15 620 m) within 39 mins, therefore an average speed of at around 25 kmh^{-1} for the whole strategy (cycle), maximal strategy length amounting to 210 s and $V_{start} = 0$ and $V_{finish} = 0$, in order to simulate operation in urban conditions (see figure 3);
- Maximal vehicle deceleration of 1 ms^{-2} during braking until stop at the end of each lap;
- Maximal operating voltage and current through the electric motors of the vehicle: U_{mot}^{max}= 44 V, I_{mot}^{max} = 9.3 A. I_{mot}^{max} is achieved in overload mode and should not be operated at for long periods of time;
- Only non-regenerative braking used;

- Maximum acceleration used in all acceleration periods of the vehicle, $I_{mot}^{max} = 9.3$ A = 100 % acceleration pedal push.

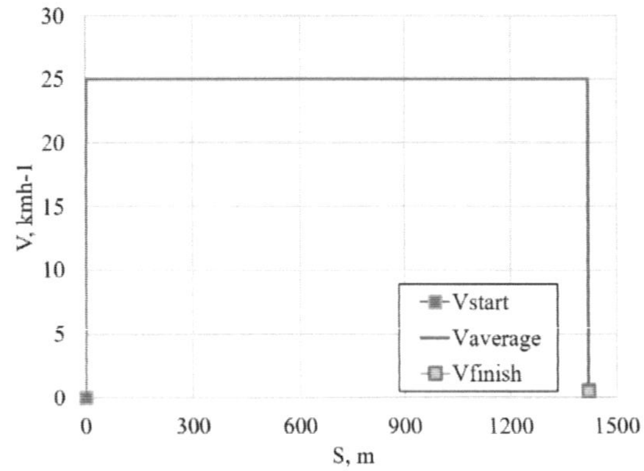

Figure 3. Main simulation limitations.

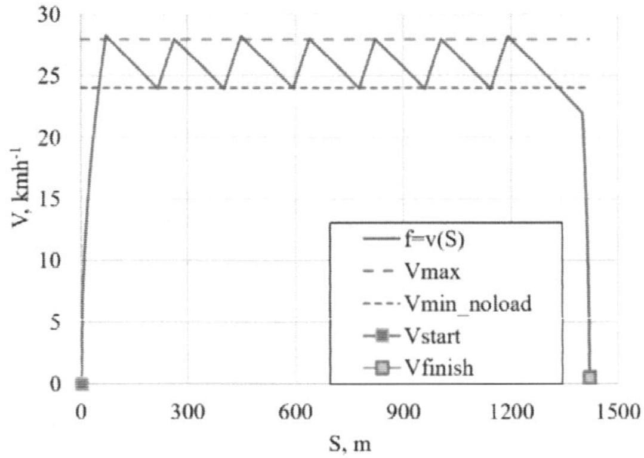

Figure 4. Proposed standard driving strategy.

3.3.2. Strategy overview

In ideal conditions the vehicle would not meet any traffic and would not have to undertake any unscheduled manoeuvres such as sudden acceleration or braking. Additionally, the power area from figure 3 could take any form and at the same time comply with the simulation limitations. Therefore, a standard driving strategy is proposed (see figure 4), consisting of the following parts:

- initial acceleration phase (IAP) from $V_{start} = 0$ up to V_{max},
- followed by an acceleration-coasting field (ACF) – plateau, in which the vehicle accelerates to V_{max}, then coasts down to the minimal speed at no-load V_{min}^{nl} and then again accelerates to the maximum speed, thus maintaining a slightly higher average speed in ACF than the one set for the whole strategy,
- followed by the last phase before the finish line which could contain either just braking, or braking and coasting to the minimal speed at the end of the strategy $V_{finish} = 0$.

In order to further define the strategy, it is important to examine each of the above-mentioned phases separately.

The vehicle could realise IAP with either 1 or 2 electric motors with the same transmission ratio. According to calculations, in terms of fulfilling the main limitations of the simulation, it is significantly better for the vehicle to accelerate with 2 motors during this period (see figure 5).

This way, not only does the vehicle reach V_{max} and ACF more quickly, while complying with the time limitation and even leaving a time reserve, but it also has an overall higher efficiency (see figure 6), by reaching more quickly the area in which the electric motors maintain their highest efficiency (the area is at high motor angular velocities – it matches with the plateau), and so lowering the energy consumption of the whole driving strategy. Therefore, IAP is realised via 2 traction motors.

ACF is where the most of the actual simulation and energy consumption comparison takes place. The optimisation is accomplished by varying certain parameters – the maximal speed V_{max}, minimal no-load speed V_{min}^{nl} and the number of motors used in the process of accelerating between the two speeds (see figure 7).

The last driving strategy period could be realised either by long and intensive braking due to the relatively high vehicle speed just before the finish line, or by a certain long coasting period after the last acceleration in ACF, followed by shorter braking. According to the calculations, the overall energy consumption of the driving strategy is better when starting to coast earlier and use braking just before the finish line, since this long coasting period at the end of each lap enables the usage of the accumulated kinetic energy during ACF (see figure 8). Moreover, it also represents a large compensative time reserve used during the whole driving strategy in cases of unexpected manoeuvres, sudden accelerations or decelerations, which are absolutely certain as the conditions during the racing are not ideal.

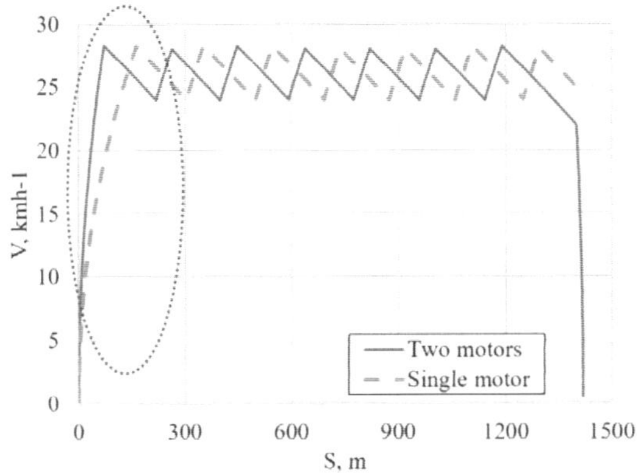

Figure 5. Comparison of IAP when using 1 or 2 traction motors (the bubble indicates IAP).

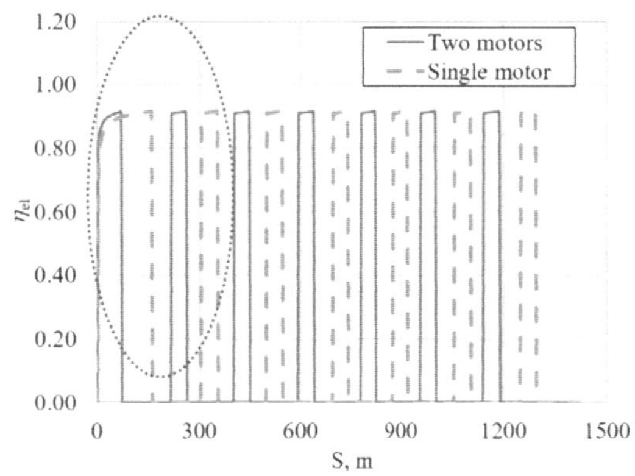

Figure 6. Comparison motor efficiency during IAP when using 1 or 2 traction motors (the bubble indicates IAP).

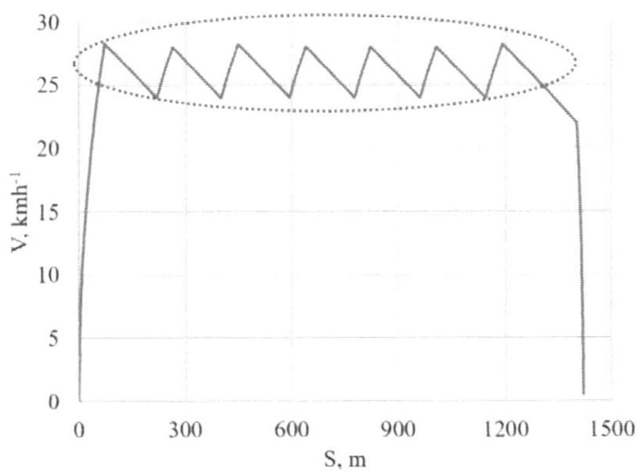

Figure 7. ACF representation – the bubble.

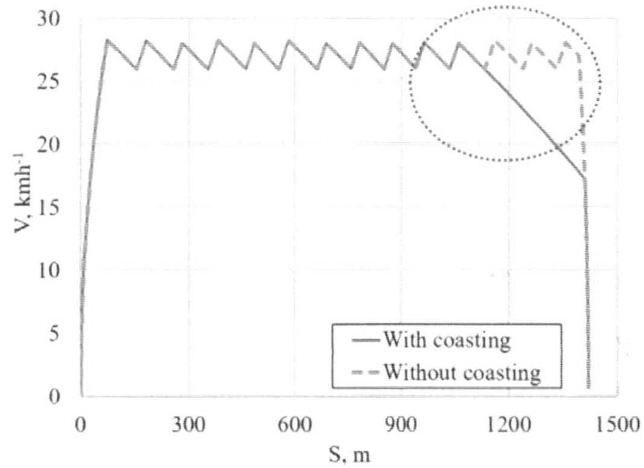

Figure 8. Using only braking in the final phase compared to using braking and coasting (the bubble indicates the final phase).

Once having the defining conditions set, the driving strategies are obtained by changing a few optimisation parameters. A total of 36 different strategies were studied in which the number of traction motors (excluding IAP), the transmission ratio (during the whole driving strategy) and the maximal V_{max} and minimal no-load V_{min}^{nl} speeds, were varied (see table 4).

4. Simulation results

In table 4 are given the case results of all 36 studied cases, that include all varying parameters, as well as the total electric motor cumulative energy used till the end of the driving strategy for each lap. All driving strategies have the set geometry of the proposed standard driving strategy (see figure 4) with a two traction motors IAP, followed by ACF and a final phase of long coasting and braking, as the cycle finishes for 210 s. In the column "comparative energy consumption" is given the energy consumption in percentage of each selected cycle compared to the one of the first one (case 1 – 100 %). In green colour are marked the results showing the lowest energy consumption.

Table 4. Results.

Case No	Motor Nominal Voltage, V	TR	Number of motors	Maximal speed, kmh^{-1}	Minimal no-load speed, kmh^{-1}	Used cumulative energy, J	Comparative energy consumption, %
1	36	200/12 =16.6	1	28	24	20920.00	100.00
2					25	20611.49	98.53
3					26	19556.57	93.48
4				29	25	19706.81	94.20
5					26	19547.83	93.44
6					27	19389.91	92.69
7				30	26	19391.86	92.70
8					27	19235.17	91.95
9					28	19225.87	91.90
10	36	200/12 =16.6	2	28	24	20913.55	99.97
11					25	20006.36	95.63
12					26	19691.54	94.13
13				29	25	19702.18	94.18
14					26	19391.23	92.69
15					27	19378.68	92.63
16				30	26	19386.15	92.67
17					27	19235.63	91.95
18					28	19224.69	91.90
19	36	200/13 =15.4	1	28	24	21347.32	102.04
20					25	20369.08	97.37
21					26	19942.95	95.33
22				29	25	19943.73	95.33
23					26	19795.36	94.62
24					27	19509.68	93.26
25				30	26	19649.53	93.93
26					27	19501.72	93.22
27					28	19355.47	92.52
28	36	200/13 =15.4	2	28	24	20787.57	99.37
29					25	20222.56	96.67
30					26	19930.35	95.27
31				29	25	19938.94	95.31
32					26	19650.86	93.93
33					27	19508.04	93.25
34				30	26	19514.17	93.28
35					27	19505.13	93.24
36					28	19346.02	92.48

On the basis of the results, the following conclusions are made:

- The optimisation in terms of TR shows that the lower TR – 200/13, leads to higher energy consumption due to the slower acceleration periods and slower reaching of the high efficiency zone of the motors, compared to TR 200/12 (see figure 9 – comparison of cases 1 and 19, 2% higher energy consumption). Despite of this, the TR 200/13 has also a few advantages – the motors operate in lower velocities, thus experiencing less stress from overloading and there is also a compensative speed reserve for possible acceleration over V_{max}.

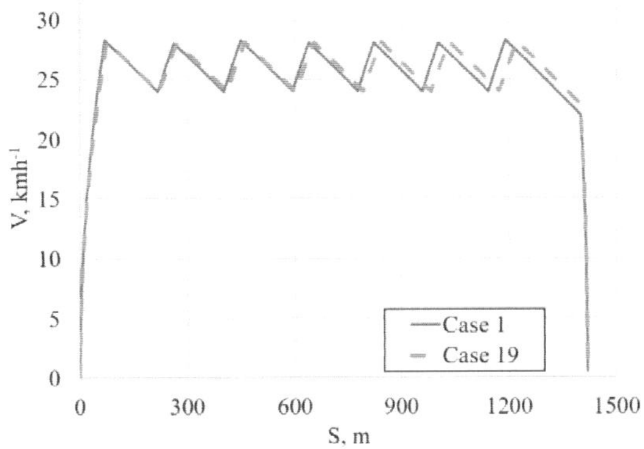

Figure 9. Comparison of cases 1 and 19, with respectively 200/12 and 200/13 TR.

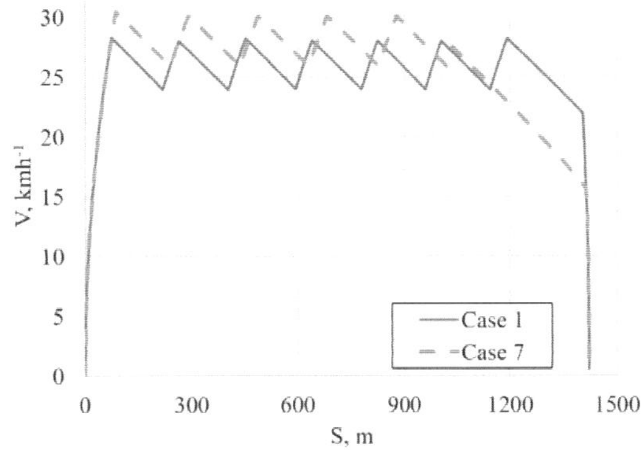

Figure 10. Comparison of cases 1 and 7, having different average speed during ACF.

- A higher average speed during the ACF – plateau, for strategies with both a single or two traction motors and all TR, leads to lower energy consumption. Additionally, a higher set V_{max} also ensures accumulation of the needed kinetic energy significantly earlier, this way also resulting in a longer coasting phase at the end of the cycle (see figure 10 – a comparison of cases 1 and 7, 7.3 % lower energy consumption). On the other hand, if the set maximal speed is too high, there is no speed reserve over V_{max}, due to technical limitations of the electric motors – U_{mot}^{max}= 44 V, I_{mot}^{max} = 9.3 A, and it also puts more stress on the motor/motors during overload mode. In real conditions it is very likely for the vehicle to need to accelerate over the set V_{max}, in order to compensate for lost time during unexpected manoeuvres.
- A smaller range between V_{max} and V_{min}^{nl} speeds and thus a higher average speed in the plateau results in less energy consumption for both a single and two traction motor strategies and all TR, yet there are more frequent accelerations during ACF, which could make it more difficult to implement the strategy by the pilot in real conditions (see figure 11– comparison of cycles case 1 and case 3, around 6.50 % lower energy consumption).
- Acceleration during the whole cycle by means of two traction motors results in approximately the same (or slightly lower) energy consumption. The advantages of such a strategy are in the higher manoeuvrability, due to the faster accelerations, and also in the more favourable conditions for the electric motors that work less time in overload regimes (see figure 12 – comparison between cases 1 and 10). Such a strategy, though, is less feasible in real conditions because of its higher ACF complexity – there are on average 1 to 2 more accelerations in ACF when the strategy adopts 2 traction motors in comparison to its 1 traction motor counterpart.

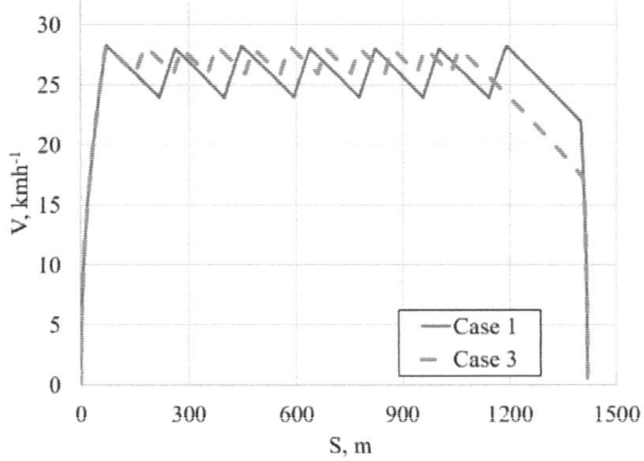

Figure 11. Comparison of cases 1 and 3, having different range between V_{max} and V_{min}^{nl} .

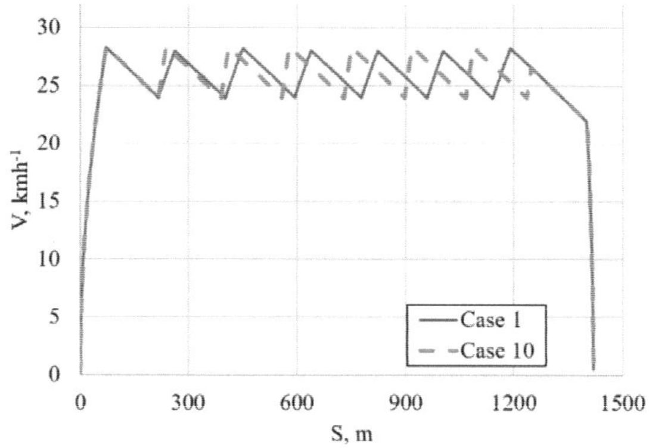

Figure 12. Comparison of cases 1 and 10, with respectively 1 and 2 traction motors.

5. Conclusions

Overall, there are many different options when choosing the correct driving strategy of an ultra-energy-efficient vehicle for this particular track in the competition. There is a well-established tendency of lowering of the energy consumption as the average plateau speed increases (up to 8 % difference in the energy consumption). Strategies with TR 200/13 consume more energy compared to these with TR 200/12, yet they offer better speed reserves over the maximal speed, and also maintain less overloading of motors during operation. Utilisation of two traction motors during acceleration in the whole cycle guarantees more manoeuvrability of the vehicle, yet such strategies become more complex to perform in real driving conditions.

When having to choose the right strategy, in terms of greater manoeuvrability, a less energy efficient strategy with lower V_{max}, two motors and TR 200/13, is suitable (see table 4, case 32, which consumes 19650.86 J of energy or is approximately 2.2 % less energy efficient than the most efficient strategy). If a lower level of complexity, yet higher efficiency is desired, a single motor, TR 200/12 and a large V_{max} - V_{min}^{nl} range, is chosen (see table 4, case 7, consuming 19391.86 J or around 0.8 % more energy than the most efficient strategy). Given that the desired cycle should purely be the most efficient one, a complex, two traction motor, with high average speed and a 200/12 TR strategy is chosen (see table 4, case 18 – consuming 19224.69 J).

Acknowledgements

This work was supported by the Bulgarian Ministry of Education and Science under the National Research Program "Low-carbon energy for transport and domestic use" approved by DCM # 577/17.08.2018.

References

[1] European Commission 2016 A European Strategy for low-emission mobility *Transp. Emiss* COM (2016) 501. https://ec.europa.eu/clima/policies/transport_en Accessed on 14.07.2020

[2] European Energy Agency 2015 Final energy consumption by sector and fuel *Indic. Assess. Data maps* ENER 016 (2020) https://www.eea.europa.eu/data-and-maps/indicators/final-energy-consumption-by-sector-10/assessment Accessed on 14.07.2020

[3] European Commission 2019 Internal Market, Industry, Entrepreneurship and SMEs *Eur. Comm.* https://ec.europa.eu/growth/sectors/automotive/ Accessed on 14.07.2020

[4] Fontaras G, Zacharof N G and Ciuffo B 2017 Fuel consumption and CO_2 emissions from passenger cars in Europe – Laboratory versus real-world emissions *Prog. Energy Combust. Sci.* **60** pp 97–131

[5] Parliament E U and E 2019 *Reducing CO2 emissions from passenger cars* Regulation (EC) No

443 (2009) https://ec.europa.eu/clima/policies/transport/vehicles/cars_en Accessed on 14.07.2020

[6] Pasaoglu G, Honselaar M and Thiel C 2012 Potential vehicle fleet CO_2 reductions and cost implications for various vehicle technology deployment scenarios in Europe *Energy Policy* **40** pp 404–21

[7] Xu R, Chou L C and Zhang W H 2019 The effect of CO_2 emissions and economic performance on hydrogen-based renewable production in 35 European Countries *Int. J. Hydrogen Energy* **44**(56) pp 29418–25

[8] Ehsani M, Gao Y and Emadi A 2017 *Modern Electric, Hybrid Electric, and Fuel Cell Vehicles* (CRC Press) p 348, p 22

[9] Wang Y, Chen K S, Mishler J, Cho S C and Adroher X C 2011 A review of polymer electrolyte membrane fuel cells: Technology, applications, and needs on fundamental research *Appl. Energy* **88**(4) pp 981–1007

[10] Bendjedia B, Rizoug N, Boukhnifer M and Bouchafaa F 2017 Hybrid Fuel Cell/Battery Source Sizing and Energy Management for Automotive Applications *IFAC-PapersOnLine* **50**(1) pp 4745–50

[11] Fathabadi H 2018 Fuel cell hybrid electric vehicle (FCHEV): Novel fuel cell/SC hybrid power generation system *Energy Convers. Manag.* **156** pp 192–201

[12] Marzougui H, Amari M, Kadri A, Bacha F and Ghouili J 2017 Energy management of fuel cell/battery/ultracapacitor in electrical hybrid vehicle *Int. J. Hydrogen Energy* **42** pp 8857–69

[13] Carignano M G, Costa-Castelló R, Roda V, Nigro N M, Junco S and Feroldi D 2017 Energy management strategy for fuel cell-supercapacitor hybrid vehicles based on prediction of energy demand *J. Power Sources* **360** pp 419–33

[14] Veneri O, Capasso C and Patalano S 2018 Experimental investigation into the effectiveness of a super-capacitor based hybrid energy storage system for urban commercial vehicles *Appl. Energy* **227** pp 312–23

[15] Gaikwad S D and Ghosh P C 2020 Sizing of a fuel cell electric vehicle: A pinch analysis-based approach *Int. J. Hydrogen Energy* **45**(15) pp 8985–93

[16] Kaya K and Hames Y 2019 Two new control strategies: For hydrogen fuel saving and extend the life cycle in the hydrogen fuel cell vehicles *Int. J. Hydrogen Energy* **44** pp 18967–80

[17] Olivier J-C, Wasselynck G, Chevalier S, Auvity B, Josset C, Trichet D, Squadrito G and Bernard N 2017 Multiphysics modeling and optimization of the driving strategy of a light duty fuel cell vehicle *Int. J. Hydrogen Energy* **42** pp 26943–55

[18] Dobrev I, Massouh F, Danlos A, Todorov M and Punov P 2017 Experimental and numerical study of the flow field around a small car *MATEC Web Conf.* **133** 02004

Electromagnetic analysis of a contactless charging station for electric vehicles

N L Hinov[1,3], N D Madzharov[2] and D N Grozdanov[1]

[1] Power electronics Department, Technical University of Sofia, 1000 Sofia, Bulgaria
[2] Electronics Department, Technical University of Gabrovo, 5300 Gabrovo, Bulgaria

[3] E-mail: hinov@tu-sofia.bg

Abstract. This report examines one of the many possible solutions of dynamic and static mode contactless power transfer, which determines the future of the automotive industry. Computer analysis of a contactless charging station for electric vehicles in certain frequency range and displacement between the transmitting and receiving coils are presented. Conceptual prototype design of 30 kW power transfer station is described in detail. Some optimization approach is also noted. The results are compared with the real experiments of prototype for a contactless charging station.

1. Introduction

A number of economic and environmental factors with increasing significance in recent years, have necessitated the search for innovative solutions in the automotive industry. There are many researches and developments in hybrid car engines - those powered by hydrogen fuel cells and many others but most of them cannot be applied in practice to compete with the internal combustion engine. Cost, technological sustainability, safety and ease of use of an innovative solution are the key factors for its market popularization. By this reason, one of the main practical factors for the mass "put on the road" of the electric car is the maximum mileage with one charge that can be realized and the method of charging with the best technical and operational parameters. Development of the element base and the methods for control of power electronic devices, the contactless transmission of electric power or wireless power transmission (WPT) is increasingly used in everyday life like domestic appliances, industrial technologies, mobile devices, medical equipment and transport. The main advantage of this technology is the elimination of power cables, the use of different standards in couplings and the depreciation of power circuits. This issue is especially serious when charging the batteries of electric vehicles, where in addition to the lack of cables and connectors, wireless technology produces a natural galvanic separation, which is extremely important to ensure safety [1-5]. The main purpose of the manuscript is to present a developed power electronic system for contactless charging of electric vehicles, operating in static and dynamic mode. Another task is to demonstrate practical aspects of the application of this relatively new technology, related to its advantages, as well as some of its disadvantages at present.

2. Topology and construction

Until recently, the main problem was the relatively low efficiency (ECE), due to both the imperfections of the semiconductor devices with which the converters are built, and the non-optimally designed electromagnetic systems for contactless transmission. Modern MOS and IGBT transistors have very good performance and thus it is possible to design high-frequency converters with high energy density. An interesting feature of the considered device is the system for transferring maximum power, which the charging station provides – over 30 kW in continuous operation. In practice, such levels of transferred power make this prototype economically adequate compared to many companies producing contactless charging systems for electric vehicles such as: *NISSAN, PRIMOVE, SIEMENS, WAMPFLER* etc. [2-6].

Figure 1 shows the topology of the power electronic system of the device. As the prototype is designed to test and demonstrate static and dynamic operation (while the vehicle is moving), it is necessary to use an advanced model of a standard full bridge converter [1, 7-11]. For this purpose, one arm of the converter operates continuously (both static and dynamic), while the other four are connected to each of the inductive power transmission (*IPT*) windings. The main advantage of this type of full bridge modification is the reduction of the total number of *IGBT* modules required for the implementation of the system – instead of four separate converters, each with two modules (eight modules in total), in this case 5 modules are used. In other words, instead of using four separate full-bridge inverters, the designed prototype uses only one modified five-link inverter.

Of course, the main unit in the *IPT* station is the power transformer with a weak magnetic connection, due to the use of an air gap. It is known from the literature that the process of designing this type of transformer is focused on the specifics of the geometry and dimensions, the range of change of the mismatch, the mechanical construction, the concentration and orientation of the electromagnetic field and others.

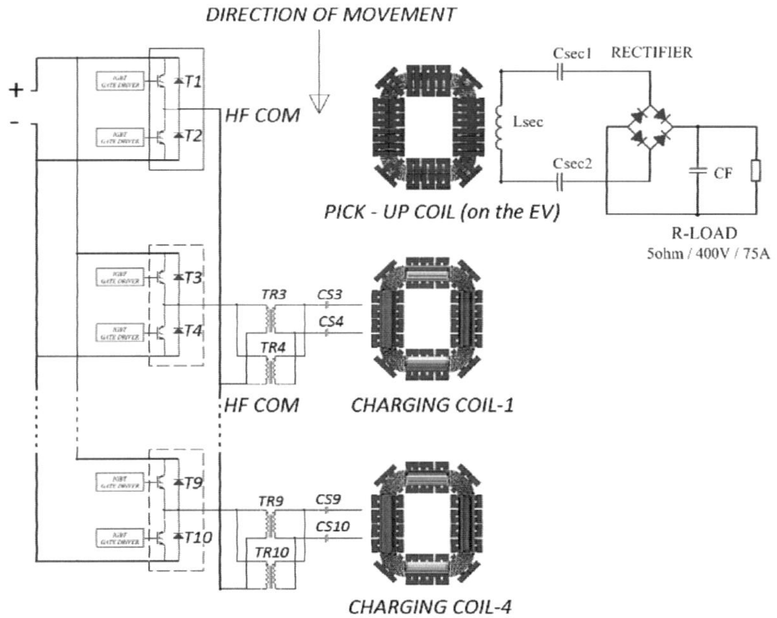

Figure 1. Selected power electronics topology of the on-route charging of EVs.

Figure 2 presents the results for evaluating the influence of the aluminum shield on the *IPT* windings obtained by modeling the electromagnetic system and computer simulations. The main objectives of this simulation are to assess the level of influence of the electromagnetic field on nearby objects, in accordance with the requirements of *ICNIRP*.

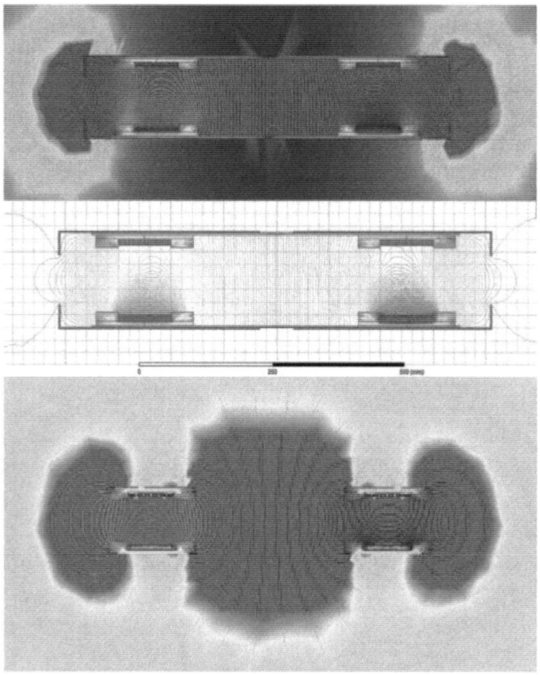

Figure 2. Electromagnetic field distribution with and without (below) AL shield.

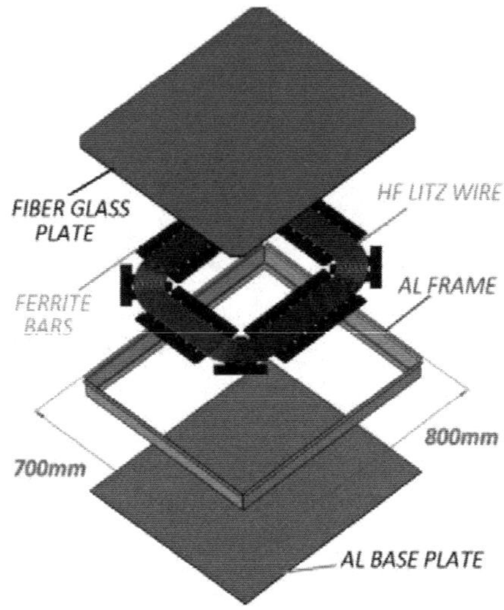

Figure 3. IPT Coil construction.

The design of the magnetic system, consisting of a transmitting coil and a model of the distribution of ferrite rods, is shown in figure 3. The main role in increasing efficiency is played by the value of the effective current density in the coil conductor (indirectly through copper losses) and the type of the material of the ferrite rods. In addition, the weight of the receiving coil (located in the *EV*) also affects the overall efficiency.

Table 1 shows the main parameters of the developed *IPT* transformer and their change for different values of the air gap and at a frequency that provides Zero Current Switching (ZCS) mode of the semiconductor devices of the inverter (*IGBTs*). The values thus determined are measured by connecting the primary and secondary windings in series - in one direction (L_{SER}) and vice versa (L_{PAR}) [9, 12].

Table 1. IPT parameters.

ZCS Frequency	14.4 kHz	13.8 kHz	13.3 kHz	13.0 kHz
Vertical Gap	55 mm	65 mm	75 mm	85 mm
N_2 / N_1	1.63	1.63	1.625	1.63
Coupling coefficient, k	0.613	0.556	0.508	0.465
M, μH	83.65	73.7	66	59.43
L_1, μH	79	76.7	75	73.6
L_2, μH	236	229.4	225	221.9
L_1+L_2 (L_{SER}), μH	482	453.2	431	413.8
L_1-L_2 (L_{PAR}), μH	147.4	158.4	167	176.1
L_m , μH	51.48	45.35	40.62	36.57
L_{11}, μH	27.52	31.35	34.38	37.03
L_{22}, μH	152.4	155.7	159	162.5

The main parameters of the magnetic system are determined by the following ratios:

$$M = L_m\left(\frac{N_2}{N_1}\right) = \frac{\left|L_{ser} - L_{par}\right|}{4}; \tag{1}$$

$$k = \frac{M}{\sqrt{L_1 L_2}}; \tag{2}$$

$$L_{11} = L_1 - L_m; \tag{3}$$

$$L_{22} = L_2 - L_m\left(\frac{N_2}{N_1}\right); \tag{4}$$

$$L_{ser} = L_1 + L_2 + 2M; \tag{5}$$

$$L_{par} = L_1 + L_2 - 2M, \tag{6}$$

where M is magnetic coupling coefficient; L_m – magnetized inductance; N_1 and N_2 – respectively the number of turns of the primary and secondary windings; k – coupling coefficient; L_1 and L_2 – inductances respectively primary and secondary windings of the transformer.

3. Output power regulation techniques

In the specific case, the frequency shift control method of the converter is used to adjust the output power. Characteristic of this method is that it allows a wide range of changes in the output parameters of the power electronic device depending on the air gap between the windings and their axial displacement. The condition for controlling the power of a continuous load circuit is defined as follows:

$$V_{HF} = \frac{Vd2\sqrt{2}}{\pi}; \tag{7}$$

$$I_{HF} = \frac{V_{HF}\cos\varphi}{R}; \tag{8}$$

$$P = \frac{{V_{HF}}^2\cos^2\varphi}{R}; \tag{9}$$

$$P = \frac{{V_{HF}}^2}{R}\frac{1}{1 + Q^2\left(\frac{\omega}{\omega_0} - \frac{\omega_0}{\omega}\right)^2}, \tag{10}$$

where V_d is DC supply voltage of the converter; V_{HF} – high frequency voltage (RMS value); I_{HF} – high frequency current (RMS value); ω – angular control frequency of the converter; ω_0 – angular resonant frequency of the AC circuit; R – is the load resistance adjusted to the output of the inverter, corrected by the corresponding efficiency; Q – the quality factor of the resonant circuit of the inverter.

In combination of simple *PWM* or Phase Shift *PWM* regulation technique the expression for power regulation is:

$$P = \frac{[V_{HF}(1 - D)]^2}{R}\frac{1}{1 + Q^2\left(\frac{\omega}{\omega_0} - \frac{\omega_0}{\omega}\right)^2}, \tag{11}$$

where D – pulse width coefficient.

Another possible option for the implementation of effective control of the output power of the system is the use of a variable connection factor between the windings by changing the parameters of the compensation components [7, 8]. Figures 4 and 5 show the achieved characteristics (figure 4) and a simulation model for changing the values of the secondary inductance L_5 or the secondary capacitor C_3, respectively. In practice, the process of switching between different sets of capacitors (C_3) or different sets of inductors (L_5) can be achieved with the help of pairs of electronic switches - *MOSFET* or *IGBT* modules [1, 2, 9 and 11].

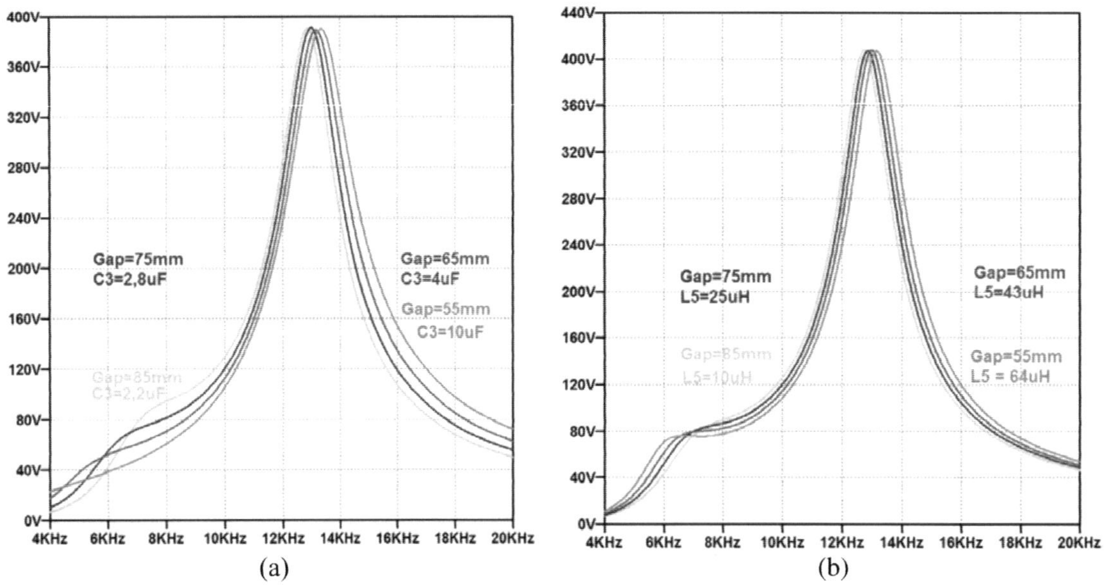

Figure 4. Output DC Voltage vs. Frequency characteristics (a) C_3 variable, L_5 = const; (b) L_5 variable, C_3 = const.

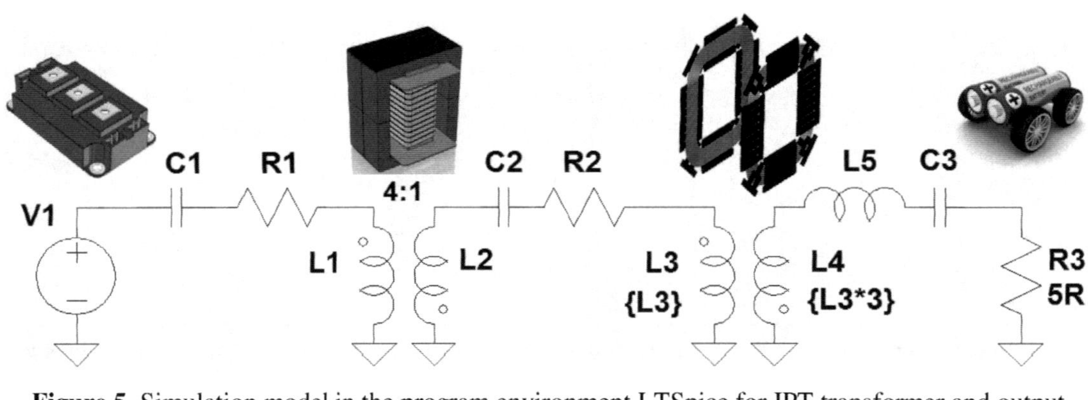

Figure 5. Simulation model in the program environment LTSpice for IPT transformer and output compensation.

During the design of the studied system it was possible to transmit energy "on-route" (charging during movement). Figure 6 shows the main parameters describing the process of dynamic charging of the battery of the electric vehicle. The length of the path along which the system does not transmit energy is determined by:

$$L_{off} = (L_{coil} - EPT) + D_{coil}\,. \tag{12}$$

Transferred energy during the movement is proportional to the speed and *EPT* area (±80 mm):

$$E_{on} = t_{on} \int_{-80}^{+80} EPT_{area}\, dx, \tag{13}$$

where *EPT* represents the field of efficient contactless energy transfer. This area is determined by the condition that the efficiency of the magnetic system is over 80%. In practice, this limits the maximum speed during dynamic charging. Another limiting factor is the time constants associated with switching the electronic converter; *PT* represents the area of energy transfer, but with significantly lower efficiency; CS_{AREA} (coil switching area) is used for starting point for the coils switching algorithm.

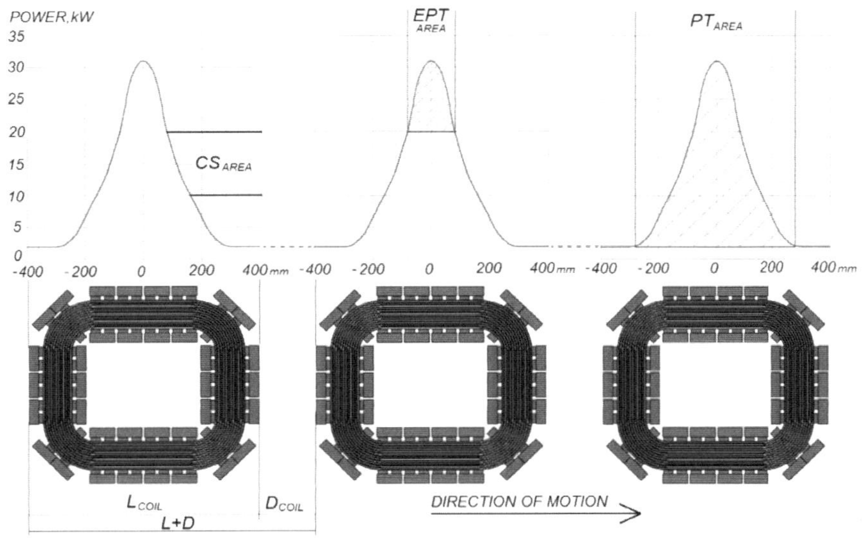

Figure 6. Defining the main dynamic charging parameters for contactless power transmission.

The presented method for determining the main dynamic parameters is very useful when it is necessary to test the actual performance of the system in different operating scenarios. On the other hand, it is not suitable for use in a real charging process, but only for testing.

Additional 2D Axisymmetric simulation is performed by using of ©COMSOL Multiphysics. Purpose of this is to find a method for optimization of energy transfer. Figure 7a represents model geometry which in our case has round shape assumed for simplification of the model. Number of coil turns, geometry shape and ferrite cores are chosen to fit with the prototype unit physical dimensions. Initial results showed that there is an optimal operating frequency in efficiency point of view (figure 7b). Choosing proper frequency, efficiency of the energy transfer can be optimized by near 2% or even more depending on the gap between the transmitting and the receiving coil [13-16].

On the other hand, the change in the efficiency of the magnetic system is not strongly dependent on the frequency, which is a good prerequisite for optimal control synthesis for the whole device. The subject of the next study is the assessment of the influence of the misalignment of the transmitting and receiving windings, which is common in practice during dynamic charging of electric vehicles.

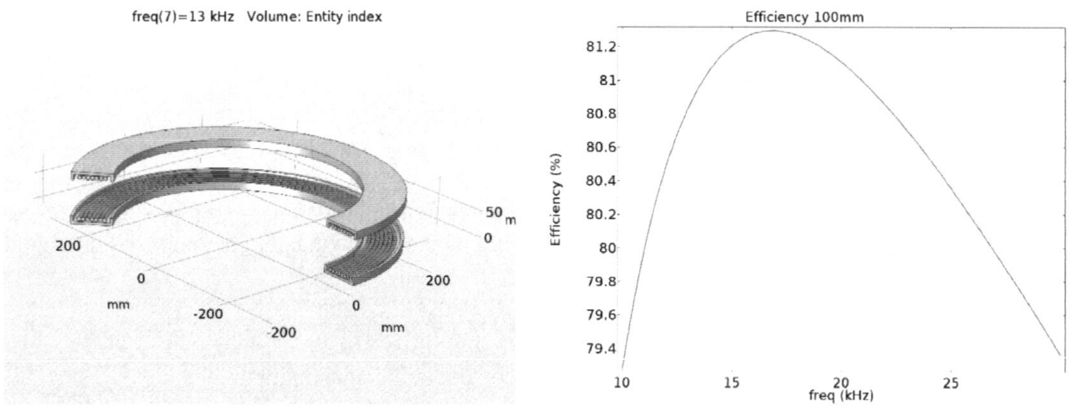

(a) (b)

Figure 7. Simulation results from the modeling of the magnetic system – (a) Model geometry; (b) Energy transfer efficiency at gap of 100 mm.

4. Converter parameters and results

Table 2 shows the main geometrical and electrical parameters of the designed IPT system. Figure 8 shows the current and voltage waveforms of the AC-Bridge diagonal at 85 mm vertical air gap.

Table 2. IPT Converter parameters.

Parameter	Value
Nominal Input Power	30kVA / 75A / 400V
Peak Input Power	45kVA @ 1min
Efficiency, %	up to 92 %
Converter Freq., kHz	20 kHz ÷ 12 kHz
Coil dimensions, mm	700 mm x 800 mm x 90 mm
Converter Control Technique	Frequency shift, Phase shift
Gap, mm	70 ÷ 90 mm
Horizontal misalignment, mm	$\Delta X=\Delta Z=\pm 100$ mm

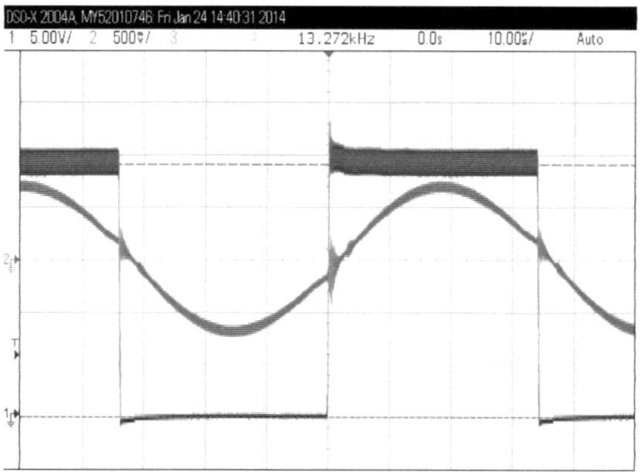

Figure 8. Current (purple) and Voltage (blue) waveforms of the AC-Bridge diagonal.

The dependence of the resonant characteristics of the converter according to the change of the air gap between the windings is shown in figure 9a and 9b. The characteristics establish the need for pre-selection in the design of the system, which is generally reduced to: changing the range of the air gap; selection of a different type of compensation circuit (with the presence of several resonant elements in order to weaken the influence of the load parameters on the mode of operation of the inverter); possibility to change any of the compensating elements in order to achieve optimal operation of the device.

From the characteristics it is established and confirmed, obtained from the modeling of the electromagnetic system, namely that the maximum output power and, accordingly, the greatest inertia of the output voltage of the inverter are obtained at a control frequency of 13 kHz.

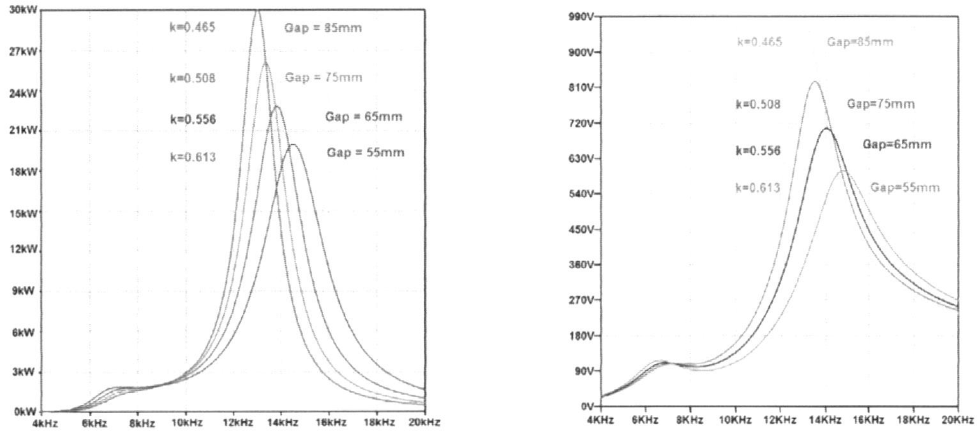

(a) (b)

Figure 9. Dependences of the main energy parameters of the system on the resonant frequency of the converter – (a) Output IPT Power vs. Resonant frequency at different vertical air gap; (b) Primary IPT coil voltage vs. Resonant frequency at different vertical air gap.

Operation with such a frequency is undesirable from the point of view of user comfort and therefore a magnetic system with other dimensions and parameters should be designed to obtain an optimal transmission frequency above 20 kHz.

The test results of the model charging station are satisfactory both in terms of the mode of operation of the *DC / AC* inverter with soft switching and good energy performance, and in terms of the influence of the change in the air gap. Further development of this work will be in the direction of improving the magnetic connection between the windings and hence the efficiency of the magnetic system based on model-based optimization of the *IPT* configuration and the parameters of its core.

5. Conclusion

The theoretical and experimental study of the studied and developed *IPT* charging station has a very large impact on the overall assessment of the feasibility of practical application. In this regard, one of the obstacles to the development of electro-mobility in general are the limitations associated with the elements for energy storage and the realization of a fast dynamic contactless charge. The content of the work done in this aspect can be summarized as follows:

- Modeling and analysis of electromagnetic processes important in the design of the IPT system.
- Creating basic relationships and interconnections in order to quickly assess and design the system.
- Obtaining characteristics for the purpose of synthesis of robust control of the system
- Prototyping and implementation of the prototype – transmission of 30 kW through an air gap of 7÷9 cm and 90% process efficiency.

Acknowledgment

The authors would like to thank to all reviewers for their advices and suggestions on improving this paper. This work was supported by the European Regional Development Fund within the Operational Programme "Science and Education for Smart Growth 2014 - 2020" under the Project CoE "National center of mechatronics and clean technologies" BG05M2OP001-1.001-0008".

References

[1] Madzharov N and Hinov N 2019 Flexibility of wireless power transfer charging station using dynamic matching and power supply with energy dosing *Applied Sciences* **9** 4767
[2] Tang J, Ma X, Gu R, Yang Z, Li S, Yang C and Yang B 2019 Cost consensus algorithm

applications for EV charging station participating in AGC of interconnected power grid *Applied Sciences* **9** 4886

[3] Pevere A, Petrella R, Mi C and Zhou S 2014 Design of a high efficiency 22 kW wireless power transfer system for EVs fast contactless charging stations *IEEE Int. Electric Vehicle Conf. (IEVC) (Florence)* pp 1–7

[4] Wu H, Gilchrist A, Sealy K, Israelsen P and Muhs J 2011 A review on inductive charging for electric vehicles *IEEE Int. Electric Machines & Drives Conf. (IEMDC)* pp 143–7

[5] Madzharov N and Ilarionov R 2011 Battery charging station for electromobiles with inverters with energy dosing *PCIM Europe Int. Exhibition and Conf. for Power Electronics, Intelligent Motion, Renewable Energy and Energy Management (Nurnberg, Germany)* pp 894–9

[6] Madzharov N, Tonchev A and Madzharov D 2012 Contactless charging system for electric vehicles *PCIM Europe Int. Exhibition and Conf. for Power Electronics, Intelligent Motion, Renewable Energy and Energy Management (Nurnberg, Germany)* pp 1400–7

[7] Budhia M, Covic G and Boys J 2010 A new IPT magnetic coupler for electric vehicle charging systems *IECON 36th Annual Conf. on IEEE Industrial Electronics Society* pp 2487–92

[8] Mohagheghi S, Parkhideh B and Bhattacharya S 2012 Inductive power transfer for electric vehicles: Potential benefits for the distribution grid *IEEE Int. Electric Vehicle Conf. (Greenville)* pp 1–8

[9] Ibrahim M et al. 2016 Inductive charger for electric vehicle: Advanced modeling and interoperability analysis *IEEE Transactions on Power Electronics* **31**(12) pp 8096–114

[10] Zhao C, Wang Z, Du J, Wu J, Zong S and He X 2014 Active resonance wireless power transfer system using phase shift control strategy *IEEE Applied Power Electronics Conf. and Exposition (APEC)* pp 1336–41

[11] Covic G and Boys J 2013 Modern trends in inductive power transfer for transportation applications *IEEE J. Emerging Selected Topics Power Electronics* **1**(1) pp 28–41

[12] Kato M, Imura T and Hori Y 2012 New characteristics analysis considering transmission distance and load variation in wireless power transfer via magnetic resonant coupling *Intelec* pp 1–5

[13] Van den Bossche A and Sergeant P 2008 Inductive coupler for contactless power transmission *IET Electr Power Appl.* pp 1–7

[14] Li S and Mi C 2015 Wireless power transfer for electric vehicle applications *IEEE J. Emerging Selected Topics Power Electronics* **3**(1) pp 4–17

[15] Brusamarello V, Blauth Y, Azambuja R, Muller I and de Sousa F 2013 Power Transfer with an inductive link and wireless tuning *IEEE Trans. Instrum. Meas.* **62**(5) pp 924–31

[16] Subudhi P and Krithiga S 2019 Wireless power transfer topologies used for static and dynamic charging of EV battery: A review *Int. J. of Emerging Electric Power Systems* **21**(1)

IOP Conf. Series: Materials Science and Engineering 1002 (2020) 012020 doi:10.1088/1757-899X/1002/1/012020

Comparative study of selected cryptocurrency

O Nakov, D Gotseva and Y Tomov [1]

Department of Computer Systems, Faculty of Computer Systems and Technologies Technical University of Sofia, 8 Kliment Ohridski blvd., 1000 Sofia, Bulgaria

[1] E-mail: yavor_tomov@tu-sofia.bg

Abstract. In the last few years, blockchain technologies have gained immense popularity both among researchers and in socio-economic life. Companies from various fields of industry implement projects based on blockchain technology. Pandemics like Covid-19 put almost all industries to the test. The work of the transport industry during the crisis is directly related to the work of the financial and banking sector. The problems that transport companies would experience from blocking the banking system during a subsequent pandemic would be solved in an innovative way using different cryptocurrencies. This article aims to make a comparative analysis of some cryptocurrencies, as well as to comment on the problems associated with them. The analysis can be used to develop blockchain-based applications in the transport industry.

1. Introduction

In 2009, S. Nakamoto [1] published an article that aroused great interest from the scientific community. A year later, the author and a group of other researchers implemented the Bitcoin project (the first cryptocurrency). Nakamoto laid down the main features that each cryptocurrency must meet: a decentralized currency independent of a third party – a bank or financial institution, anonymity or pseudo-anonymity of participants and security of transactions. Unlike traditional currencies, where a third party (bank or financial institution) is required to enable a transaction to take place between two participants, cryptocurrencies use consensus models to validate transactions. The security of the system and the anonymity of the participants are realized through cryptographic methods.

To date, there are more than 3,000 cryptocurrencies with Mагкеt Cap about $ 270 billion (figure 1). This fact speaks of a new digital economy that is evolving every day. The development of the cryptocurrency market has led to the creation of other accompanying tools from the digital industry. There is a new way for companies to raise funds for their business through Initial Coin Offering (ICO) (figure 2), digital exchanges for cryptocurrency trading, huge projects based on blockchain technologies and smart contracts.

Bitcoin aims to perform two main tasks: making transactions and storing value. Over the next few years, currencies appeared (litecoin 2011 [2], ripple 2012 [3], monero 2014 [4], dash2014 [5]) and many others that differed in certain features from Bitcoin, but they basically had the same two functionalities – performing transactions and storing value. In 2015, Ethereum (ETH) was created [6]. This new currency added new properties to those held by previous currencies. ETH is a second-generation currency based on the idea of smart contracts first proposed by N. Scebo [7].

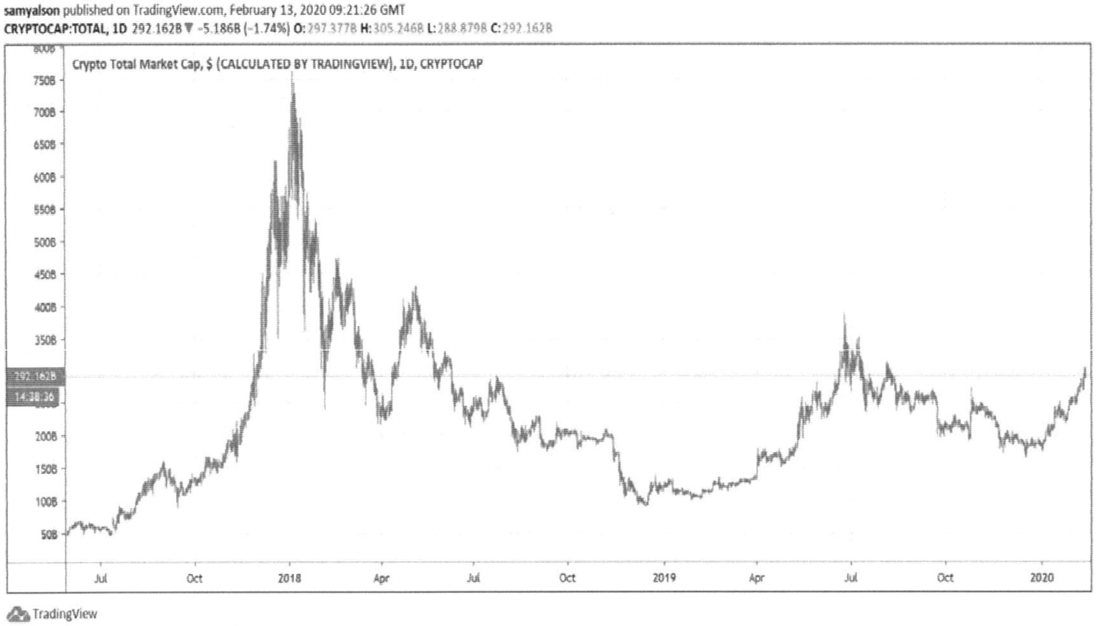

Figure 1. Cryptocurrencies Marкet Cap (tradingview.com).

Smart contracts provide an opportunity to create different applications in different industries through the use of cryptocurrencies. Cryptocurrencies and smart contracts have the ability to tweak modern relationships through a new approach to digitization.

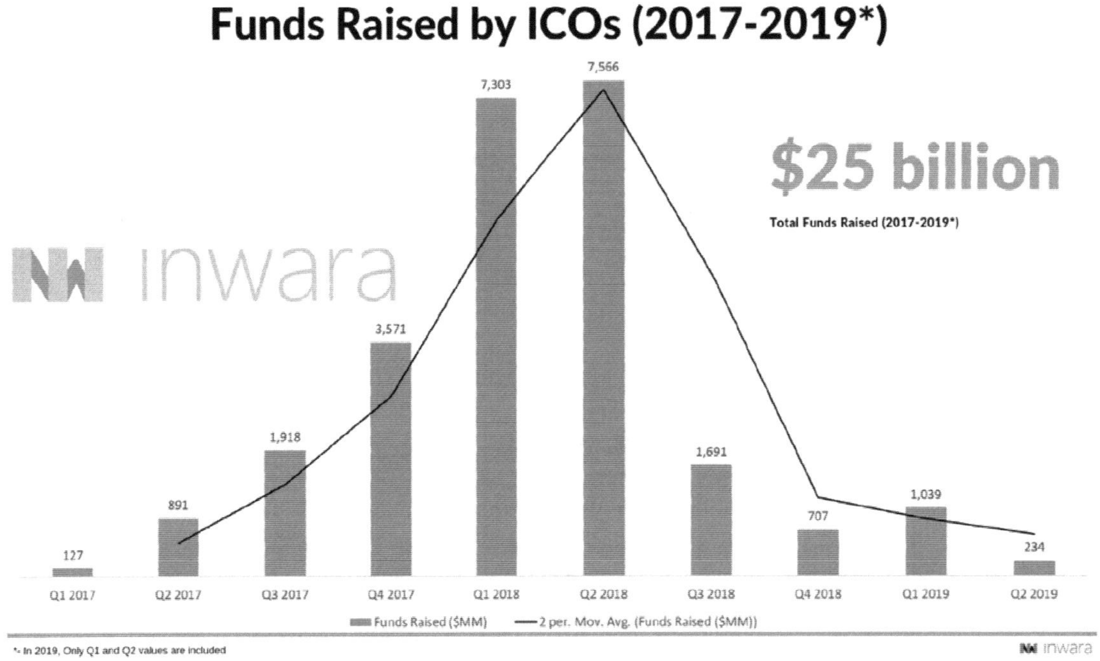

Figure 2. ICOs Marкet Cap (invara.com).

In this article, we, like other authors [8], will offer features for evaluating several cryptocurrencies and related technologies, so that these features help developers in the transportation industry to develop their applications. The cryptocurrencies that are being explored as part of this article fall into two main groups. The first one consisting of cryptocurrencies that are the most popular and well-known, while the second one involves cryptocurrencies that have huge development potential determined by the development team and applied new technology in an industry of global importance. There are different types of research that successfully integrate blockchain technology and cryptocurrencies in transport [9].

2. Background

In his article Nakamoto, describes three main technologies for the implementation of Bitcoin: blockchain, public lager, consensus model (proof of work). These technologies are the basis of most cryptocurrencies. Blockchain technology is a data structure consisting of individual blocks (figure 3).

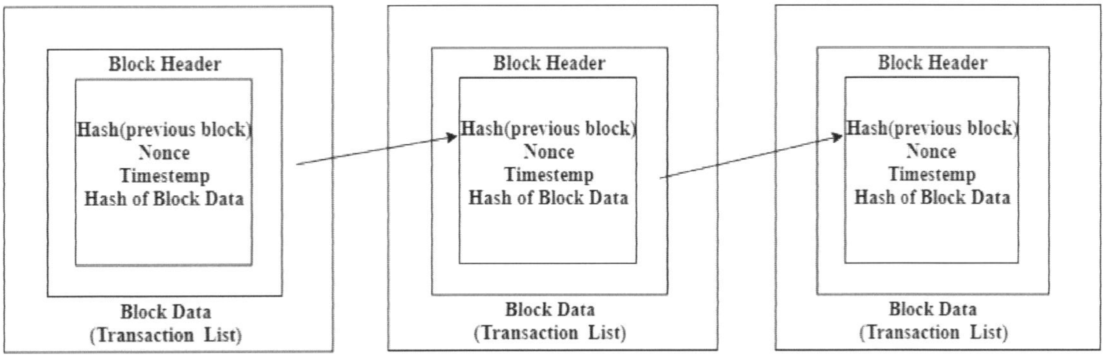

Figure 3. Blockchain data structure.

All blocks are connected to each other like a linked list. The transactions in each block are hashed and stored in a data structure called a merkle tree. The header of each block contains the previous block's hash. Using this technology, the author ensures that the information cannot be replaced. Changing a single character will generate a different hash.

Unlike fiat currencies, which use a centralized system, cryptocurrencies use decentralized, where each participant in this system has its own copy of the blockchain. This technology is known as public ledger. In centralized systems, a single participant decides on each transaction if this transaction is valid. In decentralized decision-making, all participants participate. The question arises, can't such systems be manipulated by malicious participants? Nakamoto uses asymmetric encryption techniques to solve the problem of man-in-the-middle attack (MITM) attacks. Another major contribution to security is that the participants in the system participate in proportion to their computing power, and not as the number of accounts, this protects the system from Sybil attack. The main security issue that all cryptocurrencies using the blockchain need to address is related to the problem known as double spending. Numerous algorithms have been proposed, the most popular being proof of work (PoW) [10], proof of stake (PoS) [11].

Smart contracts are computer programs uploaded to the Ethereum network and executed by the Ethereum virtual machine. These contracts, once published, cannot be changed. Ethereum provides Solidity which is a "turing complete" programming language. This means that any logic can be implemented in the Ethereum network.

3. Comparative study of selected cryptocurrencies

As mentioned above, the choice of cryptocurrencies is made on the basis of well-known currencies such as Bitcoin (Btc), Ethereum (Eth), and Ripple (Xrp) while the choice of the latter ones is based on their innovative nature and the important problems they solve – Basic Attention Token (Bat) [12], Theta (Theta) [13], and Chainlink (link) [14]. The selected evaluation criteria is shown in table 1.

Table 1. Comparative evaluation of the selected cryptocurrencies.

	Btc	Eth	Xrp	Bat	Theta	Link
intended	2009	2015	2012	2018	2018	2017
Market cap	$ 166.9 Billions	$ 25.3 Billions	$ 7.8 Billions	$ 0.36 Billions	$0.18 Billions	$ 1.6 Billions
Blockchain	yes	yes	no	yes	yes	yes
Algorithm	PoW	PoW	Ripple CA	Zero Knowledge proof	Multi-Level BFT	ChainLink algorithm
Time conformation	60 min	6 min	seconds	Such as eth 6 min	seconds	6 min
Turing complete	no	yes	no	yes	yes	yes
ledger	blockchain	blockchain	Last Closed Ledger	blockchain, Brave Micropayments Ledger	blockchain	blockchain
Coin/Token	coin	coin	coin	token	token	token
Innovation	The first crypto project	Smart contracts	Ripple algorithm	Secure Brave browser, using blockchain	Video streaming using blockchain	Blockchain connection with externel systems

The first two criteria measure the capitalization of the given cryptocurrencies since their creation. They serve as proof of their viability. Bitcoin ,being the first cryptocurrency in the world, has become the main currency for storing value (digital gold), which is evident from its capitalization. Ripple's capitalization currently puts it in third place. The main factors are the speed of confirmation of transactions, the low value of transactions, and security. Unlike Bitcoin, where the speed of transactions is very slow, Ripple was created to replace the banking system. Ethereum, with its smart contract-based functionality, has become an infrastructure for a vast number of projects and other cryptocurrencies based on the Ethereum network. An example of such cryptocurrency is Bat. Unlike currencies (coins) which have their own infrastructure, tokens work on a foreign blockchain network. This explains the influence of some currencies over others. An increase in the price of Bat would have a positive effect on Ethereum. Historically, the price of Btc affects the state of all other cryptocurrencies, just as the price of gold affects fiat currencies. Because cryptocurrencies are based on decentralized systems, they use different consensus algorithms. Bitcoin and Ethereum use the PoW algorithm, while Xrp uses the Ripple consensus algorithm. The main disadvantages of PoW are the huge amount of electricity consumed and the slow confirmation of transactions. As a matter of fact, the Btc network consumes as much energy per hour as a country like Austria. On the other hand, this algorithm is well scalable and solves the problem of double spending. Ripple is not based on blockchain technology. This cryptocurrency uses a public ledger in which the user's amount of each account is stored. Every few seconds a new ledger is created with new transactions. Last-Closed Ledger is the valid one, in which each transaction must have more than 80% of the validator servers confirming it.

Brave browser is the first browser based on blockchain. Bat is a token for the decentralized ad exchange. BAT connects advertisers, publishers, and users, creating a new, efficient marketplace. The token is based on Ethereum technology, an open source, blockchain-based distributed computing platform with smart contracts. These cryptographically secure smart contracts are stateful applications stored on the Ethereum blockchain, fully capable of enforcing performance. The token is derived from – or denominated by – user attention [12].

Unlike other browsers, where personal information is stored on servers, here the information is stored locally. This gives users much more privacy. This is realized by using the Zero Knowledge proof algorithm. The mission of Theta is to leverage blockchain technology to create the first decentralized video streaming and delivery network whereby video viewers are incentivized to share redundant computing and bandwidth resources to address today's video streaming challenges. Using the Ethereum EVM "World Compute" metaphor, the Theta Network can be viewed as the "World Cache" formed by the memory and bandwidth resources contributed by viewers [13].

One of the main problems with smart contracts is that they are only able to validate information that can be verified by the blockchain. The connection with information from the outside world can be easily manipulated. One way is for an application to make a connection between the outside world and a smart contract. However, this would lead to a breach of decentralization, which is a key feature of all blockchain technologies. The solution to this problem is the idea of oracles offered by ChainLink.

This solution will enable the creation of multiple applications based on smart contracts.

4. Conclusion

In this paper, an analysis of six cryptocurrencies was done. The choice of these currencies shows the main trends in the development of cryptocurrencies from 2009 (Bitcoin) to now. Ripple builds on top of Bitcoin by implementing an algorithm that is not based on blockchain. The main advantage is the high speed of the transactions. Until the creation of Ethereum, blockchain technologies were seen primarily as technologies capable of decentralizing the financial world. Smart contracts enable all industries to create decentralized applications for their specific needs. Proof of this are the Bat and Theta currencies, which are based on Ethereum, but solve the problems of the advertising business and video distribution. ChainLink, on the other hand, adds the ability to create smart contracts that work with data outside of the blockchain.

The analysis performed in this article can be used to develop blockchain-based applications in the transport industry. Such applications could be related to logistics problems, electronic payments and smart contracts between customers and transport companies.

References

[1] Nakamoto S 2008 *Bitcoin: A Peer-to-Peer Electronic Cash System* https://bitcoin.org/bitcoin.pdf
[2] https://litecoin.org/
[3] Schwartz D, Youngs N, Britto A *The Ripple Protocol Consensus Algorithm* https://ripple.com/files/ripple_consensus_whitepaper.pdf
[4] Saberhagen N 2013 *CryptoNote v 2.0* https://www.allcryptowhitepapers.com/wp-content/uploads/2018/05/monero-whitepaper.pdf
[5] Duffield E, Diaz D 2018 *Dash: A Payments-Focused Cryptocurrency* https://github.com/dashpay/dash/wiki/Whitepaper
[6] Buterin V *A Next-Generation Smart Contract and Decentralized Application Platform* https://ethereum.org/en/whitepaper/
[7] Szabo N 1994 *Smart contracts* http://www.fon.hum.uva.nl/rob/Courses/InformationInSpeech/CDROM/Literature/LOTwinterschool2006/szabo.best.vwh.net/smart.contracts.html
[8] Mauri L, Cimato S and Damiani E 2018 A comparative analysis of current cryptocurrencies *Proc. of the 4th Int. Conf. on Information Systems Security and Privacy ICISSP (Funchal, Madeira – Portugal)* pp 127–38
[9] Namiot D, Pokusaev O, Kupriyanovsky V and Akimov A 2017 Blockchain applications for transport industry *Int. J. of Open Information Technologies* **5**(12) (In Russian) pp 130-4
[10] Back A 2002 *Hashcash – a denial of service counter-measure*

IOP Conf. Series: Materials Science and Engineering 1002 (2020) 012020 doi:10.1088/1757-899X/1002/1/012020

http://www.hashcash.org/papers/hashcash.pdf
[11] King S, Nadal S 2012 *PPCoin: Peer-to-Peer Crypto-Currency with Proof-of-Stake* https://www.chainwhy.com/upload/default/20180619/126a057fef926dc286accb372da46955.pdf
[12] https://basicattentiontoken.org/BasicAttentionTokenWhitePaper-4.pdf
[13] https://www.thetatoken.org/
[14] Ellis S, Juels A, Nazarov S 2017 *ChainLink A Decentralized Oracle Network* https://link.smartcontract.com/whitepaper

Study of hydrogen gaseous fuel influence on the thermal condition of a diesel engine

E Dimitrov, M Peychev, Ph Michaylov and S Pantchev [1]

Department of Combustion Engines, Automotive Engineering and Transport, Technical University of Sofia, 8 Kliment Ohridski Blvd., 1756 Sofia, Bulgaria

[1] E-mail: etzd@tu-sofia.bg

Abstract. The current article presents experimental results for the thermal condition of a diesel-hydrogen powered engine. The object of the study is an air-cooled single cylinder diesel engine. A quantitative assessment of the engine thermal condition is made. It is based on four engine body part temperatures and the temperatures of exhaust gases and engine oil. The study is carried through by comparing the engine indexes via regulation characteristics as a function of the hydrogen mass content and load characteristics with constant hydrogen content. Experimental results as a function of engine speed at constant engine power are also presented. The data registered during the experiment is for two different diesel injection advance angles. A temperature for one location on the combustion chamber roof (cylinder head) is calculated. The results of the experiment are presented in charts.

1. Introduction

The internal combustion engines (ICEs) are heat engines which convert the chemical energy of the supplied fuel into mechanical work. The energy transformation takes place during the combustion and the expansion processes. The heat released during the combustion of the cyclic fuel quantity has a direct effect both on the transient temperature value of the fuel mixture in the engine cylinder and on the thermal conditions of the elements forming the inner surface of the combustion chamber. On the other hand, an optimal combustion process is directly dependent on the thermal conditions of the whole engine [1, 2]. The thermal condition of an internal combustion engine is determined by the coolant temperature, the crankcase oil temperature, the exhaust gas temperature, and the engine components' temperature.

Working on a gas-diesel cycle is one of the more effective ways to improve the emissions of diesel engines [3-6]. The gas-diesel cycle requires engine operation with two different fuel types: diesel fuel and gaseous fuels. However, the operation of a diesel engine with two fuels leads to changes of its combustion process [7-9], and therefore in its thermal conditions. Hydrogen, which is the most promising alternative fuel for internal combustion engines, has a number of properties [10,11]: high lower heating value – $H_u = 120\ \mathrm{MJ/kg}$, ignition and combustion in a wide range of fuel-air ratios – $\alpha = 0.2 \ldots 10$, high speed of flame front propagation, and other properties which directly affect the combustion process and the thermal conditions of the internal combustion engine [12,13].

2. Purpose of research

In connection with the above, the research aim is to make a quantitative assessment and analysis of the influence of hydrogen gaseous fuel on the thermal conditions of a diesel engine working on gas-diesel cycle.

3. Study subject and methodology

The subject of the current study is a single-cylinder four-stroke air-cooled diesel engine DV-550 (Bulgaria). The basic engine parameters are cylinder bore $D = 91.5$ mm; piston stroke $S = 85$ mm; compression ratio $\varepsilon = 17.5$; direct fuel injection into the cylinder, partially stratified mixture formation; nominal power $N_e = 8$ kW at engine speed $n = 3000\ \text{min}^{-1}$.

The methodology of research is based on regulating characteristics with variable gaseous fuel content [14]. These characteristics are obtained at constant engine speed – $n = const$ while varying fuel consumption of diesel fuel B_D and hydrogen B_{H_2}. Distinctive parameters are presented as a function of the hydrogen mass share in total fuel – the K_{H_2} coefficient. The K_{H_2} coefficient is described by the formula:

$$K_{H_2} = \left(\frac{B_{H_2}}{B_h}\right) \times 100, \%, \tag{1}$$

where B_h is the total fuel consumption and equals $B_h = B_D + B_{H_2}$. The load and speed engine characteristics (at constant hydrogen mass share in total fuel) can be built by obtaining full series of engine regulating characteristics.

4. Technical resources for the research

The engine subject of the current study is coupled to a *SAC 28* direct-current dynamometer, by means of which the engine performance is measured. The brake power is measured with accuracy of 0.005 kgf. An electronic frequency meter registering the pulse frequency of an inductive sensor paired with a 60-tooth wheel is used to measure engine speed. The diesel fuel consumption is measured with a volumetric flow meter for liquid fuel. The consumed diesel fuel volume is measured with accuracy of 0.2 cm^3. The hydrogen fuel consumption is determined with a G4-type volumetric flow meter for gaseous fuel. The hydrogen volume is measured with up to 0.2 dm^3 accuracy. The full test bench description is given at [15].

For the purpose of the research, six temperatures are registered: engine oil temperature – T_{oil}; exhaust gas temperature – T_{eg}, measured at a distance of 100 mm from the exhaust manifold flange; three temperatures at the cylinder's external wall – T_{cyl}^1, T_{cyl}^2, T_{cyl}^3; one temperature in the internal part of the cylinder head wall – T_{head}. All thermocouples are K-type. The measurement of engine oil temperature and exhaust gases is done with electronic devices with an error of 0.75% ± 1 °C which guarantees a maximum accuracy of 1% for the maximum measured value of 450 °C. The other four temperatures are registered with a specialised system (data acquisition device) for temperature measurements *NI 9211*, manufactured by *National Instruments*. The system works with the *LabView* software package. According to the manufacturer's datasheet, the measurement system maximum error is 2.3 °C in the 0…400 °C interval which guarantees an accuracy of 1.14% for the maximum measured value of 200 °C.

The thermocouples for measuring the cylinder external wall temperatures are mounted between the cooling ribs at distances of 15, 22, and 29 mm from the plane dividing the cylinder head from the engine cylinder. The cylinder head temperature is measured with a thermocouple mounted in a threaded collar. The collar is mounted in a threaded hole drilled in the cylinder head. During the engine indicating process, a piezoelectric sensor used to register the cylinder pressure data is mounted in the threaded hole. The thermal conduction coefficient of the threaded collar material is the same as the thermal conduction coefficient of the cylinder head material.

During the data processing, the temperatures T_{cyl}^m, calculated as an average of the three thermocouples mounted on the cylinder's external wall, and the temperature of the cylinder head surface enclosing the combustion chamber T_{head}^w are calculated by using the following equations:

$$T_{cyl}^m = \frac{T_{cyl}^1 + T_{cyl}^2 + T_{cyl}^3}{3}; \tag{2}$$

$$T_{head}^{w} = \frac{T_{head} - T_{oil}\frac{x}{\delta}}{1-\frac{x}{\delta}}, \tag{3}$$

where: x is the distance from the internal surface of the cylinder head to the thermocouple weld in the threaded collar ($x = 7$ mm) and δ is the threaded collar height ($\delta = 33.5$ mm). Equation (3) is known from the heat conduction theory [16]. It is accepted that the temperature of the upper surface of the threaded collar is equal to the temperature of the oil in the oil pan as it is situated directly under the valve rocker arms axis and is in constant contact with the engine oil under the valve cover.

The diesel fuel injection advance angle – θ, which also has a significant influence on the combustion process parameters, therefore on the engine thermal condition [2], is measured with a special device with a stroboscopic lamp and a piezoelectric sensor. The sensor is installed on the high-pressure pipeline connecting the injection pump and the nozzle. The stroboscopic lamp is activated by the voltage induced by the sensor which is proportional to the deformation of the high-pressure pipeline in the moment of fuel injection. The θ angle is recorded in numerical form in the moment when the moving and stationary marks for the piston top dead centre (TDC) match.

5. Study results

Most of the experimental results are recorded with stock engine fuel system setup where the diesel fuel injection advance angle is equal to 20 degrees (crankshaft rotation) before TDC, i.e. $\theta = 20$ deg. Full series of regulation characteristics with variable hydrogen fuel content at $\theta = 20$ deg are obtained at three engine speed modes: $n = 1500$ min^{-1}, $n = 2000$ min^{-1}, and $n = 2500$ min^{-1}. The influence of the speed on the thermal condition of the engine is also investigated with the mean effective pressure – p_e and the hydrogen mass content coefficient K_{H2} changing within narrow ranges.

In the course of the research it was established that with the stock setting of the diesel fuel ignition advance angle and engine load over 60% and percentage by mass of hydrogen over 10%, unacceptably high cylinder pressure rise rate values are reached. Therefore, a new series of regulation characteristics with variable hydrogen mass content in total fuel was obtained at speed $n = 2000$ min^{-1} and diesel fuel injection advance angle $\theta = 12$ deg. In this series however, the temperature of the cylinder head T_{head} is not registered.

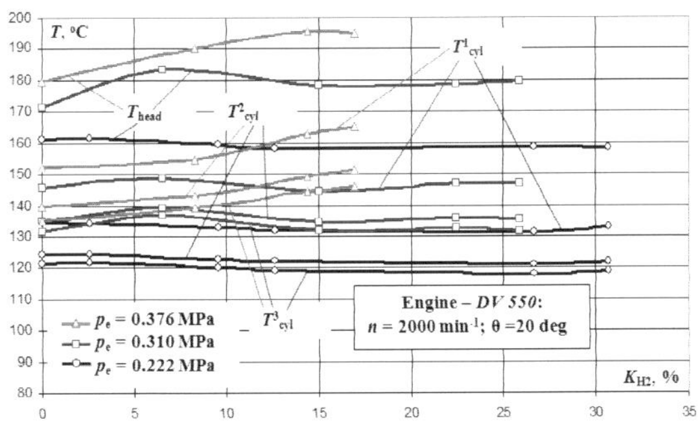

Figure 1. Body part temperatures of a DV-550 engine working on a gas-diesel cycle with variable hydrogen mass content, $n =$ 2000 min^{-1} and $\theta = 20$ deg

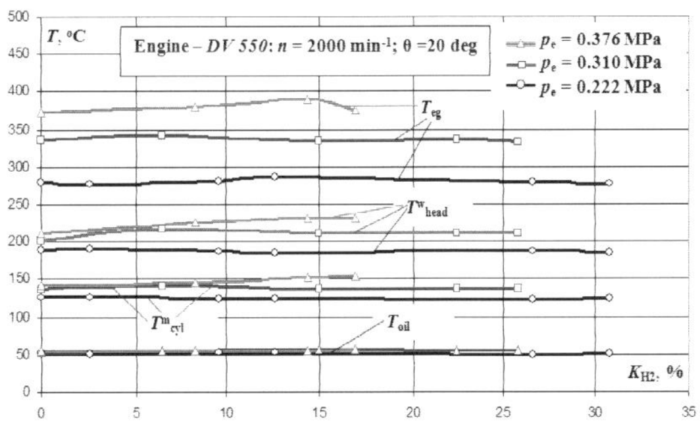

Figure 2. Body part temperatures (calculated), oil and exhaust gas temperatures of a DV-550 engine working on a gas-diesel cycle with variable hydrogen mass content, $n =$ 2000 min^{-1} and $\theta = 20$ deg

Some of the experimental results are shown on figures 1 to 7. The result analysis shows the following:

- When operating a DV-550 engine on a gas-diesel cycle and varying the hydrogen mass share in total fuel (K_{H_2}) at low and medium loads ($p_e \leq 0.3$ MPa), the maximum differences in the average cylinder temperature compared to its temperature when operating only on diesel fuel are in the range of 5.3... 8.2%. The high values of the differences are observed when working with an injection advance θ = 20 deg, and the low values – at θ = 12 deg. At the same operating modes and injection advance θ = 20 deg, the maximum differences in cylinder head temperature are 9%;

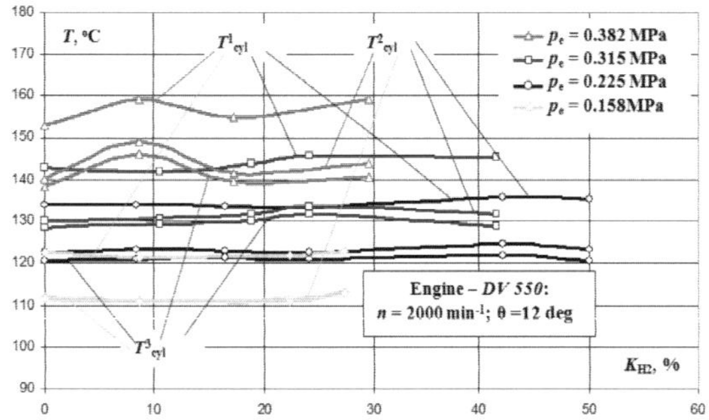

Figure 3. Body part temperatures of a DV-550 engine working on a gas-diesel cycle with variable hydrogen mass content, $n = 2000\ \text{min}^{-1}$ and $\theta = 12$ deg

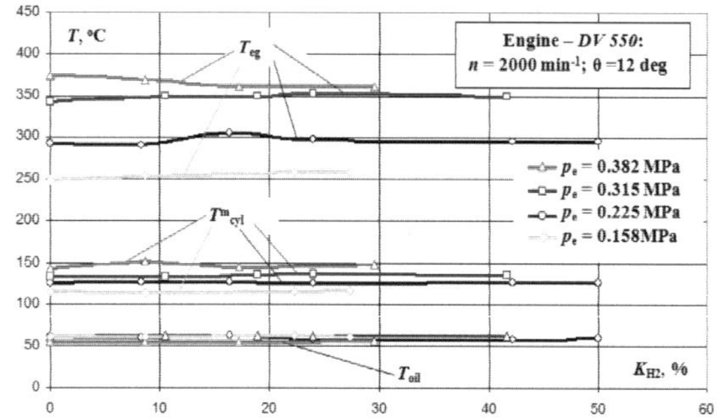

Figure 4. Body part temperatures (calculated), oil and exhaust gas temperatures of a DV-550 engine working on a gas-diesel cycle with variable hydrogen mass content, $n = 2000\ \text{min}^{-1}$ and $\theta = 12$ deg

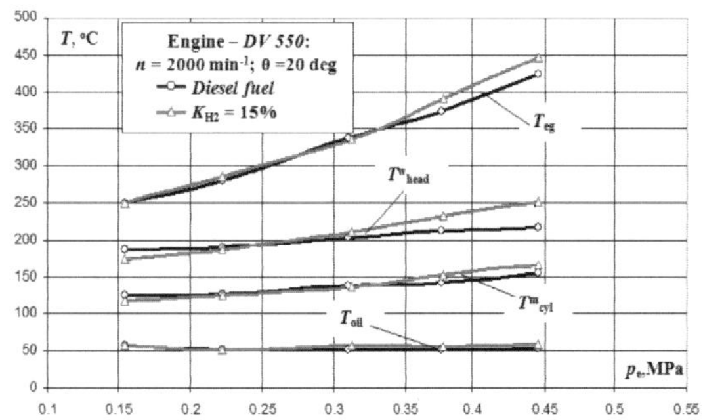

Figure 5. Comparison of distinctive temperatures as a function of the load of a DV-550 engine working on diesel fuel and on a gas-diesel cycle with hydrogen mass ratio $K_{H_2} = 15\%$, $n = 2000\ \text{min}^{-1}$ and $\theta = 20$ deg

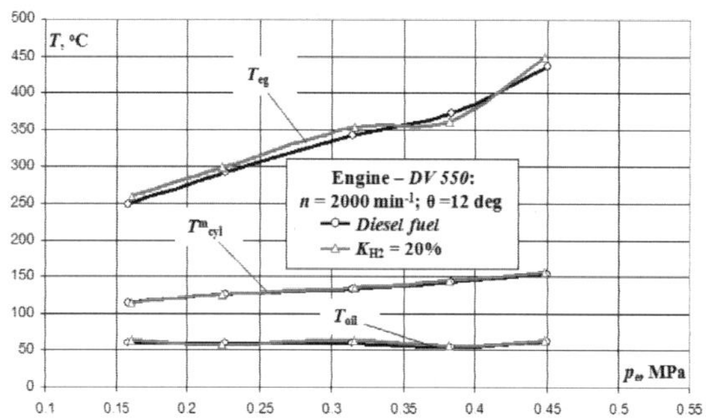

Figure 6. Comparison of distinctive temperatures as a function of the load of a DV-550 engine working on diesel fuel and on a gas-diesel cycle with hydrogen mass ratio $K_{H_2} = 15\%$, $n = 2000\ \text{min}^{-1}$ and $\theta = 12$ deg

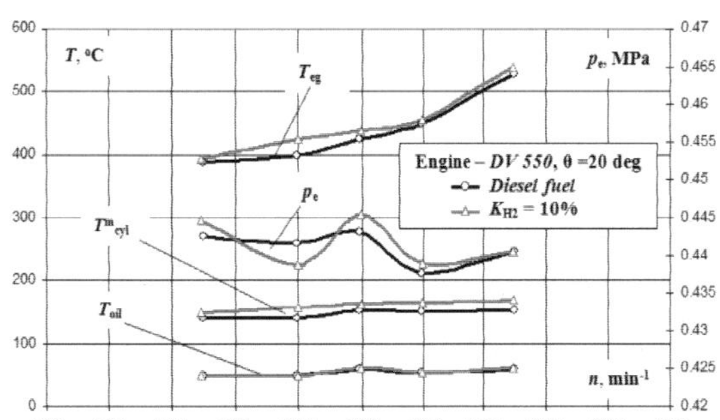

Figure 7. Comparison of distinctive temperatures as a function of the speed of a DV-550 engine working on diesel fuel and on a gas-diesel cycle with hydrogen mass ratio $K_{H_2} = 10\%$, $p_e \approx 0.44$ MPa and $\theta = 20$ deg

- When operating the engine with an injection advance $\theta = 20$ deg, the following trend is observed: with increasing engine load, the peak temperatures of the body parts shift to the higher hydrogen mass share in total fuel (K_{H_2}), and when working with an injection advance

$\theta = 12$ deg the tendency is reversed – with increasing engine load the peaks are shifted to a lower hydrogen mass share in total fuel (K_{H_2}).
- The studied operation modes do not show a significant change in the temperature of the exhaust gases when operating the engine on a gas-diesel cycle compared to the temperature when operating only on diesel fuel (the difference is up to 3%);
- According to the load characteristic at engine speed $n = 2000$ min^{-1}, the largest percentage differences in the measured temperatures of the body parts and the exhaust gases (when working on a gas-diesel cycle compared to those when working on diesel fuel only) are observed at injection advance $\theta = 20$ deg and high loads. When operating the engine with an injection advance $\theta = 20$ deg and low loads, a decrease in temperature of 5% is observed when operating on a gas-diesel cycle ($K_{H_2} = 15\%$) compared to operation only on diesel fuel. At an injection advance $\theta = 12$ deg, no significant differences in the respective temperatures are observed (up to 3%).
- In the studied speed range of the DV-550 engine with an injection advance $\theta = 20$ deg and approximately constant mean effective pressure $p_e \approx 0.44$ MPa, there is an increase in the average temperature of the cylinder (on average by 8%) when working on gas-diesel cycle compared to operation only on diesel fuel.

6. Conclusion

The results of the experimental study allow us to draw the following conclusions:

- The change of the percentage mass content of hydrogen in the total amount of fuel during the operation of a diesel engine on a gas-diesel cycle has an impact mainly on the maximum values of the temperatures, characterizing the thermal state of the engine. This effect is more pronounced at medium and high loads;
- When operating the engine on a gas-diesel cycle with hydrogen gaseous fuel, the change in the diesel fuel injection advance angle leads to a change in both the absolute values of the maximum temperature differences and the nature of their change depending on the percentage of hydrogen in the fuel;
- The change in speed under equal other conditions does not affect the differences in the engine thermal state during the gas-diesel cycle compared to the operation with diesel fuel.
- In conclusion, it can be noted that when operating the engine DV-550 on a gas-diesel cycle with hydrogen gaseous fuel in some of the modes, both in terms of load and in terms of hydrogen fuel percentage, there is a significant increase in temperatures characterizing the thermal state of the engine. When optimizing the diesel fuel injection advance angle, this negative trend can be minimized.

Acknowledgment

The authors would like to thank the Research and Development Sector at the Technical University of Sofia for the financial support.

References

[1] Heywood J 1988 *Internal Combustion Engine Fundamentals* (McGraw-Hill) pp 668–707
[2] Dimitrov P 1998 *Internal Combustion Engines (Part 1 Theory of Internal Combustion Engines)* (Technical University of Sofia) pp 94–102 (In Bulgarian)
[3] Karagöz Y, Güler I, Sandalcı T, Yüksek L and Dalkılıç A S 2016 Effect of hydrogen enrichment on combustion characteristics, emissions and performance of a diesel engine *International journal of hydrogen energy* **41** pp 656–65
[4] Nicolas C, Mario T and German Amador 2019 An experimental investigation of the performance and emissions of a hydrogen-diesel dual fuel compression ignition internal combustion engine *Applied Thermal Engineering* **156** pp 660–7

[5] Hristov R and Stefanov S 2019 Study of the potential for reducing CO_2 emissions from road transport in Bulgaria *IOP Conf. Ser.: Mater. Sci. Eng.* **614** 012009
[6] Ivanov Z, Mihaylov V, Kolev A 2012 Research system for determination ecological characteristics of automobile engines *Proc. of 18th Technical and Sci. Conf. "Transport, Ecology - Sustainable development" EKO Varna '2012* pp 511–8
[7] Nag S, Sharma P, Gupta A and Dhar A 2019 Combustion, vibration and noise analysis of hydrogen-diesel dual fueled engine *Fuel* **241** pp 488–94
[8] Bogdanov K and Dimitrov R, 2019 Research on the possibility of increasing the power output of a high - speed diesel engine by adding compressed natural gas *IOP Conf. Ser.: Mater. Sci. Eng.* **614** 012007
[9] Lata D B and Misra A 2010 Theoretical and experimental investigations on the performance of dual fuel diesel engine with hydrogen and LPG as secondary fuels *Int. J. of Hydrogen Energy* **35** pp 11918–31
[10] Wu H-W and Wu Zh-Y 2012 Investigation on combustion characteristics and emissions of diesel-hydrogen mixtures by using energy-share method in a diesel engine *Applied Thermal Engineering* **42** pp 154–62
[11] Ciniviz M and Köse H 2012 Hydrogen use in internal combustion engine: A review *Int. J. of Automotive Engineering and Technologies* **1**(1) pp 1–15
[12] Ivanov Z and Mihaylov V 2019 Transient operation of a direct injection diesel engine *IOP Conf. Ser.: Mater. Sci. Eng.* **614** 012008
[13] Burke R and Brace C 2010 The effects of engine thermal conditions on performance, emissions and fuel consumption *SAE Technical Papers*
[14] Dimitrov E, Tashev A and Pantchev S 2015 Methodology for experimental investigation of the influence of gaseous fuels on the performance of the diesel engine operating on gas-diesel cycle *BulTrans-2015 Proceedings* pp 103–7
[15] Dimitrov E, Gigov B, Pantchev S, Michaylov Ph and Peichev M 2018 A study of hydrogen fuel impact on compression ignition engine performance *MATEC Web of Conferences* **243** 03001
[16] Lienhard J IV and Lienhard J V 2019 *A Heat Transfer Textbook* 5th edition (Flogiston Press, Cambridge, US-MA) p 61

IOP Conf. Series: Materials Science and Engineering 1002 (2020) 012022 doi:10.1088/1757-899X/1002/1/012022

Operation of diesel engine with fuels treated with nanoparticle additives

Zdravko Ivanov[1], Veselin Mihaylov[1,3], Lech Sitnik[2] and Zbigniew Sroka[2]

[1] Technical University of Varna, Department "Transport Engineering and Technologies", 9010 Varna, Bulgaria
[2] Wroclaw University of Technology, Division of Vehicles and Combustion Engines, Wroclaw, Poland

[3] E-mail: v_mihaylov@tu-varna.bg

Abstract. Treatment of materials with nano-sized elements is widely used in the automotive industry. This application is associated both with surface treatment of parts and in the form of various mixtures. The use of nanoparticles as an additive to oils and fuels aims to impact the energy efficiency of the engine in order to improve mechanical losses and its environmental performance. Improving mechanical losses leads to a reduction in energy consumption and hence to a reduction in carbon emissions. Published research in this regard has shown an impact on various physical processes related to engine performance. In this paper are shown the results from performed studies on a direct injected diesel engine when working with oil and fuel additives, based on carbon nanoparticles. The results show an improvement in the operation of the fuel injection system and reduction in the mechanical losses in the engine.

1. Introduction

The aim of the present study is to investigate the impact of carbon nanoparticles dispersed in biodiesel fuel mixtures on the operating parameters of a direct injection diesel engine.

A number of studies have shown increased thermal conductivity of nanofluids compared to untreated liquids. This medium is considered to be three-component, formed by a nanoparticle as the core and surrounded by an interfacial nanolayer as a shell, which in turn is immersed in the main liquid [1]. The size of the nanoparticles and the liquid boundary layer around the particles are a major factor for this increased thermal conductivity in solutions with nanostructure sizes in the range of 1-100 nm [2]. The coefficient of thermal conductivity depends on various factors such as concentration in the solution, shape and size of the nanoparticles, the degree of homogeneity of the solution and many others [3]. Studies have shown the impact of the thickness of the interfacial layers surrounding the nanoparticles on the total thermal conductivity of the nanofluid [4].

The thermal conductivity in nanofluids also largely depends on whether the nanoparticles remain dispersed in the main liquid, forming linear chain or similar configurations [5].

The fundamental understanding of the mechanism of increased thermal conductivity is not yet clear. The effect is used in various fields of technology where forced heat dissipation is required, in particular in electrical engineering and electronics. In internal combustion engines, nanofluids are extremely efficient for use as a cooling fluid [5].

In [6] a study was performed to determine the optimal concentration of nanoparticles of diesel fuel and Jojoba methyl ester. It has been found out that the use of Al_2O_3 additive in a concentration of about 20 mg/l improves the environmental performance of the engine, most notably for NOx emissions by 70%, CO by 80%, UHC by 60% and smoke by 35%. Studies of the working process of an internal combustion engine show that when increasing the concentration of Al_2O_3 to 40 mg/l there is a decrease in specific fuel consumption by 12% and an increase in the maximum value of pressure in the cylinder – p_{max}, the maximum growth rate of the pressure – $dp/d\varphi_{max}$, and the maximum heat release rate – $dQg/d\varphi_{max}$ by 4.5%, 4% and 4% respectively. According to the authors [6], the recommended concentration of Al_2O_3 in mixtures of diesel and Jojoba methyl ester is 30 mg/l, which leads to an improvement in both power-economic and environmental performance of the engine. For a period of operation lasting about two months, no erosive effects of Al_2O_3 on the fuel system and engine exhaust system were observed. The authors [6] are of the opinion that further research is needed in the long-term operation of the engine regarding the impact of this additive on the elements of the fuel system.

The addition of nanostructured particles to the fuel and lubricants significantly reduces friction. It has been found out that the coefficient of friction continues to decrease during continuous operation of systems with nanoadditives, which can be explained by roughness equalization [7]. It has been shown that nanosized particles are deposited on the surface of the material to form a layer. The nanostructured particles are in direct contact with the metal surfaces of the elements of the fuel system and it can be assumed that they settle on the contact surface. Experiments performed with a standard dispersion of 0.5 wt % TiO_2 particles, which was tested under standard conditions (pressure: 10 MPa; velocity: 0.16 ms^{-1}), showed an increase in the titanium detected in the surface layer. It is assumed that nanoparticles tend to adhere to the contact surface between the liquid and the metal [7].

Studies in [8] show a reduction in the roughness of contact surfaces when adding nanoparticles to lubricants. The phenomenon is described as polishing and has a significant effect on reducing friction. Comparative studies of steel disks on a rotating tribometer treated with AlMgSi structures show a reduction in friction when adding nanoparticles to the fluid compared to those when no additives were added [9]. It has also been shown that the effect of reducing friction is weakly dependent on the relative speed of movement, the base oil used, the dispersion additives and the nanofluid pre-treatment method used.

The duration of the ignition delay in the diesel combustion process largely determines the behaviour of the engine [10]. It consists of two main phases – physical and chemical, the identification of which is determined by the nature of the currently prevailing physicochemical processes. The duration of the individual phases depends mainly on the properties of the fuel and the thermodynamic characteristics of the medium in which the fuel is injected [11]. For biodiesel fuels, the duration of Ignition delay increases slightly due to the higher viscosity of these mixtures and reduced volatility during the fuelling process. Some of the characteristics of nanoparticles have the potential to affect mainly the physical phase of Ignition delay. The main ones are their increased thermal conductivity reaching in Carbon nanotubes (CNT) 2000 W/mK, Diamond – 600 W/mK, Al_2O_3 – 36 W/mK, while for diesel fuel it is about 0.15 W/mK. The addition of nanoparticles reduces ignition delay by shortening mainly its physical phase. The reason for this is the higher surface-to-volume ratio and the higher thermal conductivity of Al_2O_3 nanoparticles, which increases the evaporation rate.

2. Experimental set-up

An experimental study of the operation of a high-speed diesel engine working with B7 fuel was performed on an eddy current test stand [12, 13]. The test engine is a 4 cylinder, water cooled, direct injected diesel engine with rated power 46 kW at 4500 min^{-1}, torque 122 Nm at 2500 min^{-1}, displacement 1.993 cm^3, compression ratio 18:1.

Additives of nanomodified carbon structures (diamond) with sizes from 2 to 100 nm dispersed evenly in the fuel using an ultrasonic homogenizer were used. The experimental data required for the analysis were recorded with a high-sampling rate device capable of recording experimental data at 0.1 deg per engine crankshaft rotation angle. A developed methodology was used for preliminary and subsequent processing of the experimental data registered with automated systems, in order to study the registered processes through derivatives of higher order. The workflow data are obtained by processing sequential single diagrams, each for itself, after which the characteristic result points are averaged for the specific mode of operation of the engine. The engine operates on load characteristics at maximum torque mode. To ensure sufficient time for the nanostructured particles to be in direct contact with the metal surfaces of the elements of the combustion system, for deposition on the surface of the material and for the formation of a nanolayer, the engine runs continuously with nanomodified fuel. The results of the measurements were obtained after running the engine for 13 hours.

3. Results and discussions

The duration and limits of the Ignition delay as well as the phases of the period, physical and chemical, are determined along the lines of the heat release rate.

3.1. Ignition delay

3.1.1. Physical phase. The graphs show a significant reduction in the physical phase of Ignition delay. For the shown operating modes of the engine, the average values of the duration of the physical phase when working with nanoadditives to the fuel are about 3.3 deg crankshaft rotation angle (CA) smaller than those when working without additives. Figure 1 shows a decrease in the duration of the physical phase of Ignition delay by about 28.7%. The physical processes of fuel dispersion immediately after the start of fuel injection and the evaporation of fuel droplets are probably strongly impacted by the increased thermal conductivity and accelerated heat transfer in the direction from environment with a higher temperature to the currently injected fuel. At the same thermodynamic parameters of the medium at the beginning of the fuel supply in the diesel engine there is an acceleration of the processes of fuel dispersion and evaporation.

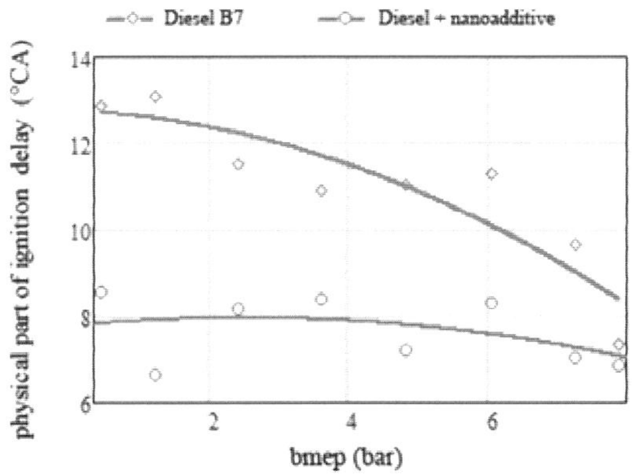

Figure 1. Duration of physical part of ignition delay vs brake mean effective pressure *bmep*.

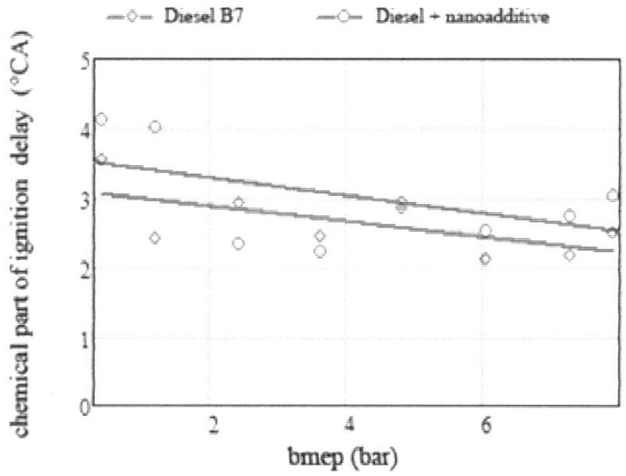

Figure 2. Duration of chemical part of ignition delay vs brake mean effective pressure *bmep*.

3.1.2. Chemical phase. The duration of the chemical phase of ignition delay is shown in figure 2. There is a certain delay in the process here. The duration of the chemical phase increases on average by about 0.4 deg, which at an average value of the order of 2.6 to 3.0 deg, is 15.4%.

3.1.3. Total ignition delay. Figure 3 shows the total duration of the ignition delay. The average decrease is of the order of 2.9 deg and represents a decrease of 13.6%. The ratio of the two phases changes. For the tested engine, the duration of the physical phase for engine operating modes by load characteristic occupies about 80.5% of the total duration of Ignition delay. After treatment of the fuel and operation of the engine with the addition of the studied nanostructures, the share of the physical phase decreased to 71.9%.

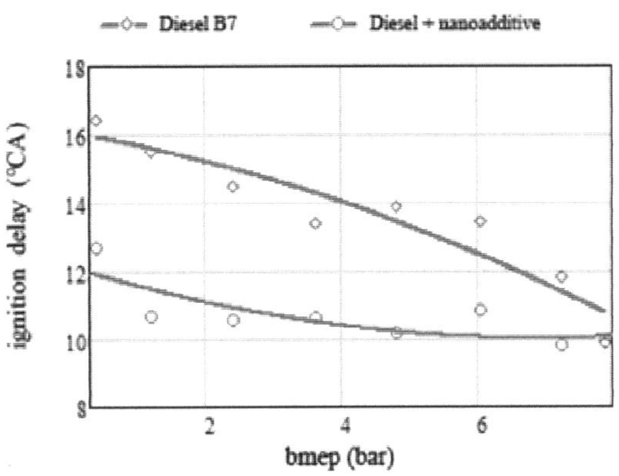

Figure 3. Ignition delay, determined by line of heat release rate, vs brake mean effective pressure *bmep*.

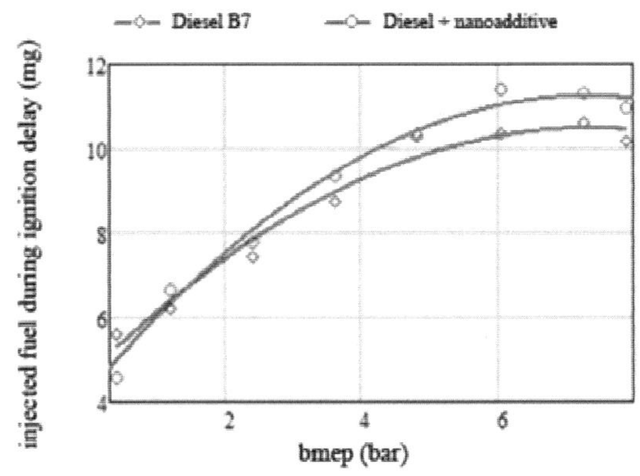

Figure 4. Quantity of injected fuel during ignition delay in mg vs brake mean effective pressure *bmep*.

3.2. Injected fuel during ignition delay
Prolonged operation of the fuel system with fuels with added carbon nanoparticles reduces the roughness of internal surfaces and this implies a reduction in hydraulic resistance. This phenomenon probably leads to a change in the differential characteristics of fuel supply, changing the amount of fuel injected during the different phases of the combustion process in diesel engines. Figure 4 shows the amount of fuel entering the engine's combustion chamber during Ignition delay. The amount of fuel injected during Ignition delay increased for all tested modes by an average of 0.347 mg, which is an increase of about 2.9% on average. The same graphs show that this increase is not the same for all engine modes. In the area of low loads such an increase is absent, even a reverse downward trend is seen.

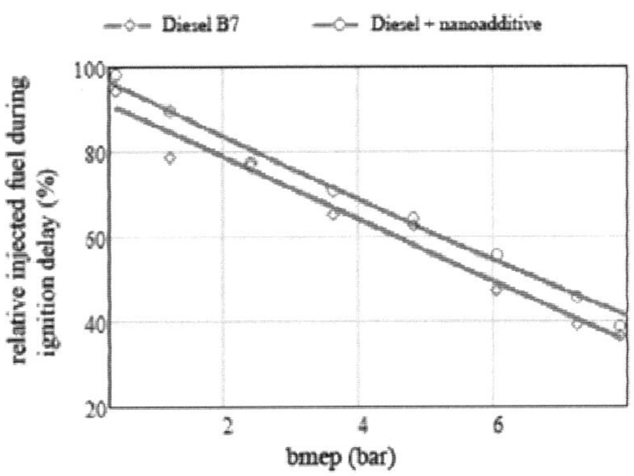

Figure 5. Relative quantity of injected fuel during ignition delay vs brake mean effective pressure *bmep*.

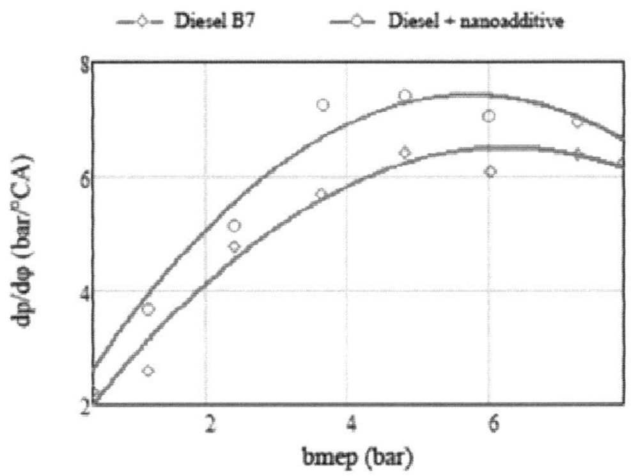

Figure 6. First derivative of in-cylinder pressure *dp/dφ* vs brake mean effective pressure *bmep*.

Figure 5 presents the results for the amount of fuel injected through the ignition delay relative to the cycle portion of fuel for the respective engine mode. Both dependences are equidistant, have the character of straight lines, although they are obtained by regression smoothing of the experimental data with a polynomial of the second degree. The average difference between them for the different operating modes of the engine is almost the same, its value is 5% and represents the increased relative amount of fuel, injected through ignition delay.

3.3. Rate of pressure rise

Reducing the duration of the Ignition delay leads to improved engine performance in indicators such as a lower maximum rate of increase in cylinder pressure. This results in reduced maximum values of pressure in the cylinder, low level of vibrations and noise, improved environmental performance and more.

The obtained values for the maximum rate of pressure increase in the engine cylinder, figure 6 shows an increase compared to the baseline. For all studied regimes, the average value of this parameter from 5.05 bar/deg CA was increased to 5.8 bar/deg CA, which is an increase of 16.5%. Despite the registered decrease in the duration of Ignition delay, by significantly shortening its physical phase, the increase in the relative share of fuel injected during this period obviously predetermines the further development of the combustion process during the premixed combustion phase.

4. Conclusion

The addition of diamond carbon nanoparticles to biodiesel fuel mixtures leads to a change in the studied parameters of the combustion process of the tested diesel engine. The increased thermal conductivity of these mixtures has a positive effect on the processes of fragmentation and evaporation, which take place at the beginning of the fuel supply to the engine cylinder. These processes have a strong impact on the duration of the physical phase of ignition delay and practically do not affect its chemical phase.

Prolonged operation of the fuel system of the engine with nanoadditives in fuel changes the differential characteristic of fuel injection. This leads to an increase in the relative amount of fuel injected into the engine cylinder during ignition delay, relative to the whole injected fuel per cycle. This change in the fuel supply characteristic has a decisive impact on the maximum values of the rate of increase of pressure in the engine cylinder.

Acknowledgment

Part of this research was supported by the Bulgarian Ministry of Education and Science through National Programme "Young scientists and postdoctoral students".

References

[1] Wang X and Mujumdar A 2008 A review on nanofluids - part I: theoretical and numerical investigations *Brazilian J. of Chemical Engineering* **25**(04) pp 613–30

[2] Machrafi H and Lebon G 2019 The role of several heat transfer mechanisms on the enhancement of thermal conductivity in nanofluids *Continuum Mech. Thermodyn.* **28** pp 1461–75

[3] Pasrija R and Srivastava S 2014 The interfacial layer effect on thermal conductivity of nano-colloidal dispersions *Int. J. of Appl. Phys. Math.* **4**(1)

[4] Rajinder P 2014 A Novel method to determine the thermal conductivity of interfacial layers surrounding the nanoparticles of a nanofluid *Nanomaterials* **4**(4) pp 844–55

[5] Eapen J, Rusconi R, Piazza R and Yip S 2010 The classical nature of thermal conduction in nanofluids *J. Heat Transfer* **132**(10) 102402

[6] El-Seesy A, Attia A and El-Batsh H 2018 The effect of Aluminum oxide nanoparticles addition with Jojoba methyl ester-diesel fuel blend on a diesel engine performance, combustion and emission characteristics *Fuel* (Elsevier) **224** pp 147–66

[7] Beel A 2019 *Tribologische Optimierung von Industrieölen durch Nanofluide* Dissertation (Dr. rer. nat.) (Fakultät für Physik, Universität Bielefeld)

[8] Demas N, Erck R, Lorenzo-Martin C, Ajayi O and Fenske G 2017 Experimental evaluation of oxide nanoparticles as friction and wear improvement additives in motor oil *J. of Nanomaterials*

[9] Tao X, Jianzheng Z and Kang X 1996 The ball-bearing effect of diamond nanoparticles as an oil

additive *J. of Physics* **29**(11) pp 2932–37
[10] Gupta H 2009 *Fundamentals of Internal Combustion Engines* (New Delhi: PHL Learning Private Limited) p 658
[11] Dimitrov E, Pantchev S, Michaylov Ph and Peychev M 2019 Economic aspects of using hydrogen in compression ignition engine operating on gas-diesel cycle *IOP Conf. Ser.: Mater. Sci. Eng.* **664** 012022
[12] Ivanov Z, Mihaylov V and Kolev A 2011 Engine indication system *Proc. of 17th Tech. Sci. Conf. ECO Varna* pp 590–7 (In Bulgarian)
[13] Ivanov Z and Mihaylov V 2019 Transient operation of a direct injection diesel engine *IOP Conf. Ser.: Mater. Sci. Eng.* **614** 012008

Experimental study of the flame propagation of gaseous fuel based mixture using optical visualisation

R Hristov[1], S Stefanov and R Dimitrov

Faculty of Mechanical Engineering and Technologies, Technical University of Varna, 1 Studentska Street, 2010 Varna, Bulgaria

[1] E-mail: rosen.hristov@tu-varna.bg

Abstract. At this article presents experimental investigations carried out to determine the flame kernel propagation of gaseous fuel-air mixture under spark-ignition engine-like conditions. Various combustion processes taking place in an experimental combustion chamber are discussed. Consecutive shots of the development of the fuel flame front in a constant volume combustion chamber under different initial conditions are presented. Direct visual recording of an image of the combustion process is performed or Schlieren photography techniques are used. Modern types of spark plugs with different gaps in the entire price range are examined. CNG and LPG are used as gaseous fuels, since they are the main alternative fuels used in internal combustion engines for vehicles. Some of the results, such as pressure, maximum pressure and electric spark indicators, are shown as graphics. To calculate the speed of the flame propagation, specialised software processing the areas of the individual photos was used. The conclusions emphasize the results of relative air/fuel ratio, most favourable type of spark plugs and their gaps and location in the combustion chamber.

1. Introduction

Road transport is one of the biggest polluters of the environment, especially in urban areas. Reduction of carbon dioxide emissions is an issue of current concern in both Europe and the rest of the world. Electric vehicles are still not widespread due to their high price and low mileage coverage, and one of the solutions is to use internal combustion engines operating on gaseous fuel. In our country, the vehicle fleet consists mainly of vehicles, which are over 15 years old and they do not meet the modern environmental requirements. Their conversion to operate on gaseous fuel would help reduce carbon dioxide emissions, which they emit into the atmosphere. The use of lean air-fuel mixtures makes it possible to further reduce HC, CO and NOx emissions. The use of gaseous fuels instead of diesel fuel significantly reduces particulate emissions [1].

Spark plugs are an important element of internal combustion engine systems and the power, economic and environmental characteristics of the engines depend on their proper selection and technical condition [2, 3]. There are studies that provide information on the impact of the number of electrodes and their shape on engine performance [4, 5].

Gaseous fuels have been used since the discovery of internal combustion engines. They have been gradually replaced by liquid ones but due to the growing energy crisis and increasing environmental requirements at present, their importance and use greatly increases. Being a gaseous fuel, methane has different burning rate and flame spread manner compared to gasoline and diesel fuels. Methane has different oxidation properties compared to other hydrocarbons. More energy is needed to break the CH chain of methane compared to that needed for higher hydrocarbons. It is hardly surprising that various experimental studies have proven that it is more difficult to ignite methane-air mixture than a mixture of air and other hydrocarbons [6]. Although various experimental methodologies have been developed and numerous previous measurements of laminar flame speed have been accomplished, small (3-5 cm/s) but still important differences exist [7].

The recording of the process provides great opportunities for studying the initial formation of the flame, its development, as well as its dependence on the turbulent motion of the fuel-air mixture and the strength of the spark [8]. As in constant volume chamber experiments can be performed on a Rapid Compression Expansion Machine (RCEM), to reproduce the operating and dynamic conditions encountered in an engine cycle, allowing for visualizations of the fuel injection and combustion processes through a transparent piston [9].

Traditionally, various methods of locating the position of the flame front have been proposed and used in the past. These mainly include the Shadowgraph, Schlieren and flame emission imaging techniques [10]. Popular optical accessible flame edges are shown in figure 1.

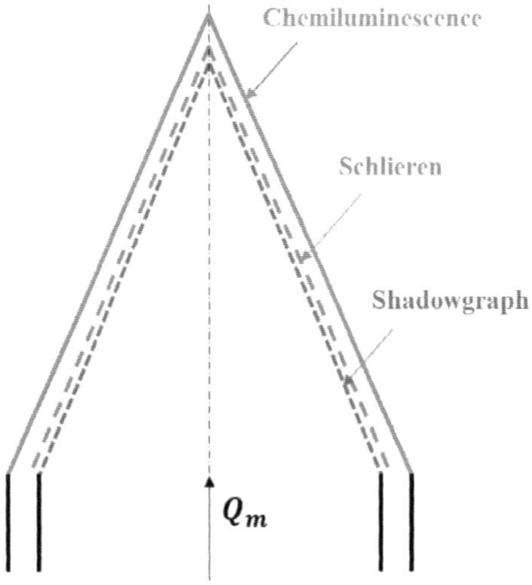

Figure 1. Popular optical accessible flame edges.

Studying the specifics of the combustion process would help reduce cycle variability in internal combustion engines, which is a significant problem, especially in single-cylinder engines [11]. When tests were performed in a constant-volume chamber to simulate exhaust gas recirculation (EGR) as in actual engine operations, it is possible to introduce different levels of inert gas [12].

Preliminary tests using the non-engine method can significantly reduce errors and facilitate the work of converting diesel engines to run entirely on gaseous fuels. Spark plugs are usually placed in place of the diesel injectors and very often it is not possible to mount them centrally in the engine's combustion chamber during this fuel change.

The aim of the present study is to perform comparative tests in a combustion chamber with specific dimensions corresponding to a real engine. Of particular importance are the data of the combustion process in lean air-fuel mixtures. The obtained results will help to properly design and adjust the fuel system of an engine converted to run on gaseous fuels.

2. Experimental setup

This report presents tests carried out in a constant volume combustion chamber. The test aims to determine the impact of the air/fuel ratio, type and location of spark plugs on the combustion process. One of the most commonly used method to study the combustion process is Schlieren photography technology. Internal combustion engines are characterised by their cyclic variability due to the random nature of the combustion process. To improve the measurement accuracy, each experiment was performed 5 times and the results were averaged.

The experimental equipment is shown in figure 2. In the centre is shown the constant volume combustion chamber. The mixture is formed in a special mixing device, which is fed with fuel and air under specific pressure. A vacuum pump is installed to better clean the combustion chamber from the exhaust gases.

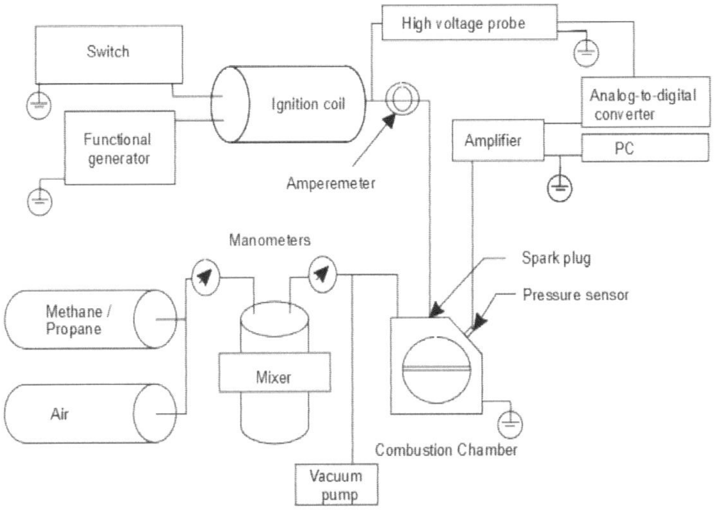

Figure 2. Schematic diagram of the experimental setup.

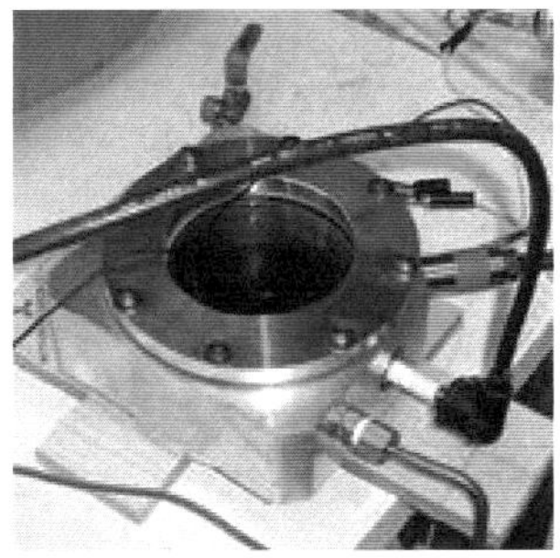

Figure 3. Combustion chamber.

Based on the control of the pressure in the chamber, the required air/fuel ratio is regulated by releasing the fuel and air separately. After the combustion chamber is filled with mixture, a spark is ignited by a spark plug to the required initial pressure. Simultaneously with the spark delivery, the photographic system and measuring devices are activated. The pressure in the combustion chamber is registered by a pressure sensor. The photographs of the flame spread are obtained through Schilieren, using high-speed CCD camera EVANS-E2. The electronic shutter works with 1/10000 s. and the shooting speed is 200 frames per second.

The combustion chamber is shown in figure 3. The chamber diameter is 98 mm, and the chamber length is 24 mm. The processes in the combustion chamber may be examined by means of quartz glasses placed on both sides. Spark plugs may be placed both centrally and laterally. Several types of spark plugs manufactured by the Japanese company NGK are used in the current tests - normal, V-Power, and iridium ones.

3. Results and discussions

Figure 4 shows photographs taken directly with the camera of the flame front spread with propane fuel. An iridium spark plug located laterally in the chamber is used.

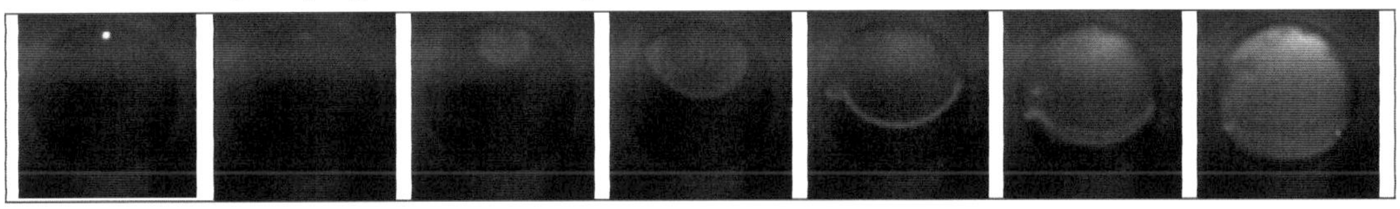

Figure 4. Serial shots of the flame propagation in a propane-air mixture.

Footage of the experimental study at different air/fuel ratios taken with the high-speed camera is shown in figure 5. The used fuel is methane. The movement of the flame front in the first 3 relative air/fuel ratios does not differ significantly, but at λ = 1.1; 1.3 and especially at λ = 1.5 the speed is visibly significantly lower. Different combustion processes are compared using area calculation software.

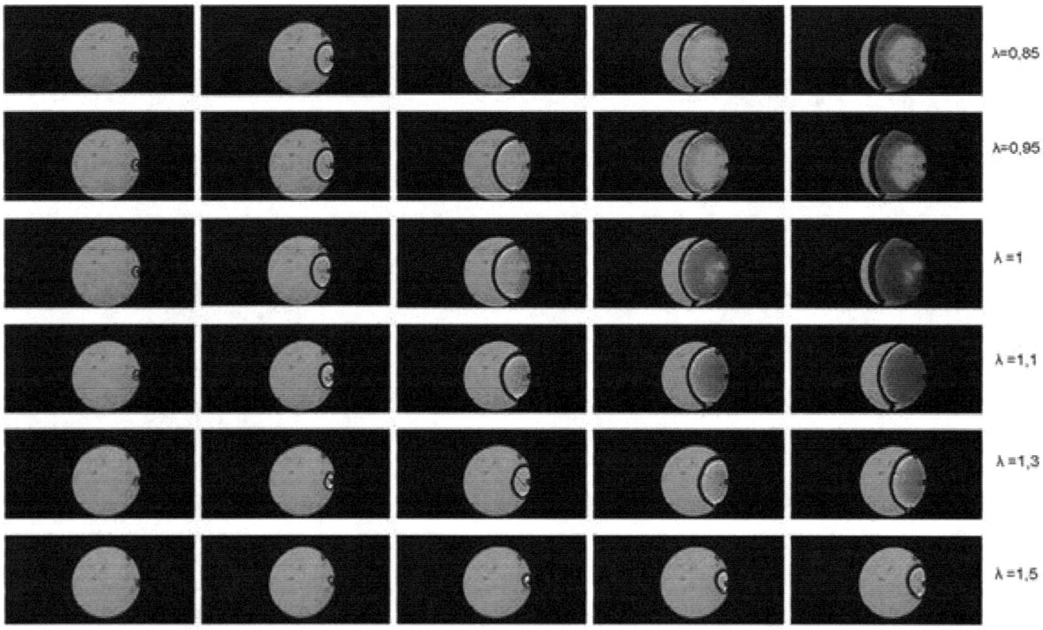

Figure 5. The development of the fuel torch in different relative air/fuel ratio λ.

During the combustion of fuel-air mixture in the combustion chamber, the highest maximum pressure is obtained with the iridium spark plug (6.46 bar), followed by "V-Power" one (6.37 bar) and the lowest pressure is obtained with the normal spark plug (6.26 bar). Figure 6 shows the maximum pressure depending on the air/fuel ratio for the three types of spark plugs, which are located laterally in the constant volume combustion chamber.

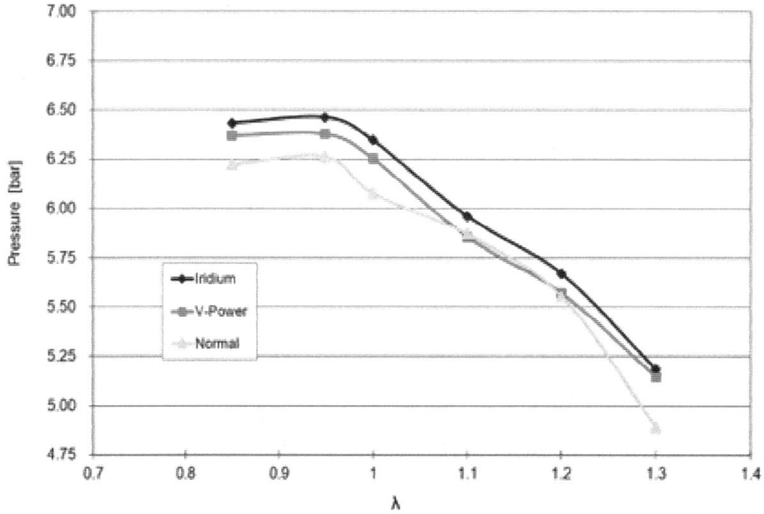

Figure 6. The maximum pressure depending on the air/fuel ratio for three types of spark plugs.

Figure 7 shows the pressure depending on the relative air/fuel ratio with an iridium spark plug on the side wall and methane fuel. It can be observed from the figure that there is big difference in the development of the combustion process when varying λ. The data from the series with the highest maximum pressure are selected.

The maximum voltage depending on the spark plug gap is shown in figure 8. When an iridium spark plug is used, the voltage is about 30% lower than that obtained in the use of normal one. The use of iridium spark plugs in lean fuel-air mixtures extend the life of the elements of the ignition system.

When the normal spark plug is located centrally, the maximum pressure (6.7 bar) is about 0.6 bar higher than in the side wall position at stoichiometric mixture. This difference is compounded if iridium spark plugs are not used.

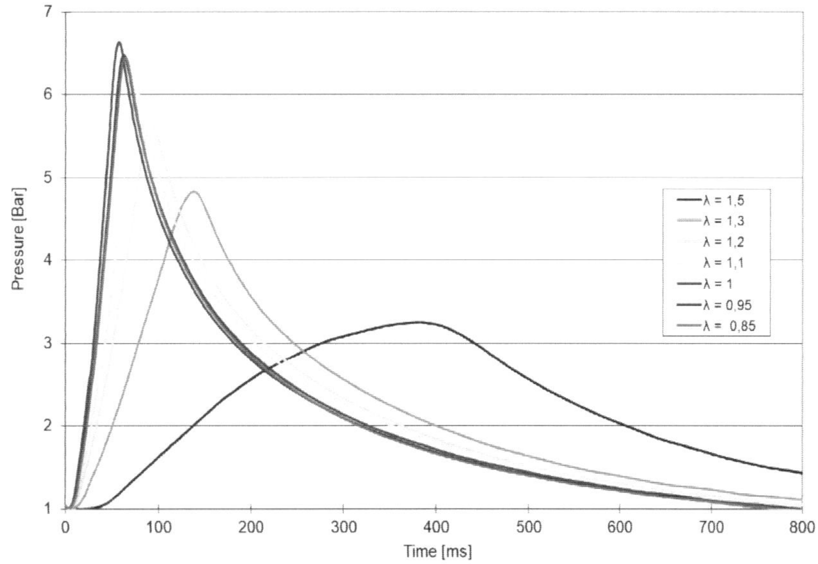

Figure 7. The pressure depending on the air/fuel ratio, iridium spark plug on side wall.

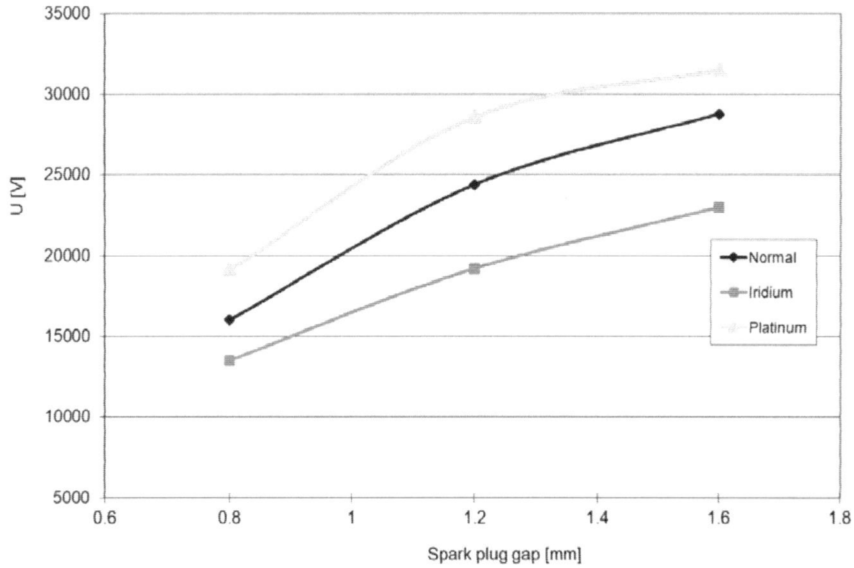

Figure 8. The maximum voltage depending on the spark plug gap.

4. Conclusion

Experimental studies were performed in a constant volume chamber for the impact of certain parameters on the combustion process such as air/fuel ratio, type and location of spark plugs. Visual examinations complement the data taken from the pressure change in the chamber and expand the possibilities for studying the combustion process.

The major observations and conclusions of this study are:

- The depletion of air/fuel mixture above 1.2 leads to significant reduction in the speed of the flame propagation. The assessment of the possibilities for depletion of the fuel-air mixture gives us a direction for the use of higher degrees of compression ratio in the reconstruction of the real engine for work with gaseous fuels.
- The central spark plug location is better and makes it possible to achieve higher pressures in the chamber. If the design does not allow for such an arrangement, the performance can be improved by using iridium spark plugs.
- The maximum pressure in a methane-air mixture is obtained in slightly rich mixtures of relative air/fuel ratio 0.95. This permits to improve fuel economy and get maximum power from the respective engine without using rich mixtures.
- Iridium spark plugs are the best in terms of the combustion process and minimum load to the ignition system. The latter is very important for the reliability and life of the ignition system. They are also suitable for use in lean fuel-air mixtures.

The study will be expanded by comparing the rate of heat transfer and real motor tests.

Acknowledgments
This work received funding from the Technical University of Varna under project KD3 with a project leader Assoc. Prof. PhD R. Hristov.

References

[1] Liu J and Dumitrescu C 2018 Flame development analysis in a diesel optical engine converted to spark ignition natural gas operation *Appl. Energy* **230** pp 1205–17
[2] Chen L et al 2019 Effects of spark plug type and ignition energy on combustion performance in an optical SI engine fueled with methane *Appl. Term. Eng.* **148** pp 188–95
[3] Gu Q, Xu M, Ye C, Hung D and Li X 2020 Investigation of the impacts of spark plug orientation on combustion stability under lean SI operation *SAE Techn. Pap*. 2020-01-1121
[4] K Ravi et al 2017 Effects of spark plug configuration on combustion and emission characteristics of a LPG fuelled lean burn SI engine *IOP Conf. Ser.: Mater. Sci. Eng.* **263** 062070
[5] Paul Hii Shu-Yi et al 2016 Analysis of spark plug gap on flame development using schlieren technique and image processing *IOP Conf. Ser.: Mater. Sci. Eng.* **160** 012044
[6] Sanal S et al 2020 A numerical study on the ignition of lean CH_4/air mixture by a pre-chamber-initiated turbulent jet *SAE Techn. Pap.* 2020-01-0820
[7] Egolfopoulos F, Hansen N, Ju Y, Kohse-Höinghaus K, Law C and Qi F 2014 Advances and challenges in laminar flame experiments and implications for combustion chemistry *Progr. in Energy and Combust. Sci.* **43** pp 36–67
[8] Sjeric M, Kozarac D and Tatschl R 2015 Modelling of early flame kernel growth towards a better understanding of cyclic combustion variability in SI engines *En. Convers. Manage.* **103** pp 895–909
[9] Salaun E, Apeloig J, Grisch F, Yvonnet C E, Nicolas B and Dionnet F 2016 Optical investigation of ignition timing and equivalence ratio in dual-fuel CNG/diesel combustion *SAE World Congress and Exhibition*
[10] Wu Y 2016 Experimental investigation of laminar flame speeds of kerosene fuel and second generation biofuels in elevated conditions of pressure and preheat temperature *Doctoral Thesis* pp 36–55

[11] Krishnasamy M, Ramachandra Babu Y, Ramesh A and Babu M 2009 Studies on reducing cycle by cycle variations and improving performance of a small carbureted gasoline engine *SAE Technical Paper* 2009-32-0097
[12] Zhang A, Naber J D and Lee S Y 2014 An experimental study of flame kernel evolution in lean and diluted methane-air mixtures at engine-like conditions *Comb. Sci. and Techn.* **186**(8) pp 988–1004
[13] Heywood J 2018 *Internal combustion engine fundamentals* (McGraw-Hill Education)

Development of a method for automotive engine control unit modification for achieving different operating modes using software correction

N Nikolov[1] and D Gotseva

Faculty of Computer Systems and Technologies, Technical University – Sofia, 8 Kliment Ohridski Blvd., 1000 Sofia, Bulgaria

[1] E-mail: n.nikolov@tu-sofia.bg

Abstract. This article describes the development of a method for automotive fuel injection engine control unit (ECU) modification for aimed at achieving operating modes such as Economy (ECO) and Sport using software alteration. The article researches the offered pattern and demonstrates the experimental staging. An automotive ECU, microcontroller and software are used for the experimental stage. The ECO mode is used to achieve fuel economy and the Sport mode are used to achieve more engine power. The microprocessor used to switch between ECU modes is PIC18F14k22 and the ROM flash memory which stores the firmware for the two modes is AM29F400BB. The ECU used in the research is Sagem s2000. In the article, both the hardware and the software modifications are described as well as the experimental results.

1. Introduction

In order to achieve operating modes such as maximum Economy and Power from the same engine, it is necessary to approach it by modification in the software of its operation. Environmental norms are an important thing but they limit the full potential of the internal combustion engine. By limiting the maximum power of a gasoline engine we achieve greater environmental friendliness, smaller amount of harmful gases emitted into the atmosphere and as a result increased economy. Maximum engine dynamics are achieved by higher fuel consumption and maintenance of richer mixtures. Through software setting – using the so called Maps tune or Chip Tuning – of engine modes such as ECO and Sport full coverage of engine capabilities, such as fuel economy and maximum developed power and dynamics, is achieved.

2. Theory and architecture

In order to achieve full fuel economy for gasoline engine, it is necessary to make the mixture lean to a certain extent. To achieve full power, the mixture has to be enriched. For the complete burning of 1 kg of gasoline requires 14.64 kg of air., the mixture must be somewhere between 15 and 16 air/fuel ratio To achieve full economy and for full power it must be about 13, 13.5 air/fuel ratio. Table 1 shows the different mixtures of gasoline and air [1]. In order to achieve fuel saving and power, maps with 16×16 values such as Lambda oxygen sensor for different modes, VE table (volumetric efficiency enrichment), ignition angle and throttle sensitivity and accelerator pedal – Driver Wish Torque are required to be changed. A 1.4-litre engine: TU3JP with 75 horse power is used for the experiment. The ECU used for fuel injection is Sagem s2000.

Table 1. Various mixtures of gasoline and air (air/fuel ratio) and their application.

Air/fuel ratio	Lambda	Comment
9.0	0.61	Black smoke/low power
11.5	0.79	Approximate rich best torque at wide open throttle
12.2	0.83	Safe best power at wide-open throttle
13.3	0.91	Approximate lean best torque
14.64	1.00	Stoichiometric air/fuel ratio (chemically ideal)
15.5	1.06	Lean cruise
16.5	1.13	Usual best economy
18.0	1.23	Carbureted lead burn limit

For the optimal operation of the gasoline engine it is necessary to increase the ignition angle by 1 degree relative to the factory settings (figure 1). Figure 1 shows the ignition angle map, which is size 16×16. This map is the same for both ECO and Sport modes.

SPARK ADVANCE AT MIDDLE THROTTLE(RPM,Load)/degrees

Load \ RPM	700	1100	1400	1800	2100	2500	2800	3200	3500	3900	4200	4600	4900	5300	5600	6000
1.00	29.88	30.60	35.10	36.00	36.18	31.14	29.34	25.34	31.68	38.70	38.34	38.34	40.32	40.86	41.22	41.76
0.90	29.34	30.24	33.66	34.38	35.28	30.78	29.34	29.34	31.68	38.70	38.88	39.34	39.06	40.86	41.58	41.76
0.80	26.82	29.16	29.88	30.24	30.42	30.78	29.34	29.34	31.68	32.40	34.56	35.64	37.26	37.98	39.06	40.14
0.75	25.38	27.00	28.80	29.34	29.88	29.70	30.24	31.14	30.78	32.22	31.50	32.76	32.94	34.38	35.28	36.54
0.70	24.12	24.48	27.00	27.72	28.26	28.80	29.52	30.24	30.24	30.96	31.14	31.50	32.40	33.84	34.56	35.64
0.65	20.88	23.58	25.56	26.28	27.00	28.44	28.80	29.52	29.34	30.60	30.42	30.96	31.50	33.30	33.66	34.56
0.60	20.16	22.32	24.84	25.38	26.46	27.72	27.90	28.62	29.52	30.06	29.88	30.24	31.14	33.66	34.38	34.74
0.55	19.08	20.52	24.30	25.02	25.74	27.54	27.72	28.26	28.80	29.52	29.34	30.06	31.68	33.84	34.38	34.20
0.50	19.08	20.16	22.14	22.86	25.38	26.82	27.36	27.50	28.44	28.98	29.16	30.24	30.78	33.30	33.84	33.48
0.45	18.72	19.44	21.78	22.68	24.48	26.46	27.00	27.72	28.44	29.16	29.88	30.06	30.60	32.58	32.94	34.02
0.40	17.64	18.36	21.78	22.68	24.48	26.28	26.46	27.64	28.26	29.16	29.52	29.88	29.34	31.32	32.58	33.30
0.35	16.20	18.36	21.42	21.96	23.76	25.92	26.74	27.36	27.90	28.44	28.26	29.16	28.44	30.60	31.14	32.40
0.30	15.12	18.36	20.70	21.42	22.68	25.02	26.74	27.36	26.46	27.54	27.72	27.72	28.26	28.44	30.24	31.68
0.25	15.12	17.64	20.16	20.70	21.96	23.58	24.48	25.38	25.56	26.64	27.36	26.82	27.54	27.90	29.34	31.14
0.20	15.12	16.56	19.80	20.16	21.42	22.68	23.58	24.48	25.02	26.10	26.46	26.28	26.82	27.36	28.44	29.52
0.10	15.12	16.56	19.44	19.80	20.88	21.96	22.50	23.76	24.66	25.56	25.56	25.56	26.28	26.64	27.54	27.72

Figure 1. Ignition angle map 16×16.

ECO and Sport modes differ in terms of Lambda oxygen sensor, VE (volumetric efficiency enrichment) and Driver Wish Torque. Figure 2 shows the Lambda oxygen sensor map for ECO mode, where the target air/fuel ratio for full load is 15. Figure 3 shows the map of Lambda Oxygen Sensor for Sport mode, where the target air/fuel ratio for full load is 13.2.

AFR - Air Fuel Ratio (Lambda Oxygen)(RPM,Load)/[Air/Fuel] Ratio

Load \ RPM	700	1100	1400	1800	2100	2500	2800	3200	3500	3900	4200	4600	4900	5300	5600	6000
100	-14.89	-14.55	-15.08	-15.03	-15.03	-15.08	-15.03	-14.97	-14.98	-15.00	-15.00	-15.03	-15.08	-15.03	-15.17	-16.14
95	-15.08	-15.10	-15.17	-15.35	-15.21	-15.23	-15.25	-15.22	-15.15	-15.19	-15.17	-15.15	-15.11	-15.11	-15.25	-15.24
85	-15.26	-15.24	-15.31	-15.41	-15.39	-15.40	-15.42	-15.41	-15.40	-15.31	-15.28	-15.26	-15.36	-15.19	-15.27	-15.34
75	-15.42	-15.38	-15.48	-15.55	-15.50	-15.53	-15.54	-15.58	-15.60	-15.62	-15.53	-15.54	-15.59	-15.53	-15.41	-15.46
70	-15.43	-15.45	-15.54	-15.49	-15.55	-15.61	-15.63	-15.66	-15.69	-15.70	-15.66	-15.62	-15.67	-15.49	-15.48	-15.54
65	-15.50	-15.52	-15.58	-15.71	-15.67	-15.69	-15.71	-15.76	-15.78	-15.79	-15.77	-15.70	-15.72	-15.61	-15.52	-15.66
60	-15.56	-15.59	-15.67	-15.79	-15.77	-15.78	-15.77	-15.84	-15.87	-15.87	-15.85	-15.78	-15.86	-15.75	-15.63	-15.82
55	-15.60	-15.66	-15.74	-15.87	-15.86	-15.79	-15.84	-15.95	-15.99	-15.96	-15.88	-15.85	-15.95	-15.98	-15.99	-15.97
50	-15.65	-15.73	-15.82	-15.93	-15.92	-15.89	-15.94	-15.99	-16.01	-16.00	-15.97	-15.91	-16.01	-16.13	-16.10	-16.04
45	-15.74	-15.77	-15.83	-15.99	-16.02	-16.02	-16.02	-16.10	-16.12	-16.02	-16.05	-15.99	-16.06	-16.12	-16.12	-16.06
40	-15.82	-15.80	-15.76	-16.00	-16.06	-16.05	-16.10	-16.03	-16.09	-16.04	-16.05	-15.99	-16.00	-16.14	-16.02	-16.06
35	-15.90	-15.85	-15.80	-16.05	-16.04	-16.12	-16.07	-16.03	-16.10	-15.96	-15.95	-15.93	-15.88	-16.05	-15.84	-15.89
30	-15.96	-16.00	-15.93	-16.27	-16.30	-16.36	-16.35	-16.36	-16.34	-16.06	-16.00	-15.94	-15.93	-15.93	-15.84	-15.84
25	-16.02	-16.12	-16.14	-16.49	-16.43	-16.54	-16.55	-16.56	-16.54	-16.32	-16.19	-15.96	-16.11	-16.07	-15.84	-15.84
20	-16.08	-16.23	-16.37	-16.61	-16.55	-16.68	-16.70	-16.71	-16.66	-16.51	-16.29	-16.07	-16.22	-16.07	-15.84	-15.84
15	-16.13	-16.33	-16.58	-16.74	-16.65	-16.88	-16.87	-16.88	-16.82	-16.67	-16.38	-16.19	-16.32	-16.07	-15.84	-15.84

Figure 2. Lambda oxygen sensor for ECO mode, the target air/fuel ratio for full load is 15.

AFR - Air Fuel Ratio (Lambda Oxygen)(RPM,Load)/[Air/Fuel] Ratio

RPM / Load	700	1100	1400	1800	2100	2500	2800	3200	3500	3900	4200	4600	4900	5300	5600	6000
100	-13.21	-13.21	-13.21	-13.21	-13.01	-13.21	-13.11	-13.21	-13.21	-13.31	-13.21	-12.90	-13.01	-13.21	-13.21	-13.21
95	-13.21	-13.21	-13.21	-13.41	-13.31	-13.21	-13.31	-13.21	-13.31	-13.41	-12.69	-13.01	-13.11	-13.21	-13.31	-13.41
85	-13.31	-13.31	-13.31	-13.60	-13.41	-13.34	-13.41	-13.11	-13.31	-13.31	-13.01	-13.21	-13.21	-13.79	-13.21	-13.51
75	-13.41	-13.41	-13.79	-13.79	-13.48	-13.48	-13.41	-13.51	-13.41	-13.31	-13.41	-13.31	-13.70	-13.79	-13.60	-13.88
70	-13.79	-13.79	-13.88	-13.60	-13.70	-13.60	-13.60	-13.70	-13.60	-13.41	-13.91	-13.41	-13.60	-13.79	-13.79	-13.79
65	-13.79	-13.79	-13.98	-13.98	-13.75	-13.88	-13.70	-13.79	-13.70	-13.41	-13.70	-13.41	-13.41	-13.79	-13.88	-14.07
60	-14.07	-14.07	-14.07	-14.07	-13.88	-13.88	-14.07	-13.88	-13.79	-13.70	-13.60	-13.58	-13.70	-13.79	-13.98	-14.15
55	-14.15	-14.15	-14.16	-14.15	-14.42	-14.15	-14.07	-14.07	-13.98	-13.79	-13.61	-13.60	-13.70	-13.98	-14.16	-14.24
50	-14.50	-14.50	-14.42	-14.50	-14.58	-14.50	-14.42	-14.33	-14.15	-13.88	-13.96	-13.75	-13.98	-14.24	-14.33	-14.42
45	-14.83	-14.83	-14.58	-14.67	-14.83	-14.67	-14.83	-14.42	-14.33	-14.33	-14.42	-14.33	-14.24	-14.33	-14.67	-14.67
40	-14.67	-14.67	-14.58	-14.83	-15.07	-15.33	-15.33	-14.67	-14.83	-14.42	-14.33	-14.50	-14.67	-14.99	-14.99	-15.14
35	-14.83	-14.83	-14.91	-15.44	-15.22	-15.30	-15.37	-14.83	-14.99	-14.67	-14.50	-14.67	-14.99	-15.14	-15.07	-15.22
30	-14.91	-14.91	-15.22	-15.73	-15.59	-15.37	-15.44	-14.99	-14.99	-14.75	-14.58	-14.67	-15.07	-15.30	-15.14	-15.07
25	-15.07	-15.07	-15.44	-15.94	-14.75	-14.76	-15.44	-14.91	-14.67	-14.67	-14.42	-14.50	-14.67	-15.07	-14.99	-14.99
20	-15.44	-15.44	-15.37	-15.80	-13.51	-15.33	-15.59	-15.22	-14.58	-14.99	-14.83	-15.07	-14.83	-14.75	-15.14	-15.22
15	-15.94	-15.94	-15.22	-15.44	-13.51	-15.33	-15.59	-13.98	-15.22	-15.37	-14.83	-15.07	-14.42	-14.42	-15.22	-15.22

Figure 3. Lambda oxygen sensor for Sport mode, the target air/fuel ratio for full load is 13.2.

Of great importance for achieving greater economy is the driver wish map, which decides how much fuel the driver wants the ECU to put into the motor. Figure 4 shows the ECO mode driver wish map and Figure shows the driver wish map 5 for Sport mode. from the data in the figures It can be concluded that the low load values for ECO mode are quite reduced and in the modes between 75 and 100 %, which correspond to WOT (Wide opening throttle), they are unchanged compared to the original ones. This is done in order for the vehicle to reach the desired power and torque in normal driving. For Sport mode, it appears that the values are well increased from original ones in order to ensure maximum power, acceleration and torque.

Torque(RPM,Load)/-

RPM / TPS %	700	1100	1400	1800	2100	2500	2800	3200	3500	3900	4200	4600	4900	5300	5600	6000
100	121	121	121	121	121	121	121	121	121	121	121	121	121	121	121	121
95	101	121	121	121	121	121	121	121	121	121	121	121	121	121	121	121
85	101	121	121	121	121	121	121	121	121	121	121	121	121	121	121	121
75	101	121	121	121	121	121	121	121	121	121	121	121	121	121	121	121
70	98	117	117	117	117	117	117	117	117	117	117	117	117	117	117	117
65	54	60	114	114	114	114	114	114	114	114	114	114	114	114	114	114
60	50	56	111	111	78	86	88	111	111	111	111	111	111	111	111	105
55	46	50	60	62	63	67	72	86	92	108	90	86	93	99	99	90
50	45	47	52	54	56	63	63	77	78	94	76	76	75	81	87	82
45	41	43	45	47	50	53	58	64	66	74	68	67	67	71	76	76
40	35	41	39	40	45	48	52	56	60	61	60	61	62	68	66	70
35	32	40	35	36	42	45	47	50	52	56	55	56	56	60	59	57
30	32	38	34	34	41	44	46	44	44	54	54	54	52	54	54	52
25	32	36	32	32	38	43	42	42	42	52	52	49	49	48	46	47
20	29	36	32	32	32	39	37	37	35	48	48	44	44	44	44	44
15	28	33	32	32	32	33	34	36	35	44	45	44	44	42	42	42

Figure 4. Driver wish map for ECO mode.

Torque(RPM,Load)/-

RPM / TPS %	700	1100	1400	1800	2100	2500	2800	3200	3500	3900	4200	4600	4900	5300	5600	6000
100	141	142	142	142	142	143	143	143	143	144	144	144	144	145	145	145
95	141	141	141	142	142	142	142	143	143	143	143	144	144	144	144	145
85	140	141	141	141	141	142	142	142	142	142	143	143	143	143	144	144
75	140	140	140	141	141	141	141	141	142	142	142	142	142	143	143	143
70	139	140	140	140	140	140	141	141	141	141	141	142	142	142	142	142
65	110	139	139	140	140	140	140	140	140	141	141	141	141	141	142	142
60	66	98	139	139	108	139	140	140	140	140	140	129	140	141	129	129
55	64	78	84	90	90	105	106	139	139	139	128	112	112	122	116	116
50	61	73	77	79	84	91	93	131	139	139	106	101	100	107	114	105
45	59	69	73	73	76	82	86	95	103	105	97	94	91	97	97	98
40	54	62	66	70	73	80	83	90	94	98	93	91	89	90	94	94
35	49	57	63	66	71	76	81	87	90	92	88	86	84	87	89	89
30	47	55	61	64	68	73	78	83	86	89	86	84	83	85	87	87
25	46	54	60	65	68	71	75	81	83	84	84	82	81	82	84	84
20	43	51	55	60	62	67	71	77	76	76	77	79	79	79	81	81
15	44	48	52	59	60	66	69	75	72	71	74	77	78	76	80	80

Figure 5. Driver wish map for Sport mode.

The following formula summarizes the work of the ECU fuel injection software, , which is responsible for the operation of the prepared fuel by injectors. The opening times of injectors are adjusted by Pulse width modulation [1]:

$$\text{Pulse width} = [(\text{MAP Voltage} / 5) \times \text{UAP} \times (\text{VE Absolute}\ \% / 100) \times \text{TPS}\ \% \times \text{CTS}\ \% \times \text{IAT}\ \% \times \text{EGO}\ \% \times \text{SE}\ \%] + \text{POT} + \text{BTO}, \quad (1)$$

where: MAP – manifold absolute pressure; UAP – user-adjustable pulse width; POT – pulse width offset time at idle; TPS – throttle position change; CTS – engine coolant temperature; IAT – intake air temperature; EGO – required air/fuel correction coefficient based on the exhaust gas oxygen; SE – starting enrichments. All correction coefficients add percentage changes to VE absolute percentage with the exception of battery voltage correction BTO.

3. Experimental stage and hardware realization

In order to switch between ECO and Sport modes, ECU Firmware must to be changed. For this purpose, two types of firmware ECO and Sport are prepared. The ECU Sagem s2000 use NOR flash memory AM29F200BB [2], which is 256 kb. In order to be able to combine both modes of operation, we need to use NOR flash memory with double the size of AM29F400BB, which is 512 kb size [3]. The switch between the two modes must be done by submitting logical low and high levels in order to select the first 256 kb block where Sport is and the second 256 kb block where ECO is stored. (figure 6).

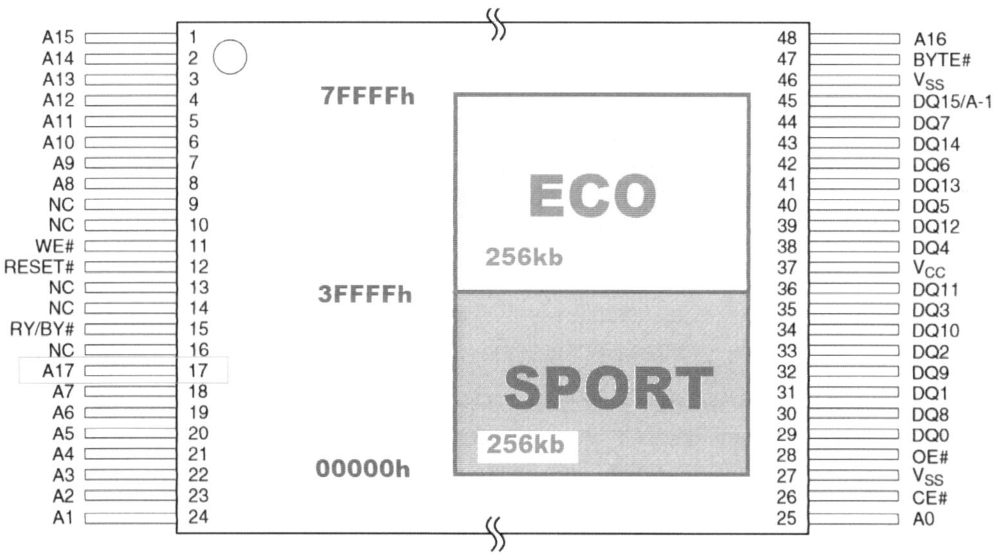

Figure 6. Memory map of firmware in nor flash memory AM29F400BB.

The two modes are selected by submitting low and high level to am29F400BB with address pin A17. But to make the mode switch, Sport and ECO, selection buttons, an indicator of the status of the mode and Reset on the ECU Sagem s2000 processor. The ECU uses a SAC c167CS microprocessor, which is 16 bit, and when switching the new mode, the SAC c167CS must be restarted and IP (instruction pointer) starts again from the reset vector [4]. The need to RESET a SAK c167CS is due to the fact that the two firmware for ECO and Sport modes have different Check Sum for data validity. Figure 7 shows the time chart to be performed to switch between engine operating modes. For this purpose, the PIC18F14K22 micro controller is used to control the switching between operating modes. Figure 8 shows the picture of the experienced staging – the ECU Sagem s2000 used with pic18F14K22 controller.

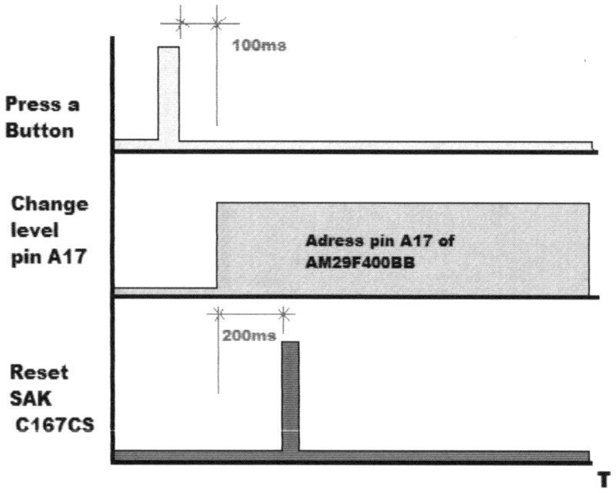

Figure 7. Time diagram for operating modes CONTROL ECO and Sport via the PIC18F14K22 micro controller.

Figure 8. The test staging – the ECU Sagem s2000 and the PIC18F14K22 micro controller.

4. Software realization

For the Firmware [5] implementation of the pic18F14K2 micro controller, the programming language is C. RTOS operating system [6-8] with cooperative scheduling Task [9, 10] are created. This RTOS has a TASK which is responsible for reading the buttons, the switching of the upper and lower 256 kbytes, and reseing the SAK C167CS controller of Sagem s2000. The main program contains the following lines:

```
Init_CPU_Freq_Clock     ( void );
Init_Hardware_GPIO      ( void );
Init_Hardware_TIMER     ( void );
Init_Hardware_IRQ       ( void );                (2)
Init_Read_EEPROM        ( void );
Scheduler_init          ( void );
for ( ; ; );
```

where the initialization of Registers, Timers, CPU Frequency Clock, EEPROM and Scheduller TASK is performed. The main TASKs are: "Task_Scan_Buttons", "Task_Switch_ECU_Mode", "Task_Reset_SAC_TC167", "Task_Control_LED_ECO_Dashboard". The chosed mode between ECO or Sport are saved in the EEPROM memory. The software executes the logic based on the time chart (figure 7). The following lines show the TASK for RESET of SAK C167CS:

```
void Task_Reset_SAC_TC167( void ){

    if( FLAG_Enable_RESET_SAC_C167CS_GPIO  == TRUE ){

      FLAG_Enable_RESET_SAC_C167CS_GPIO = FALSE;
      delay_ms( 100 );
        RESET_SAC_C167CS_GPIO = LOW ;
      delay_ms( 500 );                                   (3)
        RESET_SAC_C167CS_GPIO = HIGH ;
    }else{
        RESET_SAC_C167CS_GPIO = HIGH ;
    }
   return;
}
```

TASK itself shows how the "delay_ms(100)" function performs a software delay of 100 ms required to prepare the reset time. Cooperative scheduling in RTOS kernel is shown in these rows:

```
void Scheduler_init( void ){

 for ( ; ; ){
        Task_Scan_Buttons            ( void );
        Task_Switch_ECU_Mode         ( void );
        Task_Reset_SAC_TC167         ( void );                    (4)
        Task_Control_LED_ECO_Dashboard ( void );
   }
   return;
 }
```

5. Experimental results

Several approaches were used to study the results obtained – Dyno Jet bench which measures the power of the wheels of the car and driving on a particular route in order to compare the achieved economy. The results of Dyno Jet are shown in figure 9. The measured parameters are Torque in Nm, Power in hp and air/fuel ratio. The graph shows that the Sport mode achieves a maximum Torque 113.02 Nm and Power 64.98 hp. Comparable air/fuel ratio shows that the engine holds mixtures between 14.8÷15.20 of ECO mode for better combustion of the fuel and in Sport mode – 12 for maximum power. As a result of driving the vehicle in urban – ECO mode, the reduction in fuel consumption from 9 l/100km to 8 l/100km is detected and extra-urban from 5.4 l/100km to 5.0/100km.

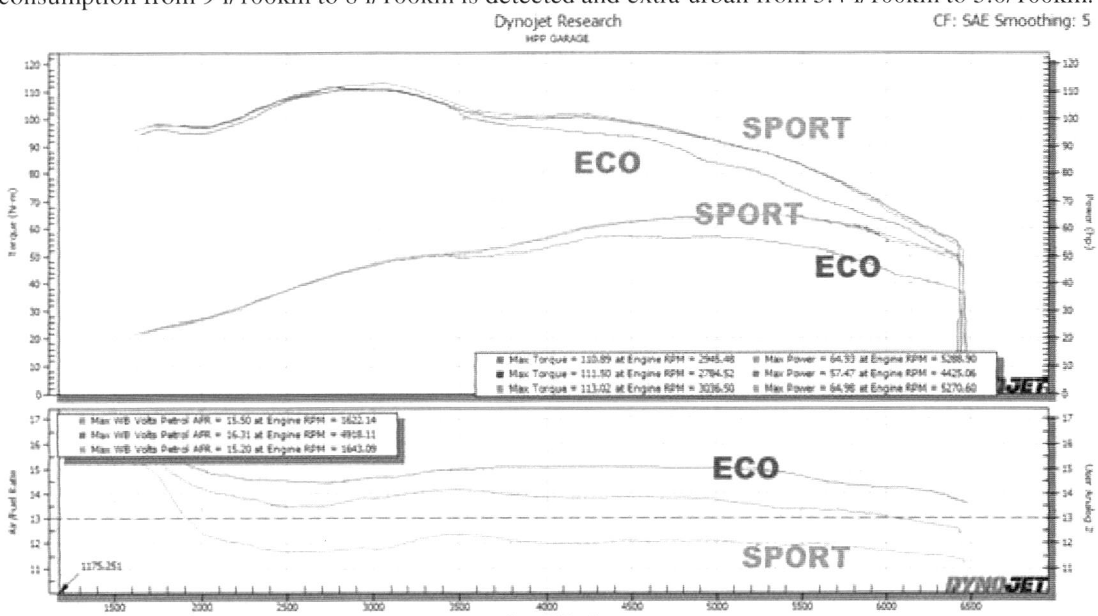

Figure 9. Dyno Jet measurements with parameter comparison such as Torque in Nm, Power in hp and air/fuel ratio.

6. Conclusion

This article describes the achievement of operating modes – maximum Economy and Power from the same engine. This was achieved through a modification in the software of its work. Two ECO and Sport modes are made, thus achieving maximum power and when it is not needed select maximum economy. This is done with ECU maps tinning. The paper describes how the hardware and software realization of the system is created. the architecture and operating model are shown. A cooperative RTOS – a real time operating system is created for controlling the process. The experimental results are shown in figure 9: the measurement parameter such as Torque in Nm, Power in hp and air/fuel ratio in both modes: ECO and Sport. The contribution of the authors consists in the realization of the software architecture and the cooperative RTOS and also in the experimental staging and the experiment.

References

[1] Hartman J 2013 *How to tune and modify automotive engine management systems* (Motorbooks Workshop)
[2] Bogdanov L and Ivanov R 2019 Flash programming low power microcontrollers over the Internet *IEEE 28th Int. Sci. Conf. Electronics (ET) (Sozopol, Bulgaria)*
[3] AM29F400B Datasheet (PDF) – Advanced Micro Devices, www.alldatasheet.com, 03.2020
[4] Colaco F and Cardoso F 1999 Flying objects: a modular real time object operating system *11th IEEE NPSS Real Time Conference (Sante Fe, NM, USA)* pp 543–6
[5] Bogdanov L V 2019 A Wi-Fi to UART bridge for firmware updates of microcontrollers *IEEE 28th Int. Sci. Conf. Electronics (ET) (Sozopol, Bulgaria)*
[6] Hoffmann M, Dietrich C and Lohmann D 2013 dOSEK: A dependable RTOS for automotive applications *IEEE 19th Pacific Rim International Symposium on Dependable Computing (Vancouver)* pp 120–1
[7] Xiao D and Zhang J 2011 A scheme and debugging of RTOS based on OSEK/VDX *Int. Conf. on Electric Information and Control Eng. (Wuhan)* pp 2924–8
[8] Shuaiqing X, Yang W, Chengjun T, Mingzhi S, Yongsheng Z and Tao X 2011 Transplantation of RTOS uC/OS-II on TriCore processor *Int. Conf. on Electric Information and Control Eng. (Wuhan)* pp 5598–601
[9] Seo SH, Lee S, Hwang SH and Jeon J W 2006 Analysis of task switching time of ECU embedded system ported to OSEK(RTOS) *SICE-ICASE Int. Joint Conf. (Busan)* pp 545–9
[10] Yang G, Zhaohui W and Long Y 2005 AlphaOS, an automotive RTOS based on OSEK/VDX: design and test *Proc. IEEE Networking, Sensing and Control (Tucson)* pp 174–9

Fuel cell electric vehicles. Investigation of the energy balance for optimal reforming process of bio-ethanol

Zlatina Dimitrova

PSA Groupe, Centre Technique de Vélizy, 78943 Vélizy-Villacoublay Cedex, France

E-mail: zlatina.dimitrova@mpsa.com

Abstract. Hydrogen ecosystem and hydrogen economy are relevant topics for the mobility. This article summarizes the ways of production of "green" hydrogen. The hydrogen storage and transportation are discussed as well. The article presents the concept of electric vehicle with serial range extender to extend the vehicle autonomy. Fuel cell system powered by hydrogen produced through reforming of liquid fuel from renewable resources such as bioethanol is considered as a range extender module. Alkaline fuel cell and solid oxide fuel cell are investigated for the use of a variety of liquid fuels instead of traditionally used hydrogen in others types of fuel cells. This article investigates the energy balance of the different types of fuel cells, powered by bio-ethanol. Three variants for ethanol conversion are investigated: variant 1: reforming of bio-ethanol in external reformer to hydrogen and its conversion in alkaline fuel cell, variant 2: reforming of bio-ethanol in external reformer to hydrogen and its conversion in a solid oxide fuel cell; variant 3: direct reforming of the bio-ethanol in the solid oxide fuel cell. The chemical processes for each variant are proposed and thermodynamic energy balance is calculated. From the results is visible that the most efficient configuration is the variant 3, the direct reforming of the bio-ethanol by the solid oxide fuel cell. The variant 3 delivers 15 MJ of energy output per kilogram of bio-ethanol, used in the fuel cell.

1. Introduction
The automotive industry is challenged by very strict emissions regulation. The CO_2 emissions levels tend to be drastically reduced to 50 g/ CO_2 km in 2030. The hydrogen question and discussion is larger than the automotive industry and the clean mobility energy services and products that the automotive sector in Europe, has to provide. The electrification of the vehicles is currently in progress in Europe, driven by the European CO_2 emissions regulation and hydrogen is seen in the energy outlooks to be used as well for vehicular propulsion, and as well as enabler to reach the CO_2 emissions targets. The current CO_2 regulation considers the emissions from tank-to-wheel perspective and thus the electric and the fuel cell vehicles are accounted as zero emissions vehicles. Introducing electricity and hydrogen as energy vectors for mobility set the questions of the global efficiency of the conversion chain, of the total energy resources to be used and the total CO_2 emissions from well-to-wheels perspective. From the hydrogen point of view the well-to-wheels perspective raises multitude of scientific, technological and economic questions.

2. Hydrogen production and storage
Renewable energies increase their part in the energy mixes due to the energy transition needs. The sustainable energy system should become clean but as well reliable and affordable. The renewable

energy production and energy demand are intermittent thus an integrated system approach is required considering energy conversion and storage. The authors propose in [1] a concept for a neighbourhood called Power-to-H3.Locally produced renewable energy is partly converted and stored in the form of heat and hydrogen. Rainwater collection, storage, purification and use is as well implemented. Hydrogen can be obtained from different sources of raw materials including water. Hydrogen production methods are required to be eco-friendly and to deliver high purity of hydrogen. Water electrolysis and especially PEM water electrolysis is considered as most promising technique for high pure efficient hydrogen production from renewable energy sources [2]. Figure 1 shows the hydrogen production methods.

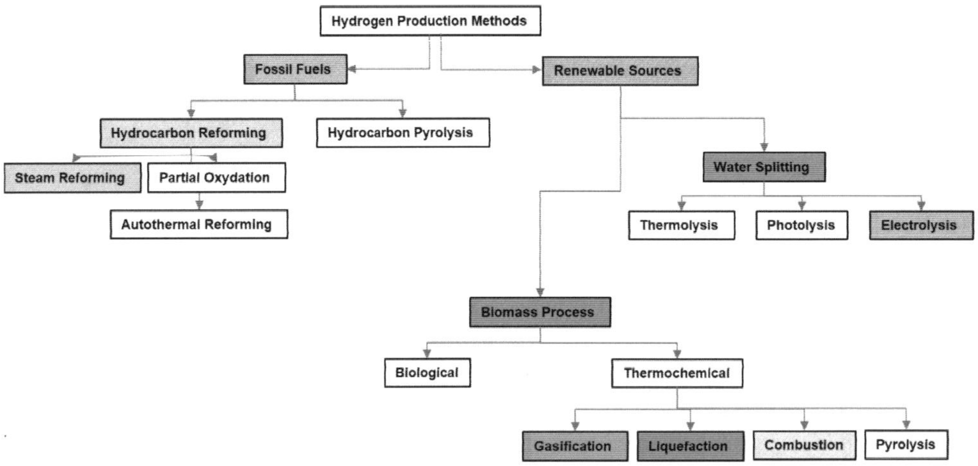

Figure 1. Hydrogen production methods.

Biomass energy contributes a major portion in global renewable energy. Biomass has ever-growing share in electricity production worldwide. H_2 is considered as a clean energy fuel having high energy density (122 kJ/g), which is three times higher than hydrocarbon fuels. Heating value of 61,100 Btu/lb of H_2 is also nearly three times higher than methane (23,879 Btu/lb). When hydrogen is produced by the reformation of fossil fuels (petroleum or coal) this involves large greenhouse-gas footprints. The authors present in [3] promising approaches applied to achieve biomass based hydrogen production by using microbes, called biohydrogen fermentation using crops straws, agri-waste; industrial waste or naturally occurring organic rich biomasses (like molasses). The hydrogen is considered as carbon free fuel. Supply of low- carbon energy for heat, balancing of electricity at national grid and application in transportation sector preferred hydrogen over other carbon based gaseous biofuels. The fertilizers and the petroleum companies are considered to be at the largest users of H_2 with respectively 50% and 37%. With the demand of the fuel cells the demand of hydrogen is increasing with 6% the last five years.

The major application of H_2 are:

- Utilization for hydration of substantial oils for fuel production, foods hydration, and alkali hydration for fertilizers production.
- As H_2 is the perfect electron donor, it is uses for diminishing nitrate, perchlorate, selenite, and a suite of other oxidized water contaminations.

In the present the industrial application of hydrogen is equivalent to only 3% of the total energy consumption. This is expected to grow significantly in the years to come [3].

For the moment the hydrogen market is with low demand. In this case, small-scale water electrolysis based hydrogen production is more cost-efficient than large-scale production. The study in [4] shows that there is a good business for local water electrolysis. The hydrogen has to supply captive fleets of trucks.

The life-cycle environmental performance of hydrogen as a transportation fuel focuses on its production [5]. The steam reforming of natural gas still arises as the key hydrogen production technology in the short term. Water electrolysis is the main technology in the medium and long term. In scenarios with very restrictive carbon footprint limits, biomass gasification also appears as a key hydrogen production technology in the long term.

Metal hydrides are known as option for high-density hydrogen storage [6]. In addition, they give the opportunity to develop a new hydrogen compression technology due to their thermodynamic possibilities.

Calcium carbonate ($CaCO_3$) and Mg-substituted $CaCO_3$ were selected as adsorbents for hydrogen with high amount of release after high-pressure hydrogen loading. [7]. Their crystal structures, specific surface areas, hydrogen content at high pressure, and the amount of hydrogen released in water were investigated.

Hydrogen can be considered as balancing and storage grid for intermittent energies [8]. Two onsite storage technologies, i.e., power-to-hydrogen-to-power and lithium battery, are investigated to size the storage system and to estimate its profitability [8].

Among several candidates of hydrogen storage, liquid hydrogen, methylcyclohexane (MCH), and ammonia are considered as potential hydrogen carriers, in terms of their characteristics, application feasibility, and economic performance [9]. Liquid hydrogen faces challenges in huge energy consumption during liquefaction and boil-off during storage. MCH has main obstacles in largely required energy in dehydrogenation. Ammonia encounters high energy demand in both synthesis and decomposition (if required). In terms of energy efficiency, ammonia is predicted to have the highest total energy efficiency (34–37%), followed by liquid hydrogen (30–33%) and MCH (about 25%). In addition, from cost calculation, ammonia with direct utilization (without decomposition) is considered to have the highest feasibility for being massively adopted, as it shows the lowest cost (20–22 JPY/Nm3-H_2 in 2050).

The authors outline in [10] the four biggest measurement challenges that are faced by the hydrogen industry including flow metering, quality assurance, quality control and sampling.

This article [11] presents a detailed breakdown of the energy conversion chains from intermittent electricity to a vehicle. Battery electric vehicles (BEVs) and Fuel cell electric vehicles (FCEVs) are considered. The traditional well-to-wheel analysis is used to study the effective coupling of renewable electricity sources and associated storage needs.

The recent literature review states the art of the hydrogen technologies by giving orders of magnitude of processes, costs, technologies, efficiencies, about the technical and the economic viability of the "hydrogen economy" concept. The main idea is to cover the whole "life cycle" of the hydrogen as energy vector. The main categories are: hydrogen production, hydrogen storage, hydrogen transport and distribution and hydrogen conversion on vehicles or energy grid conversion. From the literature review it comes out that the hydrogen is seen as energy carrier and connector for different energy grids: electricity, natural gas and as a storage medium for renewable energies. The leading spots in term of hydrogen research are in North of Europe and in Japan. Japan has the ambition to demonstrate the progress of hydrogen energy on real scale during the Olympic games of Tokyo in 2021. Recent studies from Nederland and Norway consider the hydrogen with production from renewable energy and highlight the economic benefit of stand-alone installations with small production capacity. The literature review summarizes the ways of hydrogen production and classifies them. The main need is to produce clean, "green" hydrogen. Clean hydrogen is seen to be produced by renewables: water electrolysis powered by renewable electricity and biomass. The current production of hydrogen is coming from the natural gas steam reforming, but water electrolysis and biomass are seen as future ways (technologies) for clean hydrogen production, from 2030. Other question concerning the hydrogen production is the sizing of the capacity of the installations. This question is directly related to the storage of the hydrogen. Hydrogen is very reactive molecule and its storage in pure state is a complex process. The literature review describes the compressions levels for the compressed gas hydrogen. The compression process presents important operating costs. The classification of the hydrogen storage ways is proposed. According to the classification, hydrogen can be stored in physical, adsorption or chemical forms. The physical forms of storage are the compressed gas and the liquefied hydrogen. They deliver pure hydrogen but present high operating costs related to the high-energy consumption to maintain them is their respective physical state. The chemical storage concerns the metal hydrides and the chemical hydrides. Metal hydrides present different opportunities for storage of hydrogen in a solid state. They need to be investigated to improve efficiency, storage capacity, to reduce cost and energy demand. The adsorption of hydrogen is done in porous materials: MOFs or carbon based materials (calcite). This way of storage is as well under scientific investigation. Japan has the leading position with the hydrogen storage in calcite. The chemical hydrides are the most promising way of hydrogen storage and transportation in liquid form, in the future. Ammonia is the best chemical hydride candidate, presenting the highest hydrogen content. A research direction related to the ammonia is how to convert it directly in adapted converters (fuel cells) without dehydrogenation process, which is costly, and need purification. Japan is as well leading in the ammonia research area. The literature review points out lack of regulation and difficulties by the distribution of the hydrogen at the tank stations to the customers. Main challenges are the measures of the quantities, the control of the quality and the sampling of the hydrogen for purity control. These difficulties are related to the lack of measure instruments for large utilization, calibrated to work at 700-900 bar pressure and the purity control capacity available locally. Hydrogen is clearly seen as an energy carrier for extended energy grids, as stocker for renewable electricity and depending of the installation size can be competitive to Li-Ion stationary battery for local grid storage. Hydrogen production from renewables and Li-Ion stationary batteries demonstrate benefit to balance the global electricity grid and to reduce losses related to intermittence of the renewables.

3. Fuel cell range extender concept

Several solutions have been developed in order to improve the competitiveness of the electric vehicles. The range extender concept (serial hybrid) is one of the concepts proposing an additional range for the electric vehicles. A second advantage is the reforming of a liquid fuel on board what avoiding the storage of hydrogen under elevated pressure. This proposal suggests the use of fuel cell system powered by hydrogen produced through reforming of liquid fuel from renewable resources such as bioethanol. The fuel cell installation has an integrated fuel reformer. This article considers alkaline fuel cell and solid oxide fuel cell (SOFC). The SOFC is a fuel cell operating at high temperature range 800°C to 1000°C. Figure 2 presents the overall layout system based on SOFC. This kind of fuel cell can use a variety of fuels instead of traditionally used hydrogen in others types of fuel cells. It can be powered by CH_4, C_2H_5OH, CO or others fossil fuels what is very advantageous from the application point of view of such fuel cells on board of automobiles. In this case we have not any problems with the purity of the produced hydrogen which is strongly required to power PEMC stack system. That is why we need high active and selective catalytic system exhibited 100% of fuel conversion, which will be applied in the micro reactor directly before the SOFC fuel cells stack system (figure 2).

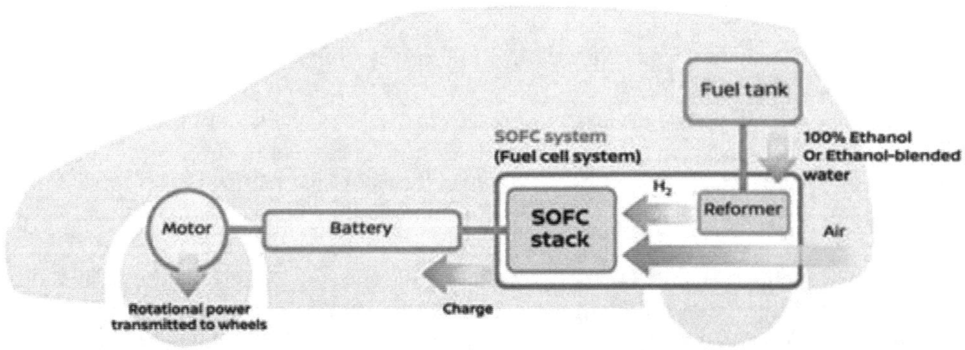

Figure 2. Vehicle propulsion system with an innovative range extender system [12].

During the research tasks the most suitable liquid fuel will be selected and will be used as a substrate for hydrogen generation and finally to produce electricity using a SOFC or alkaline fuel cell. One of the best candidates for fuel is the bio-ethanol, which can be obtained from sugar beet, sugar cane and other plants. Ethanol and especially bio-ethanol is convenient for reforming in the reactor before the fuel cell and has the advantage to be a renewable fuel, produced from biomass, as illustrated from figure 3. The main advantage of the concept is to obtain a long range, with easily supply of liquid fuel that has a carbon neutral life cycle.

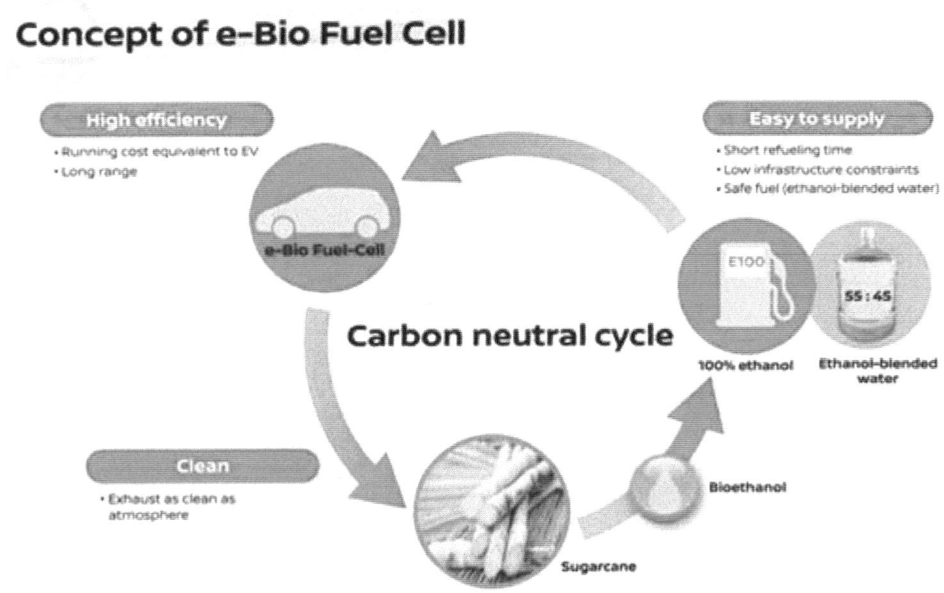

Figure 3. Carbon neutral cycle of the e-Bio Fuel Cell vehicle [12].

4. Energy balance of the fuel cell system, production and conversion of hydrogen

The choice of bio- ethanol is favorable by the fact that this non- toxic compared to methanol. It is easily available. It is easily converted under atmospheric pressure and it is a large possibility to obtain it from biomass. Bioethanol has the property to be stored on easy way under liquid form, and to be used on a safety way. Bio ethanol is as well a renewable energy source.

It is possible to obtain different configurations of energy vectors (fuels) and types of fuel cell. According of the type of fuel, the processes that the fuels pass through are different (figure 4).

One can observe from Figure that to use the fuels for the alkaline fuel cell the fuels have to pass through different steps in order to reduce the content of CO after the reforming stage. The alkaline fuel cell needs pure hydrogen, because of risk of contamination of the electrodes by CO.

The following processes are applied:

Variant 1:

Reforming (Steam reforming) – A process in which reacts steam at high temperature with the fossil fuel, as noted in equation (1) and in equation (2):

$$CH_3CH_2OH + 3H_2O \Rightarrow 2CO_2 + 6H_2 \qquad \Delta H = 347 \text{ kJ mol}^{-1} \qquad (1)$$

$$CH_3CH_2OH + H_2O \Rightarrow 2CO + 4H_2 \qquad \Delta H = 298 \text{ kJ mol}^{-1} \qquad (2)$$

IOP Conf. Series: Materials Science and Engineering 1002 (2020) 012025 doi:10.1088/1757-899X/1002/1/012025

WGS (Water-gas shift reaction WGSR) – describes the reaction of carbon monoxide and water vapour to form carbon dioxide and hydrogen. It is one of the most important reactions used to balance the H_2/CO ratio, as given in equation (3):

$$CO + H_2O \Rightarrow CO_2 + H_2 \qquad \Delta H = -41.16 \text{ kJ mol}^{-1} \tag{3}$$

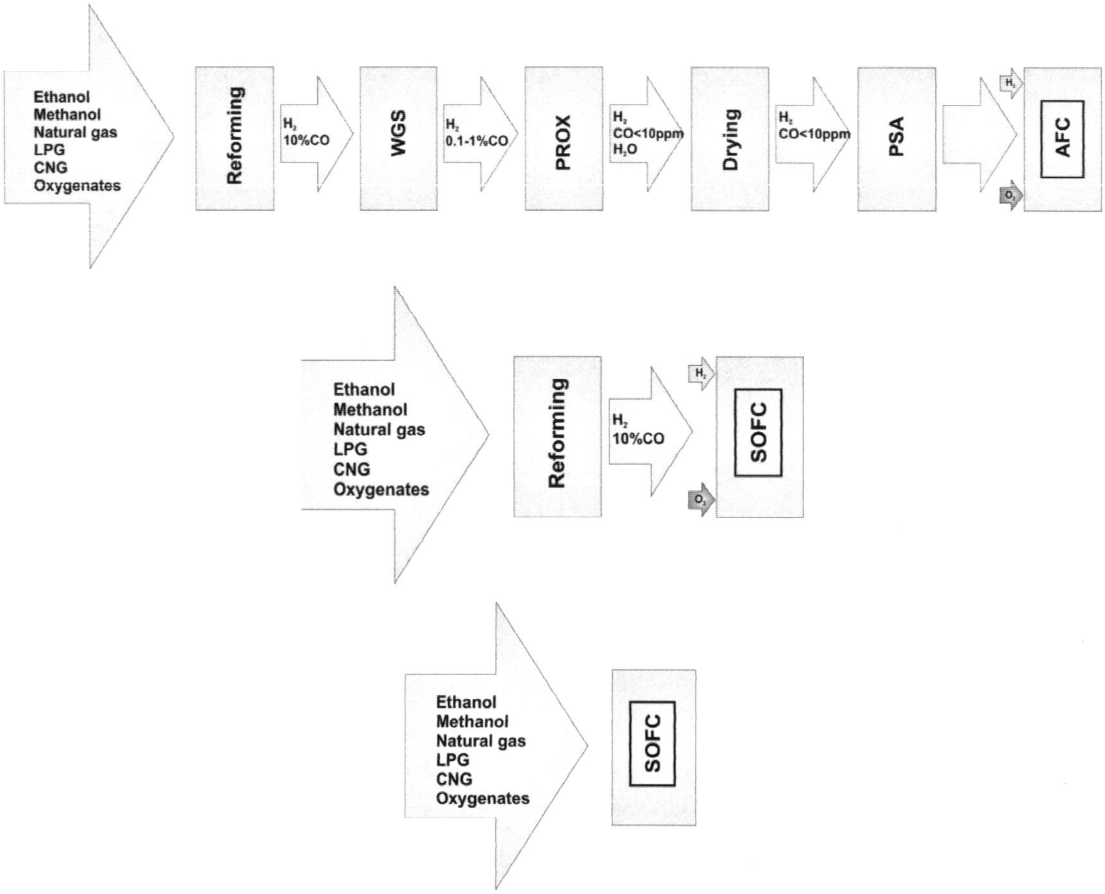

Figure 4. Schematic diagram of the various type of fuel cell powered by various feedstocks.

PROX (Preferential Oxidation) – the catalyst preferentially oxidises CO. The PROX process allows for the reaction of CO with oxygen, reducing CO concentration from approximately 0.5–1.5% in the feed gas to less than 10 ppm, as given in equation (4):

$$2CO + O_2 \Rightarrow CO_2 \qquad \Delta H = -172.35 \text{ kJ mol}^{-1} \text{ neglected} \tag{4}$$

Drying. The drying process serves to evaporate water.

PSA (Pressure Switching Adsorption) is a technology used to separate CO_2 species from a mixture.

Gibbs free energy of formation ΔG_f is the energy that accompanies the formation of 1 mole of a substance from its constituent elements.

Standard Gibbs free energy of formation ΔG_0 is the energy that accompanies the formation of 1 mole of a substance in its standard state (absolute pressure p = 1 atm (101.3kPa) and temperature T=273.15+25 K) from its constituent elements in their standard state. The values of ΔG_0 (and ΔH_0) are given in tables of thermodynamic properties.

Gibbs free energy of formation $\Delta G_f = 0$ or $\Delta G_f \cong 0$ for chemical elements (C, H, O, etc.) is expressed by equation (5).

$$\Delta G^0_{reaction} = \left(\sum v_i \Delta G^0_i\right)_{products} - \left(\sum v_j \Delta G^0_j\right)_{reactants} \quad (5)$$

The following definition is applied:

$\Delta G^0_{reaction} < 0 - Heat\ is\ released\ from\ the\ reaction (exothermic\ reaction)$

$\Delta G^0_{reaction} > 0 - Heat\ is\ consumed\ by\ the\ reaction (endothermic\ reaction)$

The SOFC fuel cell has the advantage to accept the hydrogen from after the reforming of different fuels or to be directly power by the different fuels, without any transformation of the fuels.

Variant 2:

This variant considers SOFC fuel cell with external reformer, where the liquid fuel is decomposed to hydrogen and other substances. The hydrogen is thus injected to the SOFC. This variant is as well illustrated in figure 2.

Variant 3:

Variant 3 considers the case where the liquid fuel is injected directly to the SOFC, without external reforming. The bio-ethanol is reformed in internally integrated reformer.

Figure presents the working principle of the e-Bio Fuel Cell system. The system produces hydrogen from ethanol- blended water through reformer. The main reaction is $C_2H_5OH + 3H_2O \rightarrow 6H_2 + 2CO_2$. The H_2 produced in the reformer reacts with the oxygen from the air in the SOFC and the produced electricity and heat. The electricity is used to charge the high voltage battery of the vehicle. The heat is reused in the reformer to help the reforming of bio-ethanol to water. Thus, the system has a high efficiency of 55%.

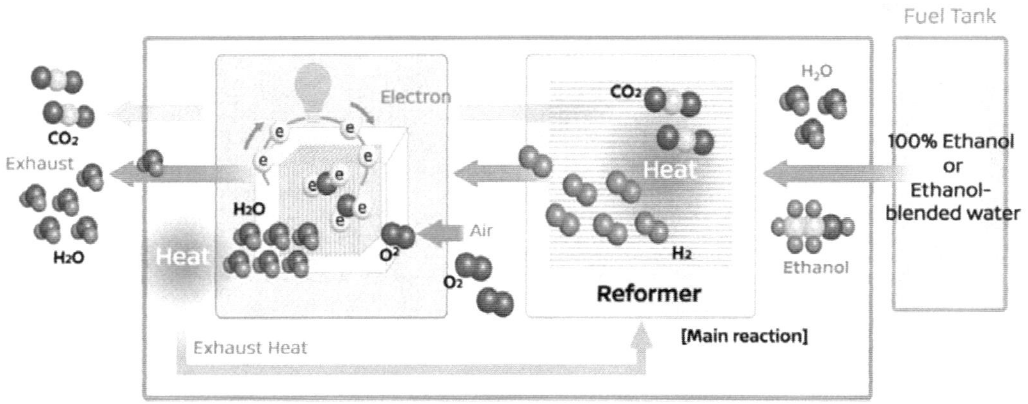

Figure 5. Working principle of the e-Bio Fuel Cell system.

Figure defines three different energy systems between fuels and converters. The energy balance of these three systems will be investigated to understand and to compare the efficiency of conversion and the global energy efficiency of the three scenarios. For the energy balance the thermodynamics based on the enthalpies of the reactions is used.

Table 1 presents the comparisons of the values of the energy balance for the different configurations (variants) presented in Figure . From the results is the visible that the most efficient configuration is Variant 3, the direct reforming of the bio-ethanol by the SOFC Fuel Cell. The energy consumption of the different processes are presented in table 1. The criteria to be observed is to output power of the fuel cell for the same quantity of bio-ethanol injected in the fuel converter. Variant 3 presents 15.15 MJ/kg of injected ethanol in the system.

Table 1. Comparison of the values of the energy balance for the different configurations.

	Vaporiser kJ/mol	Reformer kJ/mol	Dryer kJ/mol	PSA (CO_2 separation) kJ/mol	Process of H_2 generation MJ/kg_{Efuel}	Fuel Cell output $MJ_{electric}/kg_{EtOH}$
Variant 1 ($EtOH+H_2O$)	-466.45	-337.98	39.31	-21.85	-7.78	11.04 (PEM) 12.04 (AFC)
Variant 2 ($EtOH+H_2O$)	-466.45	-342.10	–	–	-8.08	12.61 (SOFC)
Variant 3 (EtOH)	–	–	–	–	0.00	15.15 (SOFC)

Error! Reference source not found. summarizes the advantages and the disadvantages of the different conversion configuration, presented in Figure .

Table 2. Advantages and disadvantages of the different configurations (variants).

	Advantages	Disadvantages
Variant 1	Low cost fuel cell. Compatible to all types of FC.	The lowest overall system efficiency. Multiple chemical processes. Three catalytic converters (at least one of them contain Pt catalyst). System load can not be adjusted on-line. Overall large dimensions difficult to use in a small size vehicles.
Variant 2	Overall system efficiency is slightly higher in comparison to multistep process. System load can be adjusted on-line in the limited range. It is not necessary to remove CO from the H_2 stream. Only one catalytic reactor is used. Low requirements on the type of fuel.	Only SOFC cell could be used. Limited number of start/stop cycles. Limited lifetime of SOFC system.
Variant 3	The highest overall system efficiency. The most compact configuration. System load can be adjusted on-line in the limited range. It is not necessary to remove CO from the H_2 stream. Only one catalytic reactor is used. Low requirements on the type of fuel.	Only SOFC cell could be used. Limited number of start/stop cycles. Limited lifetime of SOFC system.

5. Conclusion

The variant 3 presenting direct reforming of bio-ethanol is the most efficient combination. In conclusion, is proposed the working principle of the bio ethanol fuel cell for range extender unit for vehicle application with 15.15 MJ of electric energy per kilogram of converted bio-ethanol. SOFC with integrated reformer of the bioethanol is the best fuel cell for the concept for a range extender unit. It presents the best energetic balance, from energy conversion point of view.

It presents the following advantages to be used to power electric vehicles:

- Long lifetime.
- Low - costs of the electrodes.
- Gases produced in SRE process can directly be used to power fuel cell stack.
- High fuel efficiency.
- Various compounds could be used to power fuel stack even natural gas.
- Chemical stable of the solid electrolyte.
- Low cost of raw materials for the electrolyte production.

Disadvantages of using SOFC to power electric vehicles:

- High operating temperature.
- Long start-up.
- Mechanical and chemical compatibility issues.

Next research will be the investigation for the best catalyst system to integrate to the reformer to maximize the hydrogen yield and the stability of the conversion reaction from bioethanol to hydrogen.

References

[1] van der Roest E, Snip L, Fens T and van Wijk A 2020 Introducing Power-to-H3: Combining renewable electricity with heat, water and hydrogen production and storage in a neighbourhood *Applied Energy* **257** 114024
[2] Shiva Kumar S, Himabindu V 2019 Hydrogen production by PEM water electrolysis – A review *Materials Science for Energy Technologies* **2** pp 442–54
[3] Mishra P, Krishnan S, Rana S, Singh L, Sakinah M, Ab Wahid Z 2019 Outlook of fermentative hydrogen production techniques An overview of dark, photo and integrated dark-photo fermentative approach to biomass *Energy Strategy Reviews* **24** pp 27–37
[4] Ulleberg Ø and Hancke R 2020 Techno-economic calculations of small-scale hydrogen supply systems for zero emission transport in Norway *Int. J. of Hydrogen Energy* **45** pp 1201–11
[5] Navas-Anguita Z, García-Gusano D, Dufour J, Iribarren D 2020 Prospective techno-economic and environmental assessment of a national hydrogen production mix for road transport *Applied Energy* **259** 114121
[6] von Colbe J B et al. 2019 Application of hydrides in hydrogen storage and compression: Achievements, outlook and perspectives *Int. J. of Hydrogen Energy* **44** pp 7780–808
[7] Okuda R, Komatsu K, Nakamura A, Ito O, Nambu K, Saitoh H 2019 Evaluation of released amount of hydrogen after high pressure hydrogen loading in carbonate *Results in Engineering* **4** 100047
[8] Zhang Y et al. 2019 Balancing wind-power fluctuation via onsite storage under uncertainty: Power-to-hydrogen-to-power versus lithium battery *Renewable and Sustainable Energy Reviews* **116** 109465
[9] Aziz M, Oda T, Kashiwagi T 2019 Comparison of liquid hydrogen, methylcyclohexane and ammonia on energy efficiency and economy *Energy Procedia* **158** pp 4086–91
[10] Murugan A et al. 2019 Measurement challenges for hydrogen vehicles *Int. J. of Hydrogen Energy* **44** pp 19326–33
[11] Ligen Y, Vrubel H, Girault H 2018 Mobility from renewable electricity: Infrastructure comparison for battery and hydrogen fuel cell vehicles *World Electric Vehicle J.* **9** 10003
[12] Nissan unveils world's first Solid-Oxide Fuel Cell vehicles Nissan News USA, 04-août-2016. https://usa.nissannews.com/en-US/releases/nissan-unveils-world-s-first-solid-oxide-fuel-cell-vehicle Accessed on: 02/02/2020

IOP Conf. Series: Materials Science and Engineering 1002 (2020) 012026 doi:10.1088/1757-899X/1002/1/012026

Energy demand assessment for long term operation of hybrid electric vehicles

Lech J Sitnik[1,3], Zdravko D Ivanov[2] and Zbigniew J Sroka[1]

[1] Wroclaw University of Science and Technology, Department of Automotive Engineering, Wyb. Wyspianskiego 27, 50-370 Wroclaw Poland
[2] Technical University Varna, Transport Engineering and Technologies Dpt., Studentska 1, 9000 Varna, Bulgaria

[3] E-mail: lech.sitnik@pwr.edu.pl

Abstract. This publication deals with the assessment of fuel demand for long-term operation of hybrid vehicles, because only with this kind of analysis it could be possible to clearly and unequivocally assess the actual progress in achieving the intended goals. The long-term test is the opposite of short-term tests – currently WLTP and RDE or other tests developed mostly by specialized magazines. Long-term car use means mileage measured in thousands of kilometers over a period of several years. The assessment of fuel consumption during long-term operation of HEVs from the presented point of view is new, in which the theory of demand for cumulated energy (as fuel) proved to be very useful. Mathematical models of fuel consumption along with an assessment of its intensity and a description of fuel economy constitute the vehicle's "energy footprint". Cumulative fuel consumption increases as mileage increases, but the increase is not a linear function. The discrepancy between fuel consumption in a group of vehicles of the same make and type is significant and increases as mileage increases. The average fuel consumption of the analyzed HEVs during their long-term operation significantly exceeds what is determined by the short tests, e.g. WLTP (and earlier NEDC). The actual fuel consumption is much higher than the one from the short tests.

1. Introduction

The average emission of all cars sold in the European Union in 2018 was 120.5 g CO_2/km. From 1 January 2020, 95 percent of cars sold within the European Union must meet the average maximum emission limit of 95 g CO_2/km. The European Commission proposes that in 2025 this value should be reduced by 15 % and in 2035 by 30 %. Targets are expressed in percentages because the new standard will have to be recalculated according to the new (more rigorous) WLTP emissions test [1].

The hopes associated with the WLTP test are greater because the discrepancies between the results obtained in the NEDC test compared to the real ones, were not respectively satisfying (were too large) [2] – shown in figure 1.

In 2001, the average CO_2 emissions, determined in the test (NEDC) were 170 g/km with a spread about of 20 g/km. In real life emissions were by 10 % higher.

In 2015, in the test (NEDC), is achieved reduction in the average values of emissions below 120 g/km. The dispersion also decreased to about 10 g/km. On the other hand, the average values from real operation differed from the test ones by over 40 % – which means that the actual progress in reducing emissions was unsatisfactory.

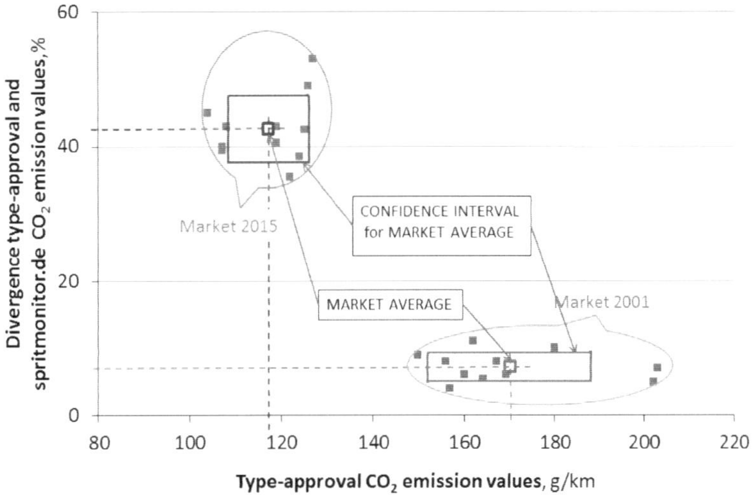

Figure 1. CO_2 emissions in test and real drive conditions over years (on the basis of data in [2]).

Despite this, especially in the opinion of the car industry, the European Commission proposals of reduction of CO_2 emissions targets for cars and vans for years 2025 and 2030, are criticized because they fail in 3 key respects:

1) The 30 % reduction from 2021 to 2030 is far below the trajectory needed to achieve the "Paris Climate goals".
2) The regulation fails to require the supply of zero emission vehicles – instead this is only incentivized.
3) There is no mechanism to ensure emissions reductions are delivered on the road – not just in the laboratory – or that the 2021 baseline will be not manipulated.

The conclusion of 3) is very important for this publication. Such organizations as T&E [3], for example, are therefore proposing the following goals:

a) A 20 % (instead of EC 15 %) average CO_2 emissions reduction from new fleet of cars of each carmaker from 2021 to 2025, to apply to both cars and vans.
b) A 50-60 % (instead of EC 30 %) reduction of CO_2 emissions by 2030 – the target to be reviewed in 2022 to allow for sufficient lead time.
c) A 0g CO_2/km (zero g CO_2/km) target for 2035 – in order to have the time to indicate the required direction and speed relevant for improvement of industry solutions.

The inclusion of c) target is essential since it doubles the lowering of the anticipated carbon emissions by 2030. In the absence of target c), much less improvement is envisaged in new car and van CO_2 emissions between 2021 and 2025 and, as a result, the fleet will use more fuel (it will emit more CO_2) as if it were possible if the 2025 limit was introduced.

We could and should ask a question – is electricity really clean, i.e. is it CO_2 free? And, is the short test relevant for assessing energy (fuel) demand?

The answer to the last question is the key for all efforts to really reduce CO_2 emissions from transport.

For this reason, the paper is about energy demand assessment of long term operation of real hybrid vehicles in everyday conditions. The mileage is in thousands of kilometers over a period of several years measured. That is in contrast to short-term tests – such as WLTP [1] and RDC [4, 5] or various specialist tests [6-8].

The assessment of long term energy demand as a cumulative fuel consumption of HEV, from the presented point of view, is brand new concept. Therefore, there are no papers with direct discussion or comparison of results. Also the earlier works of the authors are concerned with the use of the theory of cumulative fuel consumption for assessing fuel consumption of standard motor vehicles (e. g. [9-13]) but not HEVs. The authors' experience lead to the conclusion that a description of fuel consumption in long term operation of HEV using the theory of cumulative fuel consumption makes sense.

Very good adequacy of the mathematical models was obtained. Mathematical models of cumulative fuel consumption, the intensity of this consumption and specific fuel economy together form the vehicle's "energy footprint".

In chapter 3 the analysis of fuel consumption in the long term operation of HEVs of the same make and type are presented. The obtained analysis results seem to be interesting.

Chapter 4 contains the conclusions. The main one is that the theory of cumulative fuel consumption in long term operation is a very good tool for energy consumption analysis. That is only beginning. The analysis with the use of the theory, lead to much more interesting and unexpected next conclusions.

By assessing fuel consumption in the long term operation of vehicles over the years and thousands of kilometers then the question is, what methods for those assessments are sensible to use.

First of all it must be clearly defined – is the operational fuel consumption a linear function to the mileage. Maybe in some cases it is true, but there is not a base for this assumption for all cases in long term operation of vehicles.

Secondly, it should be clarified whether the fuel consumption determined in the test (e.g. WLTP) is appropriate to assess such consumption, but in standard long-term operation. In particular the question is – can the test results be overstated or understated in relation to real life.

A typical, most used, factor for assessing the fuel consumption of cars during standard operation is e.g. fuel economy *FE*. This is calculated as a $dm^3/100km$ ratio (or in Anglo-Saxons countries as MPG – miles per gallon). An example of *FE* values are given in figure 2.

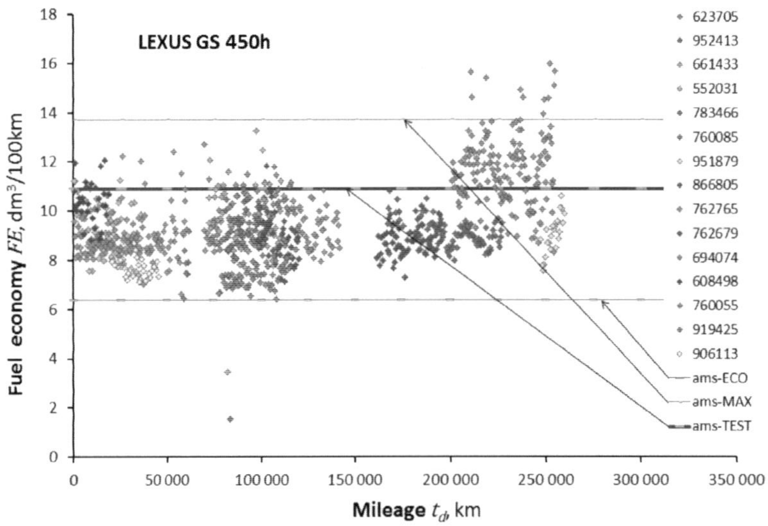

Figure 2. Typical results of fuel economy between refuelling HEV LEXUS GS 450h. Numbers on the figure right side are dedicated to the separate cars ams-test of fuel economy [14].

The presented results of fuel economy between refueling are for HEV LEXUS GS 450h calculated.

The fuel consumption and trip distance data are from the German website spritmonitor.de [14]. The data are publicly available for car with given code number. The technical data of the studied HEV Lexus GS 450h can be found in [15].

The results are from long term operation of cars, including urban, peri-urban and high way drive. When these results are compared with the short test, for example with ams-tests result for LEXUS GS 450h, the extensive discrepancies are visible.

The ams-ECO result is too low. Practically no driver (here analyzed) achieves such low fuel economy.

The ams-TEST average is too high vs. other values. Without statistical dispersion analyses it can be seen that the density of fuel economy (shown in figure 2) is between ams-ECO and ams-TEST placed.

The ams-MAX is on the one hand too low – they is higher fuel economy noticed. On the other hand it is perhaps too high because only some drivers achieved such high fuel economy. Figure 2 demonstrates that most results of fuel economy reach 9 liters per 100 km.

From these results, it is not possible to form an opinion about the real changes of fuel economy vs. mileage. The same results as in figure 2 are in figure 3, but with trends line (predictions) presented. All the trends are equivalent. It is not possible to decide which line is to be taken in account.

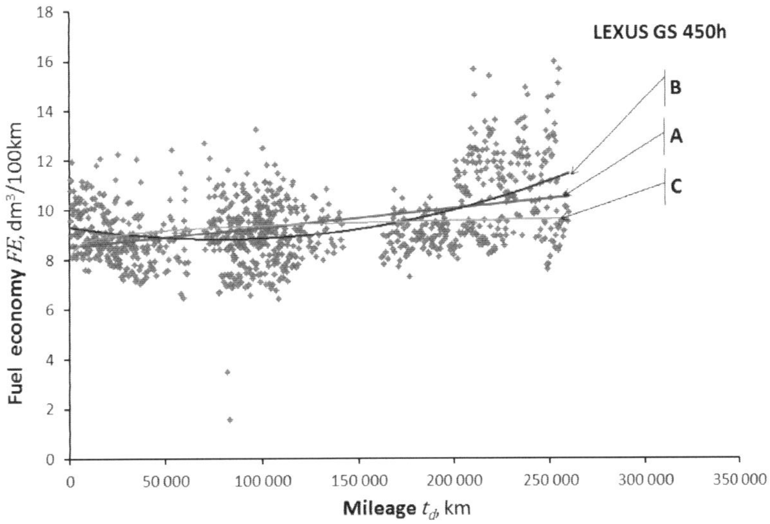

Figure 3. Fuel economy and its trends over mileage.

However, it seems that with increased mileage fuel economy also increase. That is very good to show by the linear trend ("A" in figure 3). In "B" case, fuel economy at the beginning decreases and then increases. In case "C", it is the opposite. There is not any clarity in assessing this possibility. Such clarity can be achieved if the theory of cumulative fuel consumption is used.

2. Theory of cumulative fuel consumption

As mentioned above, traditionally, if the energy demand of ICEV is analyzed, then the analyzed fuel consumption is expressed as fuel economy (*FE*) in $dm^3/100km$ or MPG. It is the same for HEV because the energy to drive a vehicle comes from the combustion of fuel in the internal combustion engine.

Through the assumptions that F_i – *i*-th refueling; t_{di} – mileage to F_i, then the fuel economy can be expressed as:

$$FE_i = 100F_i(t_{di} - t_{di-1}) \tag{1}$$

By *n* refueling the average fuel economy (*FEA*) can be expressed as

$$FEA_n = \frac{1}{n}\sum_{i=1}^{n} FE_i \tag{2}$$

The cumulative fuel consumption to achieve mileage t_{dn} can be described as

$$CFC(t_{dn}) = \sum_{i=1}^{n} F_i \tag{3}$$

A sample presentation of FEV data ([spritmonitor.de], vehicle code 783466) is shown in figure 4. The mileage of the car was in the range 181758 to 227713 of kilometers, in the time period 09.09.2016 to 02.03.2019.

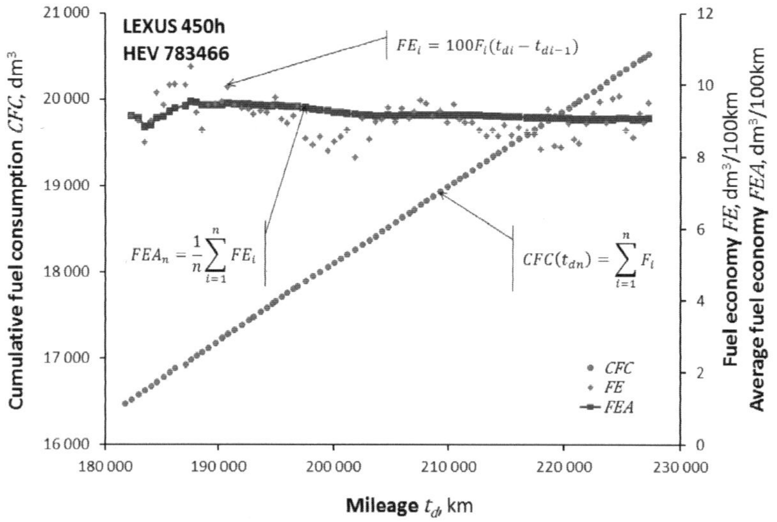

Figure 4. Fuel consumption of analyzed middle class of HEV.

The fuel economy (*FE*) in operation time changes between 8 and 11 dm^3/100km.

Figure 4 also presents the values of the average fuel economies (FEA). They are in the narrower range of 9 to 10 dm^3/100km. As the car mileage increases, the average fuel economy generally decreases (in the case of the data of this car).

The cumulative fuel consumption changes (quasi linear) from 16472 to 20532 dm^3.

Without knowledge of any functional relationship describing cumulative fuel consumption as a function of time, (and traditionally as a function of vehicle mileage), it is not possible to answer such a simple question as: Is it really a quasi-linear function or another relationship?

A function is therefore sought

$$CFC = f(Mileage) \tag{4}$$

The answer to the above question is given by the theory of cumulative fuel consumption.

The consumption of fuel by a car, in its natural operation, is a random process. It is described in a quantum model. Fuel quanta has a random size. Random size has also time between the quanta of fuel consumption. The ratio of the total amount of quanta of fuel supplied by the car engine in its operating period to the time *t* is called cumulative fuel consumption and can be determined as follows:

$$CFC(t) = \sum_{i=1}^{n(t)} f_i = n(t)\bar{f}(t) \tag{5}$$

where f_i– *i*-th quantum of fuel, $\bar{f}(t)$ – the average size of the quantum of fuel supplied for time *t*, $n(t)$ – fuel quantum number supplied for time *t*, $CFC(t)$ –cumulative fuel consumption for time *t*.

After many transformations, described in detail in [3], and assuming that the time between successive quantum of fuel is described by the Poisson distribution [7] there is a relationship

$$CFC(t) = ct^{a+1} \tag{6}$$

where: *c*, *a* – constants.

This is a relatively simple relationship describing cumulative fuel consumption as a function of time. on the assumption that mileage t_d is proportional to operation time t is, also ($t \rightarrow t_d$), the relationship (6) can be shown in the form (7)

$$CFCc(t_d) = ct_d^{a+1} \tag{7}$$

the intensity of the cumulative fuel consumption is Derivative of the cumulative energy consumption based on the vehicle engine run time t_d .

$$ICFC(t_d) = \frac{dCFC(t_d)}{dt_d} = c(a+1)t_d^a \tag{8}$$

The intensity of the cumulative fuel consumption takes infinitely large values, when $a<0$ and at the same time where $t_d \rightarrow 0$ and almost immediately after the beginning of the trip (but it decreases rapidly when the travelled distance increases). By $t_d=0$ the ICFC not exist.

To analyze the specific cumulative fuel consumption there exists a new relationship

$$SCFC(t_d) = \frac{CFC(t_d)}{t_d} = ct_d^a \tag{9}$$

The *SCFC(t_d)* can be expressed in dm^3/km or parallel to fuel economy as *FEC* in dm^3/100km or MPG.

$$FEC(t_d) = 100\frac{dCFC(t_d)}{t_d} = 100ct_d^a \tag{10}$$

Constants c and a can be derived from the data obtained from the use of vehicles in natural operation.

With the data presented earlier for car of the HEV 783466 type the achieved coefficients c and a are the following: $c = 0.107235$, $a = -0.013887$.

Knowing coefficients c and a, it is possible to present all above mentioned indicators regarding fuel consumption to present. Graphic illustration of their course is as follows (figure 5).

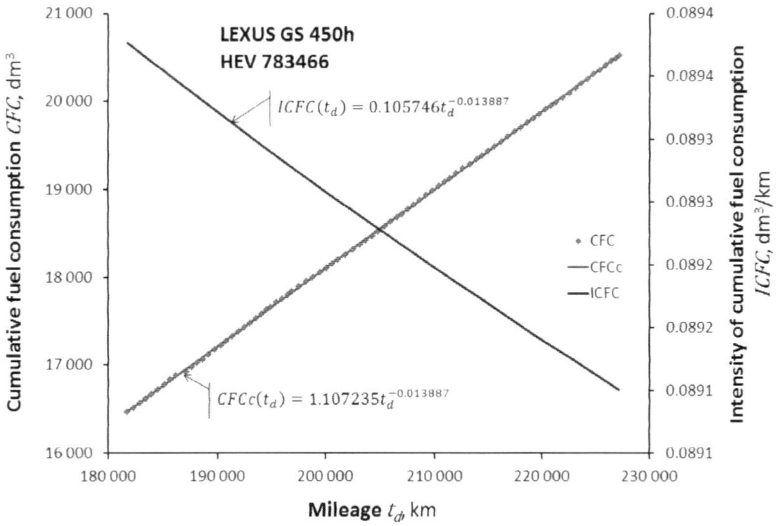

Figure 5. Graphic illustration of course of cumulative fuel consumption and its intensity for HEV 783466.

The correlation between the measurement points and mathematical model points (for *CFC*) gets unexpected high values. That is also confirmed by the statistical coefficients (table 1).

Table 1. Results of statistical analysis of the correlation of mathematical model (7) for cumulative fuel consumption of HEV 783466.

Multiple R	0.999968
R Square	0.999936
Adjusted R Square	0.999935
Standard Error	0.000524
Observations	82

The data presented can be considered sensational. This is not an exception – as was presented earlier in [9-12].

It is worth noting the relatively steep decrease in the intensity of cumulative fuel consumption (*ICFC*), however it is worth noting the axle values, which are contained within very narrow limits. However, based on these data, it can be concluded that this intensity although slightly decreases as the mileage increases.

It was mentioned earlier that the presented indicators are a kind of "footprint" regarding the fuel consumption of a vehicle. For HEV 783466 this "footprint" is shown in figure 6.

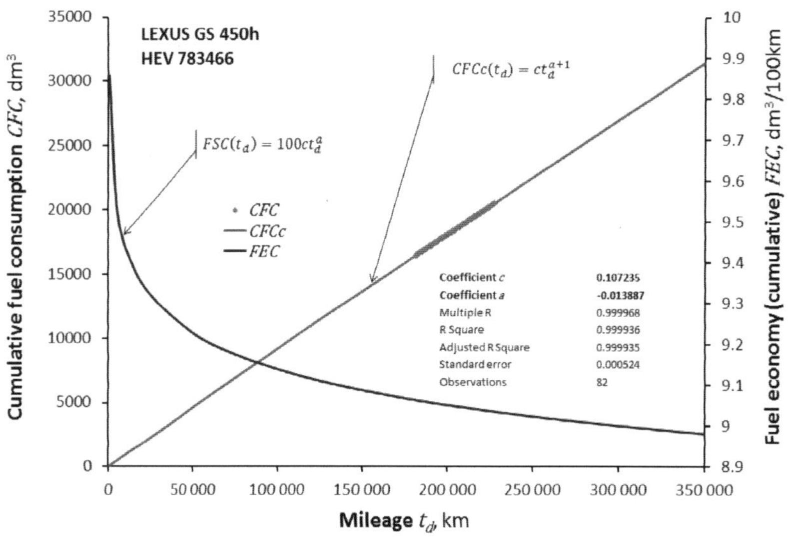

Figure 6. Energetic (fuel consumption) footprint of HEV 783466.

The cumulative fuel consumption for a mileage to 350,000 km gets a quasi-linear function. In reality it is not a linear relation. Also fuel economy decreases with the increase in mileage. However, this change does not exceed 1 $dm^3/100$ km.

3. Results and discussion

The data for fifteen HEVs were used for further analysis to determine the energy footprint of each analyzed vehicle and on this basis also for brand and type of analyzed cars.

The calculations results are presented in figure 7.

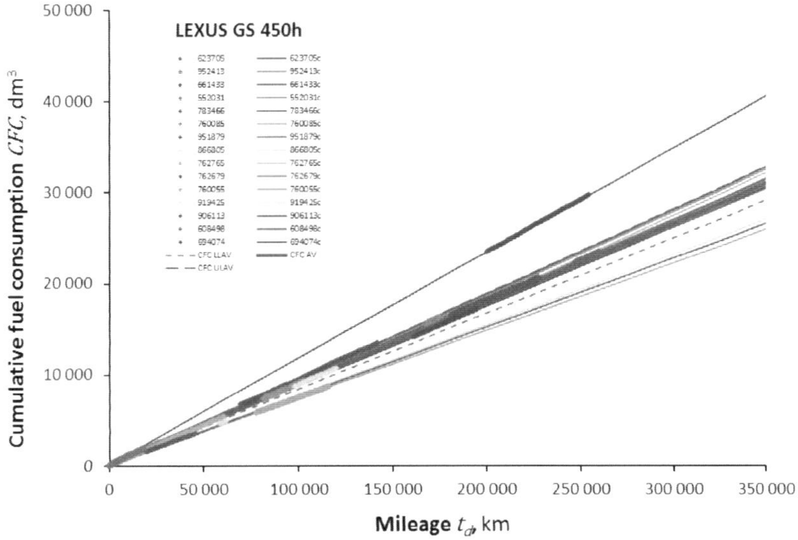

Figure 7. Cumulative fuel consumption of analyzed HEV's.

The points lying on the cumulative fuel consumption curves correspond to the individual refueling of the vehicle.

There are values of coefficients *c* and *a* in table 2 in addition to the results of the statistical analyses of the model (7) adequacy for all analyzed HEVs.

Table 2. Results of the statistical analysis of the correlation of mathematical model (7) for cumulative fuel consumption of the analyzed HEVs.

N° of probe	HEV code	Coefficients		Statistics				Numbers of Observations
		c	*a*	Multiple R	R-Square	Adjusted R-Square	Standard Error	
1	623705	0.110896	-0.013202	0.999791	0.999582	0.999579	0.004213	152
2	952413	0.095743	-0.002292	0.999795	0.999590	0.999561	0.000306	16
3	661433	0.136929	-0.033048	0.999567	0.999135	0.999124	0.003724	86
4	552031	0.114006	-0.021083	0.999649	0.999299	0.999290	0.030652	80
5	783466	0.107235	-0.013887	0.999968	0.999936	0.999935	0.000524	82
6	760085	0.075480	0.015431	0.999407	0.998814	0.998781	0.002653	38
7	951879	0.098589	-0.007028	0.999387	0.998775	0.998722	0.000544	25
8	866805	0.093372	-0.004373	0.999620	0.999240	0.999224	0.001724	48
9	762765	0.089745	-0.001666	0.999951	0.999902	0.999901	0.009157	113
10	762679	0.091714	-0.003779	0.999963	0.999925	0.999924	0.000487	65
11	760055	0.078709	-0.004642	0.999964	0.999928	0.999926	0.000993	66
12	919425	0.083097	-0.005624	0.999541	0.999082	0.998929	0.000643	8
13	906113	0.082063	-0.006025	0.999907	0.999814	0.999808	0.003360	37
14	608498	0.196559	-0.064007	0.998123	0.996250	0.996136	0.066573	35
15	694074	0.147210	-0.018636	0.999872	0.999745	0.999743	0.001095	160

The adequacy of model (7) is very high in all cases.

The course of *CFC* is quasi linear (it would be linear if factor $a = 0$). It is interesting that in case of the HEV 760085 the course has got another character as for the rest of analyzed cars have.

Some waveforms (e.g. HEV 694074) clearly differ from others. However, this is neither a mistake in entering the data nor a calculation error. These data reflect the operating conditions.

Knowing the mathematical model (vehicle energy footprint), we can approximate virtually the fuel amount for any mileage, because the prediction of the mathematical model gets very high.

The presented data shows a large dispersion of cumulative fuel consumption courses (although the vehicles of the same brand and type analyzed). This dispersion increases with the increase of the vehicle mileage (figure 8).

With the data on the vehicle group, various statistical analyzes can be carried out. Figure 8 presents not only the averages values of *CFC* and *FEC* but also the confidence intervals for these averages.

It is clear that the cumulative fuel consumption is not constant over the long term of use of the vehicle. Generally, there is an opinion that that fuel economy increases as the vehicle mileage increases, while in most cases presented here it decreases. Therefore without clear analysis with mathematically based on-line methods, assessment of real fuel consumption is difficult.

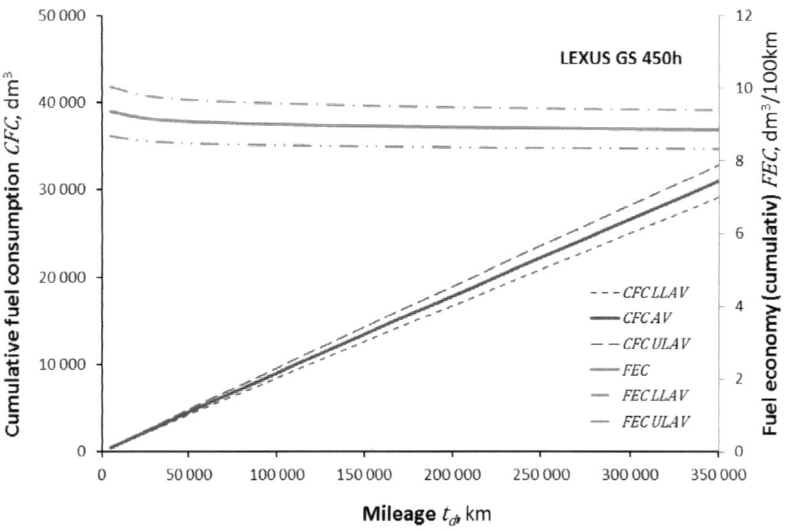

Figure 8. CFC and FEC of analyzed HEVs.

Knowing the long term fuel economy, for a specific make and type of cars, broader analyses are possible. An example of data (for LEXUS GS 450h) for further analysis is given in figure 9.

Figure 9 shows the results of fuel economy from long and short term operation.

The data of *FEC* are the long term result. Other data are for the short term. They are presented so because it is believed that the results of short tests (for example WLTP) are valid for the whole life of the vehicle.

The *FEC* data in figures 8 and 9 differ slightly. Figure 8 gives the average *FEC* and confidence interval for this mean. Figure 9 shows the average and the range of volatility. The volatility range is slightly wider than the confidence interval.

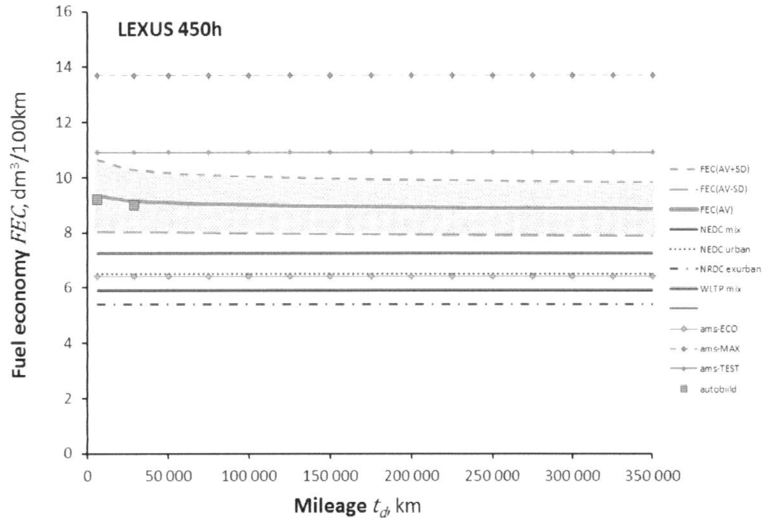

Figure 9. Data of fuel economy of LEXUS DS. 450h for analysis.

It is interesting that the results are very close to the average ones determined for a randomly selected group of 15 vehicles. This is of course accidental. In principle, however, this confirms the sense of using the presented methods of fuel consumption assessment.

The introduction of WLTP slightly improves the situation, but still the test results are lower in relation to the values obtained in natural exploitation of cars. They do not even fall into a wide range of volatility. If the results of other analyses confirm this, it will mean that work on the WLTP test should be carried out permanently.

4. Conclusions

One of the many important challenges of the modern world is the reduction of CO_2 concentration in the ambient air. Various actions are taken, including in the area of transport, in which the internal combustion engine still occupies a leading place among the sources of propulsion.

To eliminate the effects of internal combustion engines, hybrid vehicle technologies are developing relatively quickly, in which CO_2 emissions are reduced by reducing fuel consumption. However, the question arises how to assess the energy efficiency of hybrid vehicles?

The authors of this publication have undertaken to indicate a new approach to the energy assessment of the use of HEVs.

The method refers to the assessment of fuel consumption during the long-term operation of vehicles, which will ensure that only then will it be possible to properly assess the progress in achieving the intended goal of reducing CO_2.

This assumption is in opposition to the use of short-term tests – currently WLTP and RDE or various tests created by the editors of popular car magazines. The presented point of view is relatively new.

The work concerns the method of assessing fuel consumption in the long-term operation of hybrid vehicles. The assessment method is based on the theory of cumulative fuel consumption. The theory of cumulative fuel has proven to be very useful.

Data from long-term operation of 15 randomly selected HEVs of the middle class were taken for calculations. Very good adequacy of mathematical models was obtained. The R-Square ratio ranged from 0.99625 to 0.99994, which was an unexpectedly positive high value.

Mathematical models of cumulative fuel consumption, intensity of this fuel consumption and fuel economy together form the vehicle's "energy footprint". The footprint is individual for each car.

The cumulative fuel consumption, as the average for specific make and type of cars, increases as mileage increases, but the increase is not linear. The discrepancies between the fuel consumption in the vehicles group (the same brand and type) are large and increase as mileage increases.

The average fuel consumption of analyzed HEVs in their long-term operation, exceeds significantly what is determined in short tests, e.g. WLTP (and earlier NEDC).

The real fuel consumption is significantly higher than the one obtained by short tests. On the other hand the results are positive –fuel consumption is lower in long term operation in comparison to the average one – published in magazines, from their short tests.

Use of the presented methodology is possible to compare the fuel consumption of the same make and type cars with and without a hybrid dives. The result should be given as information on the real sense of HEV implementation, especially in the aspect of CO_2 reduction. The relevant operation data is collected. A comparison work will be presented in Authors' near future publications.

References

[1] Younes Z, Boudet L, Suard F, Gerard M and Rioux R 2013 Analysis of the main factors influencing the energy consumption of electric vehicles *Electric Machines & Drives Conf.* pp 247–53

[2] Tietge U, Díaz S, Mock P, German J, Bandivadekar A and Ligterink N 2016 From laboratory to road. A 2016 update of official and real-world fuel consumption and CO_2 values for passenger cars in Europe *ICCT* White Paper

[3] Earl T, Ambel C, Poliscanova J, Murphy A, Abasov F, Nix J and Buffet L 2020 How European transport can contribute to an EU -55% GHG emissions target in 2030 *Transport & Environment*

[4] Holmén B and Sentoff K 2015 Hybrid-electric passenger car carbon dioxide and fuel consumption benefits based on real-world driving *Environ. Sci. Technol.* **49**(16) 10199–208

[5] Mock P 2017 Real-driving emissions test procedure for exhaust gas pollutant emissions of cars and light commercial vehicles in Europe *Int. Counc. Clean Transp.* pp 1–10

[6] Bennion K and Thornton M 2009 Fuel savings from hybrid electric vehicles *Technical Report NREL/TP-540-42681*

[7] Heejung J 2020 Fuel Economy of plug-in hybrid electric and hybrid electric vehicles: effects of vehicle weight, hybridization ratio and ambient temperature *World Electr. Veh. J.* **11**(2)

[8] Kyoungho A and Hesham A 2019 A simple hybrid electric vehicle fuel consumption model for transportation applications (IntechOpen)

[9] Sitnik L 2004 Cumulated fuel consumption *Archiwum Motoryzacji* **7**(3) pp 227–54

[10] Sitnik L 2009 Cumulated LPG consumption suplied of cars enginees *J. of KONES* **16**(4) pp 429–34

[11] Sitnik L 2014 Theory of cumulative fuel consumption and example for its application *22nd Int. Sci. Tech. Conf. Trans&Motauto* pp 17–20

[12] Sitnik L 2015 Theory of cumulative fuel consumption by LPG powered cars *J. of KONES* **22**(4) pp 275–80

[13] Sitnik L 2004 Ekopaliwa silnikowe *Oficyna Wydawnicza PWr* pp 0–336

[14] https://www.spritmonitor.de Accesseded on: 15.02.2020

[15] https://www.lexus.com Accesseded on: 15.02.2020

IOP Conf. Series: Materials Science and Engineering 1002 (2020) 012027 doi:10.1088/1757-899X/1002/1/012027

Random vibration endurance test of automotive component using virtual prototyping

Georgi Todorov, Konstantin Kamberov[1]

Laboratory CAD/CAM/CAE in Industry, FIT, Technical University – Sofia, 8 Kliment Ohridski Blvd., 1797 Sofia, Bulgaria

[1] E-mail: kkamberov@tu-sofia.bg

Abstract. Automotive component manufacturers are required to validate their product reliability by assessing their mechanical endurance through predefined by the car manufacturer test procedures. A possible way to avoid (or to decrease their number) failures is to use Virtual Prototyping (VP) techniques. This study aims to present a methodology for random vibration endurance test of automotive component using virtual prototyping, based on common engineering analysis tools. There is a certain number of specialized tools for this type of testing but they require directed investment that are not accessible for every automotive industry subcontractor. This is demonstrated by a case study from the automotive industry – a cluster assembly (dashboard).

1. Introduction

The automotive industry aims to ensure certain reliability and safety parameters that are highly dependent on endurance assessment of car structures. Vehicles are subjected to complex random variable amplitude loadings, leading to faults in the components [1]. Automotive component manufacturers are required to validate their products reliability by assessing their mechanical endurance through predefined by the car manufacturer test procedures. These test procedures define vibration profiles (Power Spectrum Density of acceleration) along each axis and mono-axial dynamic load is applied subsequently for all axes, in predefined work environment (temperature, humidity, etc.). Existing studies of vibration testing in the automotive industry [2, 3] propose the following classification of the types of vibration signals used:

1) Fixed sine: sinusoidal cycle loadings, applied at a given frequency and at a certain acceleration value.
2) Swept sine: sine function that gradually changes frequency and/or amplitude over time.
3) Shocks: pulsed signals, based on an excitation of single type half-sine to represent a shock input on both directions along tested axis.
4) Random (PSD): very close to real component loads. It simulates loads by PSD, or represents random vibration environment, typical for driving conditions. This type of testing is the most realistic one for automotive component validation [2, 3].

The random vibration endurance test is quite typical and commonly used method. This test evaluates the function and strength of a particular component when vibration is applied by simulating vehicle vibration conditions, and also checks for deterioration or predicts failure in its parts. All parts and materials must be functional without any permanent deformation. Mechanical damage such as deformation or cracks is forbidden. This test is performed over a physical prototype of the product.

Thus, it is performed, usually, when the product is in the stage of prototyping and any more serious problem that is found during testing means there is the necessity for design change and new physical prototype. Failures found during testing means expenses – in terms of time and money – and they should be avoided.

A possible way to avoid (or to decrease their number) failures is to use Virtual Prototyping (VP) techniques. Under fierce development globally, approach of incorporating VP in Product Development Process (PDP) to validate products is becoming common practice. It has enhanced capabilities, offered to the engineers, for reducing the design cycle times of product and costly prototypes [4-6]. The virtual prototype is based on simulation models, but taking into account the connectivity and mutual influence of the effects and induced effects in studying the behavior of the object, thus getting as close as possible to the needed final physical object in terms of shape, size, functions and elements of behavior. It is possible to "reproduce" physical testing also, and provides opportunities to explore the product behavior in detail by analyzing its "internal" parameters (stress, deformations, etc.) and to understand the process and even the complex reasons for certain failure.

Virtual Prototyping is a promising technology that could decrease time-to-market and expenses, but requires an appropriate approach (methodology) for a modeled physical process. Usually, it is based on an existing hypothesis or theory.

Damages and failures, caused by random vibration endurance testing, are typically caused by fatigue. Fatigue assessment uses SN curves of critical structural details, in addition to the spectra, i.e. it represents material behaviour locally. This requires a known hypothesis for damage accumulation. There is a wide variety of hypotheses that examine this type of calculations - cumulative damage. A widely used method for calculation of cumulative damage is the Palmgren-Miner-Hypothesis. It gives certain simplicity, and it is based on SN curve. This hypothesis could use a modification according to Haibach, where there is a change after knee-point (k' = 2k-i; k' = k for materials with corrosion) in order to also include the damage by stress amplitudes below the knee-point (refer to figure 1) [7-9].

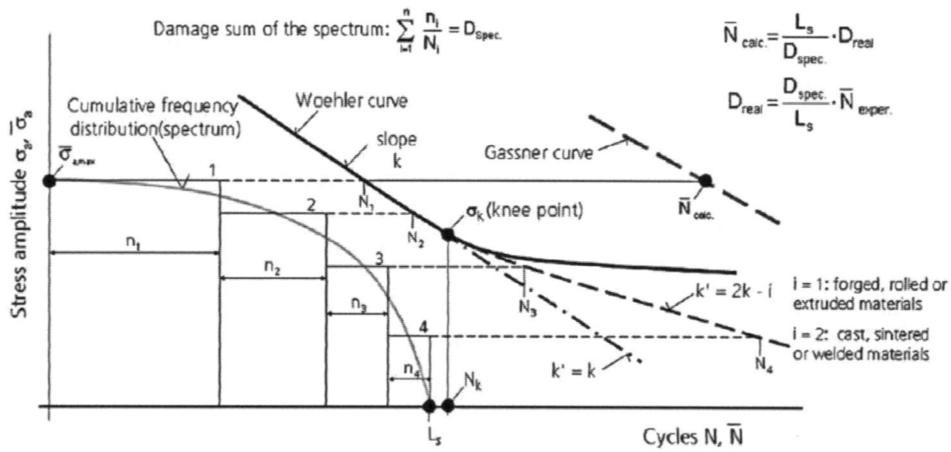

Figure 1. Most commonly used cumulative damage calculation methods [5].

This study aims to present a methodology for random vibration endurance test of automotive component using virtual prototyping, based on common engineering analysis tools. There is a certain number of specialized tools for this type of testing but they require directed investment that are not accessible for every automotive industry subcontractor.

2. Methodology

The methodology is based on Palmgren-Miner-Hypothesis. It uses data for base excitations (Power Spectrum Density – PSD) along each of the three axes, as specified by car manufacturer. The methodology is step-based, and uses available types of analyses in commercial conventional Finite

Element Analysis (FEA) software – modal and random vibration (with applied PSD). Further, the results are additionally combined and calculated to assess fatigue of the examined component.

The methodology is presented graphically in figure 2 below.

Random vibration analysis is usually performed over a large range of frequencies — from 5 Hz to 1000 Hz. This is repeated in all 3 directions of application of the base excitation and is combined in a final result. The steps, as shown in figure 2, are described below:

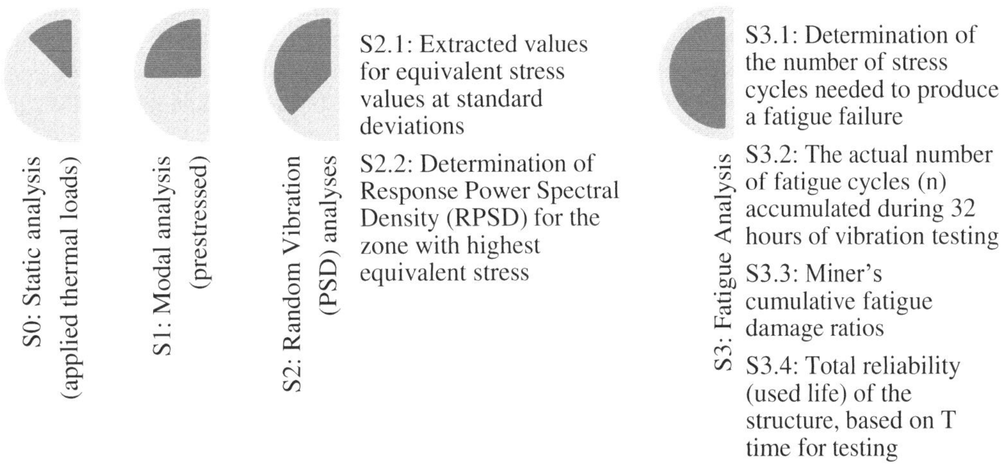

Figure 2. Step-based methodology of random vibration endurance test using virtual prototyping.

- **S0: Static analysis:** Thermal loads are applied (temperature of environment) to determine prestressed condition of the entire structure and to be used in the next step;
- **S1: Modal analysis:** to obtain modal frequencies and use them as a basis for subsequent PSD analyses;
- **S2: Random Vibration (PSD) analyses:**
  - 3 analyses – each for base excitation at X, Y and Z. Excitation is by acceleration PSD, measured during test (defined by car manufacturer) with values as shown in table 1.
  - S2.1: Extracted values for equivalent stress values at standard deviations 1σ, 2 σ, 3 σ – for each of the above mentioned PSD analyses.
  - S2.2: Determination of Response Power Spectral Density (RPSD) for zone with highest equivalent stress – to determine dominant frequency (natural) for the proper direction: $f_{d(X/Y/Z)}$.
- **S3**: Fatigue Analysis: root mean square (RMS) stress quantities are used in conjunction with the standard fatigue analysis procedure – Three-Band Technique using Miner's Cumulative Damage Ratio:
  - **S3.1:** Determination of number of stress cycles needed to produce a fatigue failure.

The approximate number of stress cycles N_1 required to produce a fatigue failure for the 1σ, 2σ and 3σ stresses can be obtained from the following equation:

$$N_{1(1\sigma/2\sigma/3\sigma)} = N_2\left(\frac{\sigma_2}{\sigma_{1(1\sigma/2\sigma/3\sigma)}}\right)^b, \qquad (1)$$

where: $N_{1(1\sigma/2\sigma/3\sigma)}$ – number of cycles to fail @ dynamic stresses of 1σ, 2σ and 3σ; $\sigma_{1(1\sigma/2\sigma/3\sigma)}$ – RPSD stress of 1σ, 2σ and 3σ.

Needed material data – $N_{2(material)}$, $\sigma_{2(material)}$ and $b_{(material)}$, are derived from material characteristics.

Table 1. PSD acceleration data.

f (Hz)	PSD G acc. X g^2/Hz	f (Hz)	PSD G acc. Y g^2/Hz	f (Hz)	PSD G acc. Z g^2/Hz
5	6.1×10^{-3}	5	2.43×10^{-2}	5	3.13×10^{-2}
10	6.1×10^{-3}	10	2.43×10^{-2}	35	3.13×10^{-2}
11	1.22×10^{-2}	11	1.38×10^{-2}	36	9.15×10^{-2}
15	1.22×10^{-2}	30	1.38×10^{-2}	40	9.15×10^{-2}
16	2.31×10^{-2}	31	1.14×10^{-2}	41	5.33×10^{-2}
25	2.31×10^{-2}	40	1.14×10^{-2}	45	5.33×10^{-2}
26	4.22×10^{-2}	41	1.81×10^{-2}	46	2.41×10^{-1}
45	4.22×10^{-2}	50	1.81×10^{-2}	52	2.41×10^{-1}
46	6.62×10^{-2}	51	9.77×10^{-3}	53	4.76×10^{-2}
54	6.62×10^{-2}	60	9.77×10^{-3}	60	4.76×10^{-2}
55	1.92×10^{-2}	61	1.34×10^{-2}	61	1.64×10^{-2}
70	1.92×10^{-2}	70	1.34×10^{-2}	80	1.64×10^{-2}
71	8.4×10^{-3}	71	8.65×10^{-3}	81	6.31×10^{-3}
85	8.4×10^{-3}	80	8.65×10^{-3}	90	6.31×10^{-3}
86	3.06×10^{-3}	81	2.47×10^{-3}	91	4.91×10^{-3}
100	3.06×10^{-3}	100	2.47×10^{-3}	100	4.91×10^{-3}
101	6.1×10^{-3}	101	7.44×10^{-4}	101	2.88×10^{-3}
120	6.1×10^{-3}	170	7.44×10^{-4}	170	2.88×10^{-3}
121	9.12×10^{-4}	171	7.56×10^{-4}	171	1.62×10^{-3}
250	9.12×10^{-4}	250	7.56×10^{-4}	250	1.62×10^{-3}

- **S3.2:** The actual number of fatigue cycles (n) accumulated during 24 hours of vibration testing for one direction, @ one of the examined temperatures (t_{max} = 85°C, t_{amb} = 20°C and t_{min} = -40°C) can be obtained from the percentage of time exposure for the 1σ, 2σ and 3σ values:

$$n_{(1\sigma/2\sigma/3\sigma).(85^\circ C/20^\circ C/-40^\circ C)} = f_{d(X/Y/Z)} \times T_{temp(85^\circ C/20^\circ C/-40^\circ C)} \times P_{(1\sigma\,/\,2\sigma\,/\,3\sigma)}, \tag{2}$$

where: $T_{temp(85^\circ C/20^\circ C/-40^\circ C)} = T^*\Delta_{(85^\circ C/20^\circ C/-40^\circ C)}$ and T = 24h = 86400, sec – time for testing at each of directions; $\Delta_{(85^\circ C)} = 0.603$; $\Delta_{(20^\circ C)} = 0.087$; $\Delta_{(-40^\circ C)} = 0.310$ – share of test duration at proper temperature (refer to figure 3); $P_{(1\sigma)} = 0.683$; $P_{(2\sigma)} = 0.273$; $P_{(3\sigma)} = 0.042$ (refer to figure 4);

- **S3.3:** Miner's cumulative fatigue damage ratios (every stress cycle uses up part of the fatigue life of a structure) for each direction, @ each of examined three temperatures are calculated, based on $T_{temp(85^\circ C/20^\circ C/-40^\circ C)}$ time for testing:

$$R_{(85^o C/20^o C/-40^o C)} = \sum_{1\sigma}^{3\sigma} \frac{n_{(1\sigma/2\sigma/3\sigma)}}{N_{1(1\sigma/2\sigma/3\sigma)}} = \frac{n_{(1\sigma)}}{N_{1(1\sigma)}} + \frac{n_{(2\sigma)}}{N_{1(2\sigma)}} + \frac{n_{(3\sigma)}}{N_{1(3\sigma)}}, \tag{3}$$

– reliability of the structure for $T_{temp(85^\circ C/20^\circ C/-40^\circ C)}$ testing with excitation along the proper axis (X or Y or Z) at proper temperature – 3 values in total for each direction of base excitation;

- **S3.4**: Reliability (used life) of the structure, in the proper direction is calculated as follows:

$$\begin{aligned} R_X &= R_{X.85^\circ C} + R_{X.20^\circ C} + R_{X.-40^\circ C} \\ R_Y &= R_{Y.85^\circ C} + R_{Y.20^\circ C} + R_{Y.-40^\circ C} \\ R_Z &= R_{Z.85^\circ C} + R_{Z.20^\circ C} + R_{Z.-40^\circ C} \end{aligned} \tag{4}$$

- **S3.4**: Total reliability (used life) of the structure, based on 3*T = 3*24h = 72h time for testing:

$$R_\Sigma = R_X + R_Y + R_Z \tag{5}$$

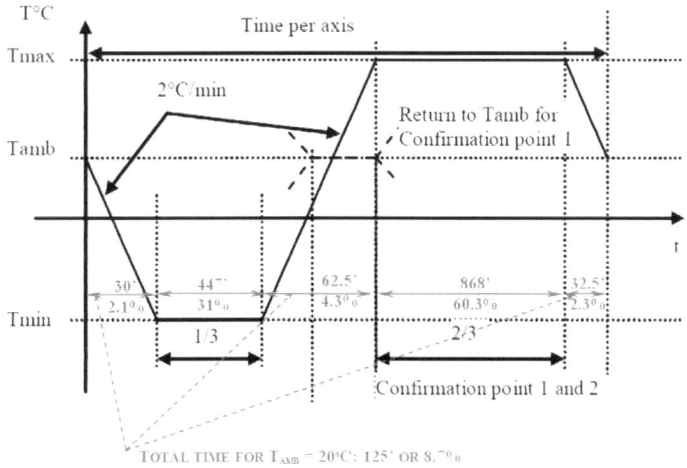

Figure 3. Thermal profile during vibrations as provided by car manufacturer.

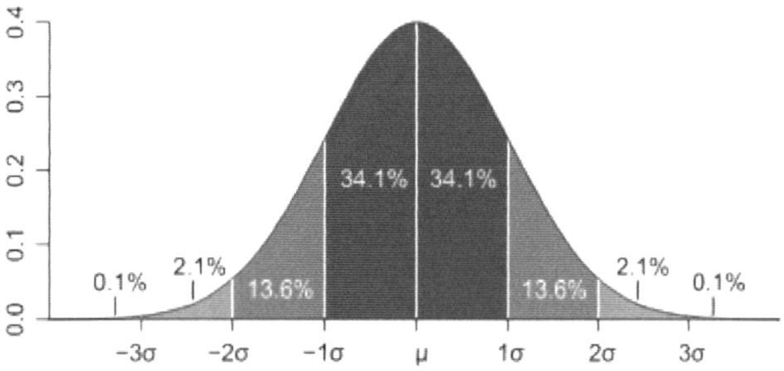

Figure 4. Normal distribution diagram: 1σ→2×34.15=68.3%; 2σ→2×13.65=27.3%; 3σ→2×2.1=4.2%.

3. Virtual prototyping demonstration by use case

The herein presented methodology is illustrated through a demonstration of a case case of an automotive component – dashboard (cluster assembly). It is a relatively complex assembly and it is shown in figure 5, together with materials specification for each of modelled part.

A meshed structure is built, based on the above mentioned virtual prototype. This numerical model contains 833 000 nodes and 433 000 elements and is shown in figure 6 to present the density of the mesh. All contacts between both modelled parts are presented as "Bonded" type.

3.1. Material properties

Six different materials are used in the examined model and their parameters are based on data defined by material suppliers and additional internet available resources. All major parameters used in subsequent analyses are shown in table 2 and in figure 7.

Table 2. Material properties for components of the numerical model (material designations according to figure 5).

Parameter	ABS	PC	PMMA	PP T30	PP T40	FR4
Elasticity modulus, E, MPa	defined as function of temperature – see figure 7					
Poisson's ratio, μ[1]	0.35	0.37	0.38	0.36	0.36	0.2
Density, ρ, kg/m^3	1050	1200	1190	1140	1250	2800
Coef. of therm. exp., α, 10^{-6}m/m×K	95	65	80	68	55	11

[1] - Used data source – common value for plastics

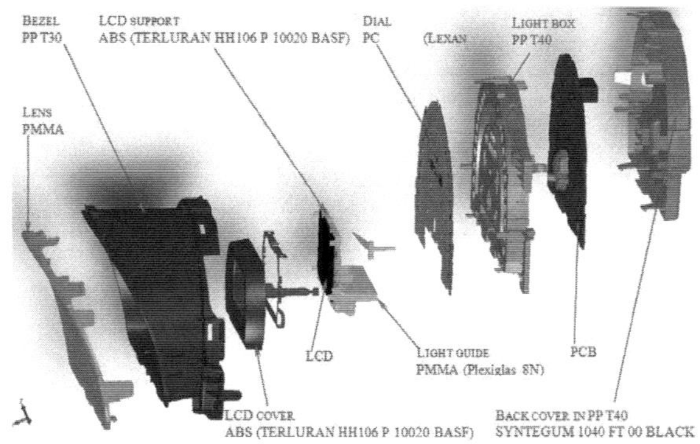

Figure 5. Virtual prototype of a cluster assembly, used for demonstration.

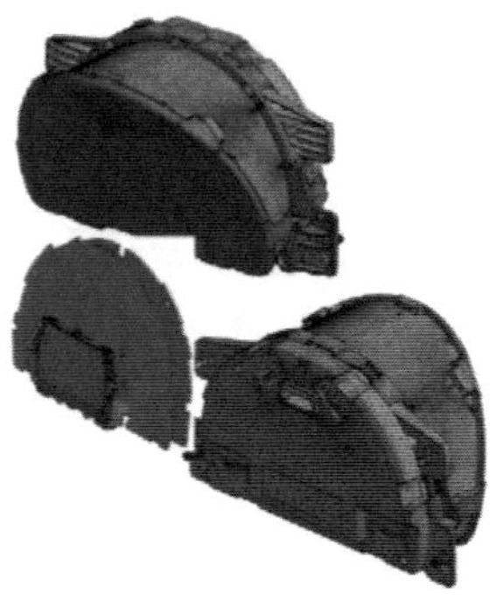

Figure 6. Meshed model of examined cluster assembly.

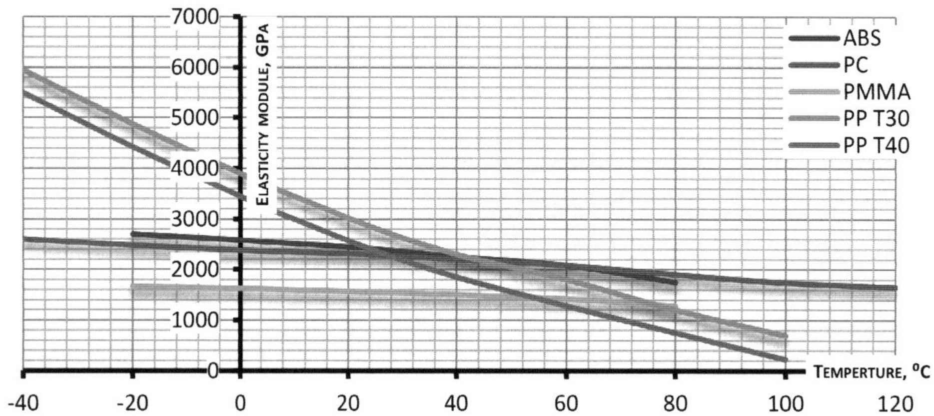

Figure 7. Elasticity module values versus temperature – for all used materials.

3.2. Boundary conditions

The applied boundary conditions – as base excitations – application surfaces (supports) and PSD used – are shown in figure 8. They represent mounting conditions of the examined cluster assembly. Three different temperatures: -40°C, 20°C and 85°C – are used for static analyses and subsequent modal analyses. The steps for each analysis include initial steady state analysis, analysis under thermal loads, and subsequent prestressed modal analysis that uses static analysis results. Next, random vibration analysis is performed.

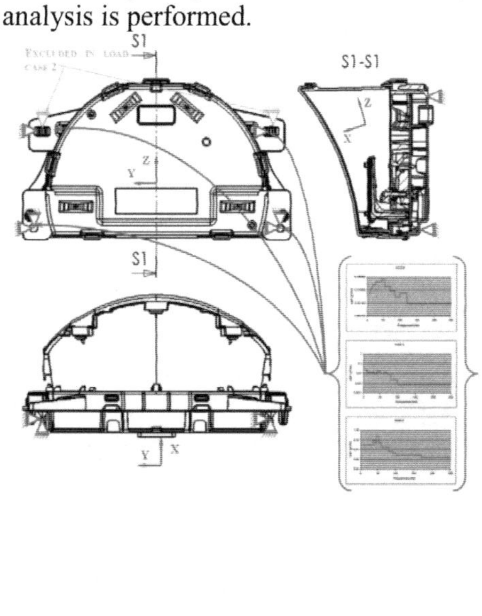

Figure 8. Applied boundary conditions.

3.3. S0 and S1: Modal analyses results

The first three natural frequencies for each load case (-40°C, 20°C and 85°C) are compared graphically in figure 9.

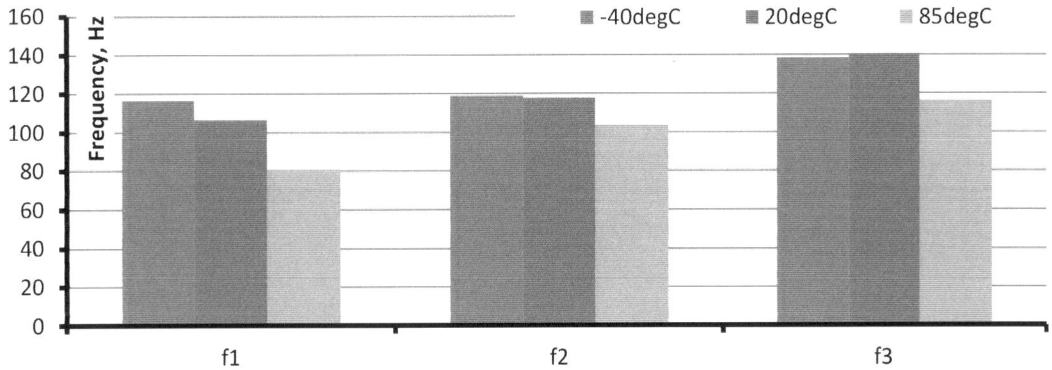

Figure 9. S1: Modal analysis results.

3.4. S2: Random Vibration (PSD) analyses – S2.1: Extracted values for equivalent stress values at standard deviations 1σ, 2 σ, 3 σ

The results from step S2, at three different temperatures (3 separate analyses), and three directions of applied base excitation – are shown by samples of equivalent (von Mises) distribution fields for 1σ standard deviation in figures 10a, 10b and 10c below.

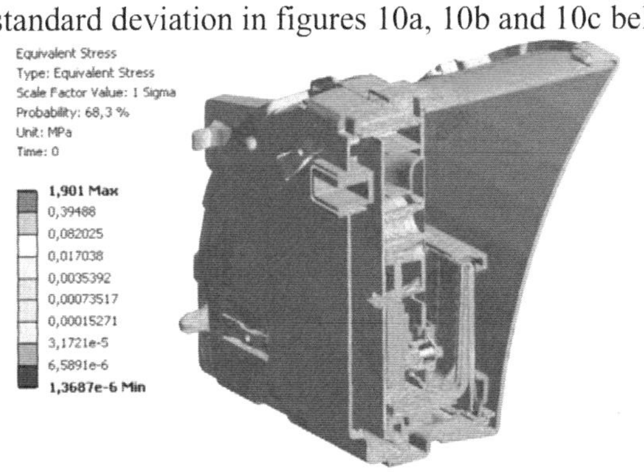

a) excitation along X axis, @85°C

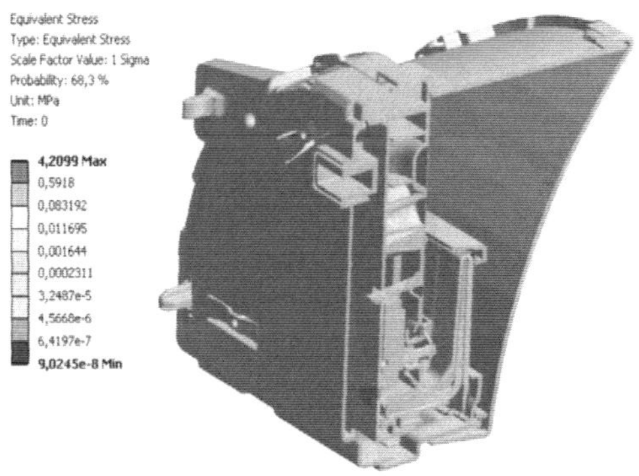

b) excitation along Y axis, @20°C

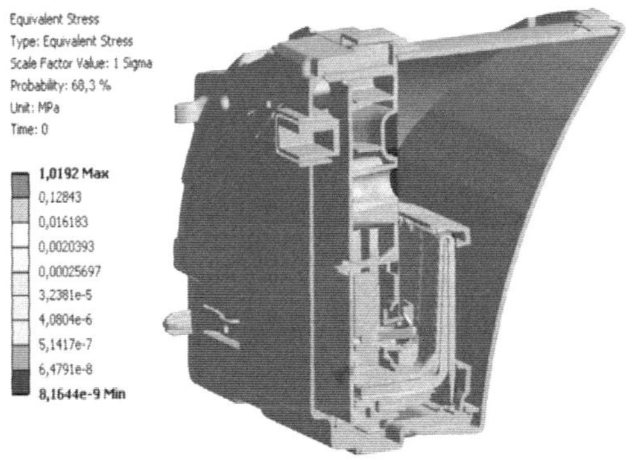

c) excitation along Z axis, @-40°C

Figure 10. Equivalent (von Mises) distribution fields at different excitations and temperatures, MPa.

All needed results, for each temperature range and each excitation direction, are summarized in table 3 below by percentage of occurrence (1s, 2s, 3s) extracted from the performed analyses – for each component.

Table 3. Random vibration analyses results – extracted values for equivalent stress values at standard deviations 1σ, 2 σ, 3 σ.

Max eq stress, MPa		Components	Back cover	Light box	LCD support	LCD cover	Bezel	Lens	Dial	PCB
X	1σ (68.3%)	85°C	0.071	0.330	0.180	0.093	0.180	0.360	0.120	0.350
		20°C	0.050	0.500	1.340	0.770	0.350	0.920	0.160	0.290
		-40°C	0.002	0.280	0.290	0.130	0.060	0.008	0.420	0.040
	2σ (27.3%)	85°C	0.142	0.660	0.360	0.186	0.360	0.720	0.240	0.700
		20°C	0.100	1.000	2.680	1.540	0.700	1.840	0.320	0.580
		-40°C	0.004	0.560	0.580	0.260	0.120	0.016	0.840	0.080
	3σ (4.2%)	85°C	0.213	0.990	0.540	0.279	0.540	1.080	0.360	1.050
		20°C	0.150	1.500	4.020	2.310	1.050	2.760	0.480	0.870
		-40°C	0.006	0.840	0.870	0.390	0.180	0.024	1.260	0.120
Y	1σ (68.3%)	85°C	0.071	0.330	0.180	0.093	0.180	0.360	0.120	0.350
		20°C	0.050	0.500	1.340	0.770	0.350	0.920	0.160	0.290
		-40°C	0.002	0.280	0.290	0.130	0.060	0.008	0.420	0.040
	2σ (27.3%)	85°C	0.142	0.660	0.360	0.186	0.360	0.720	0.240	0.700
		20°C	0.100	1.000	2.680	1.540	0.700	1.840	0.320	0.580
		-40°C	0.004	0.560	0.580	0.260	0.120	0.016	0.840	0.080
	3σ (4.2%)	85°C	0.213	0.990	0.540	0.279	0.540	1.080	0.360	1.050
		20°C	0.150	1.500	4.020	2.310	1.050	2.760	0.480	0.870
		-40°C	0.006	0.840	0.870	0.390	0.180	0.024	1.260	0.120
Z	1σ (68.3%)	85°C	0.071	0.330	0.180	0.093	0.180	0.360	0.120	0.350
		20°C	0.050	0.500	1.340	0.770	0.350	0.920	0.160	0.290
		-40°C	0.002	0.280	0.290	0.130	0.060	0.008	0.420	0.040
	2σ (27.3%)	85°C	0.142	0.660	0.360	0.186	0.360	0.720	0.240	0.700
		20°C	0.100	1.000	2.680	1.540	0.700	1.840	0.320	0.580
		-40°C	0.004	0.560	0.580	0.260	0.120	0.016	0.840	0.080
	3σ (4.2%)	85°C	0.213	0.990	0.540	0.279	0.540	1.080	0.360	1.050
		20°C	0.150	1.500	4.020	2.310	1.050	2.760	0.480	0.870
		-40°C	0.006	0.840	0.870	0.390	0.180	0.024	1.260	0.120

3.5. Random Vibration (PSD) analyses – S2.2: Determination of Response Power Spectral Density (RPSD) for zone with highest equivalent stress

The results from the analyses are used also to determine the dominant frequencies by direction and are listed below:

- along X axis: $f_{X(85^o)} = 88$Hz, $f_{X(20^o)} = 108$Hz and $f_{X(-40^o)} = 115$Hz.
- along Y axis: $f_{Y(85^o)} = 134$Hz, $f_{Y(20^o)} = 170$Hz and $f_{Y(-40^o)} = 119$Hz.
- along Z axis: $f_{Z(85^o)} = 130$Hz, $f_{Z(20^o)} = 174$Hz and $f_{Z(-40^o)} = 120$Hz.

3.6. S3: Fatigue Analysis

Next, fatigue calculations are performed, using the formula shown in the previous chapter. The results are shown in table 4 below – for all examined directions of support excitation.

The maximal time used for testing of each component is less than 0.01% of its resource. There is no component with reliability parameter for life prediction less than 1×106h. Nevertheless, the component results are compared in figure 11 below by reciprocal value of their used time (R_Σ) during tests. Thus, it is seen that most loaded components (with less R that corresponds to less remaining life) are LCD support, LCD cover and the dial.

Nevertheless, there are no critical components and total life prediction for the examined component – cluster assembly – fully corresponds to the car manufacturer requirements. This result could be used for design approval and reason for proceeding further with the required physical prototyping.

Table 4. Fatigue analysis calculations.

	Parameters / Components		Back cover	Light box	LCD support	LCD cover	Bezel	Lens	Dial	PCB
X	$N_{1(1\sigma)}$ (68.3%)	85°C	1.8E+90	5.1E+73	1.3E+20	1.6E+22	3.0E+83	5.8E+39	7.6E+21	2.8E+24
		20°C	4.0E+95	1.0E+67	5.7E+13	3.2E+15	5.7E+72	8.6E+32	9.9E+20	1.5E+25
		-40°C	4.0E+144	2.3E+76	4.0E+18	1.4E+21	1.5E+101	3.0E+67	1.1E+18	9.5E+32
	$N_{1(2\sigma)}$ (27.3%)	85°C	5.1E+79	3.4E+62	8.2E+17	1.0E+20	2.0E+72	5.2E+34	5.6E+19	5.2E+21
		20°C	1.1E+85	6.8E+55	3.6E+11	2.1E+13	3.8E+61	7.8E+27	7.3E+18	2.9E+22
		-40°C	1.1E+134	1.5E+65	2.5E+16	8.8E+18	1.0E+90	2.7E+62	7.9E+15	1.8E+30
	$N_{1(3\sigma)}$ (4.2%)	85°C	3.5E+73	9.8E+55	4.3E+16	5.3E+18	5.8E+65	5.9E+31	3.2E+18	1.3E+20
		20°C	7.5E+78	2.0E+49	1.9E+10	1.1E+12	1.1E+55	8.7E+24	4.2E+17	7.2E+20
		-40°C	7.5E+127	4.4E+58	1.3E+15	4.6E+17	3.0E+83	3.0E+59	4.5E+14	4.5E+28
	$n_{1(85°C)}$	1σ	3.1E+06	3.1E+06	3.1E+06	3.1E+06	3.1E+06	3.1E+06	3.1E+06	3.1E+06
		2σ	1.3E+06	1.3E+06	1.3E+06	1.3E+06	1.3E+06	1.3E+06	1.3E+06	1.3E+06
		3σ	1.9E+05	1.9E+05	1.9E+05	1.9E+05	1.9E+05	1.9E+05	1.9E+05	1.9E+05
	$n_{1(20°C)}$	1σ	5.5E+05	5.5E+05	5.5E+05	5.5E+05	5.5E+05	5.5E+05	5.5E+05	5.5E+05
		2σ	2.2E+05	2.2E+05	2.2E+05	2.2E+05	2.2E+05	2.2E+05	2.2E+05	2.2E+05
		3σ	3.4E+04	3.4E+04	3.4E+04	3.4E+04	3.4E+04	3.4E+04	3.4E+04	3.4E+04
	$n_{1(-40°C)}$	1σ	2.1E+06	2.1E+06	2.1E+06	2.1E+06	2.1E+06	2.1E+06	2.1E+06	2.1E+06
		2σ	8.4E+05	8.4E+05	8.4E+05	8.4E+05	8.4E+05	8.4E+05	8.4E+05	8.4E+05
		3σ	1.3E+05	1.3E+05	1.3E+05	1.3E+05	1.3E+05	1.3E+05	1.3E+05	1.3E+05
	$R_{X.(85°C)}$		5.6E-69	2.0E-51	6.1E-12	4.9E-14	3.3E-61	3.3E-27	8.3E-14	1.7E-15
	$R_{X.(20°C)}$		4.5E-75	1.7E-45	2.4E-06	4.3E-08	3.1E-51	4.0E-21	1.1E-13	5.5E-17
	$R_{X.(-40°C)}$		1.7E-123	3.0E-54	1.3E-10	3.8E-13	4.3E-79	4.4E-55	4.0E-10	3.3E-24
	$R_{X.\Sigma}$		**5.6E-69**	**1.7E-45**	**2.4E-06**	**4.3E-08**	**3.1E-51**	**4.0E-21**	**4.0E-10**	**1.8E-15**
Y	$N_{1(1\sigma)}$ (68.3%)	85°C	4.0E+100	2.6E+86	4.6E+21	1.1E+24	3.0E+98	4.6E+45	2.5E+24	1.1E+28
		20°C	2.1E+110	9.8E+80	4.9E+16	2.4E+18	2.6E+91	4.2E+40	1.0E+24	4.5E+28
		-40°C	2.4E+152	5.3E+88	4.2E+19	1.9E+23	8.1E+112	3.3E+72	1.8E+20	5.1E+35
	$N_{1(2\sigma)}$ (27.3%)	85°C	1.1E+90	1.7E+75	3.0E+19	7.0E+21	2.0E+87	4.2E+40	1.8E+22	2.1E+25
		20°C	6.0E+99	6.5E+69	3.1E+14	1.6E+16	1.7E+80	3.8E+35	7.6E+21	8.5E+25
		-40°C	6.6E+141	3.5E+77	2.7E+17	1.2E+21	5.4E+101	3.0E+67	1.4E+18	9.5E+32
	$N_{1(3\sigma)}$ (4.2%)	85°C	7.5E+83	5.1E+68	1.5E+18	3.6E+20	5.8E+80	4.7E+37	1.0E+21	5.3E+23
		20°C	4.0E+93	1.9E+63	1.6E+13	8.1E+14	5.1E+73	4.2E+32	4.3E+20	2.1E+24
		-40°C	4.5E+135	1.0E+71	1.4E+16	6.4E+19	1.6E+95	3.3E+64	7.7E+16	2.4E+31
	$n_{1(85°C)}$	1σ	4.6E+06	4.6E+06	4.6E+06	4.6E+06	4.6E+06	4.6E+06	4.6E+06	4.6E+06
		2σ	1.8E+06	1.8E+06	1.8E+06	1.8E+06	1.8E+06	1.8E+06	1.8E+06	1.8E+06
		3σ	2.8E+05	2.8E+05	2.8E+05	2.8E+05	2.8E+05	2.8E+05	2.8E+05	2.8E+05
	$n_{1(20°C)}$	1σ	8.7E+05	8.7E+05	8.7E+05	8.7E+05	8.7E+05	8.7E+05	8.7E+05	8.7E+05
		2σ	3.5E+05	3.5E+05	3.5E+05	3.5E+05	3.5E+05	3.5E+05	3.5E+05	3.5E+05
		3σ	5.4E+04	5.4E+04	5.4E+04	5.4E+04	5.4E+04	5.4E+04	5.4E+04	5.4E+04
	$n_{1(-40°C)}$	1σ	2.2E+06	2.2E+06	2.2E+06	2.2E+06	2.2E+06	2.2E+06	2.2E+06	2.2E+06
		2σ	8.7E+05	8.7E+05	8.7E+05	8.7E+05	8.7E+05	8.7E+05	8.7E+05	8.7E+05
		3σ	1.3E+05	1.3E+05	1.3E+05	1.3E+05	1.3E+05	1.3E+05	1.3E+05	1.3E+05
	$R_{Y.(85°C)}$		3.8E-79	5.6E-64	2.5E-13	1.0E-15	4.9E-76	6.1E-33	3.8E-16	6.2E-19
	$R_{Y.(20°C)}$		1.3E-89	2.8E-59	4.4E-09	8.9E-11	1.1E-69	1.3E-28	1.7E-16	2.9E-20
	$R_{Y.(-40°C)}$		3.0E-131	1.3E-66	1.3E-11	2.8E-15	8.5E-91	4.1E-60	2.4E-12	6.5E-27
	$R_{Y.\Sigma}$		**3.8E-79**	**2.8E-59**	**4.5E-09**	**8.9E-11**	**1.1E-69**	**1.3E-28**	**2.4E-12**	**6.5E-19**
Z	$N_{1(1\sigma)}$ (68.3%)	85°C	6.8E+88	1.6E+74	4.8E+20	5.7E+22	2.4E+85	2.5E+40	1.4E+22	1.1E+25
		20°C	3.5E+109	4.0E+82	8.7E+16	4.0E+18	2.6E+91	2.2E+41	1.0E+24	2.4E+29
		-40°C	1.4E+155	4.0E+82	4.2E+19	6.9E+22	6.7E+110	7.8E+70	1.5E+19	5.1E+35
	$N_{1(2\sigma)}$ (27.3%)	85°C	1.9E+78	1.1E+63	3.1E+18	3.6E+20	1.6E+74	2.3E+35	1.0E+20	2.1E+22
		20°C	9.9E+98	2.7E+71	5.5E+14	2.5E+16	1.7E+80	2.0E+36	7.6E+21	4.4E+26
		-40°C	4.0E+144	2.7E+71	2.7E+17	4.4E+20	4.5E+99	7.1E+65	1.1E+17	9.5E+32
	$N_{1(3\sigma)}$ (4.2%)	85°C	1.3E+72	3.1E+56	1.6E+17	1.9E+19	4.6E+67	2.5E+32	5.9E+18	5.3E+20
		20°C	6.7E+92	7.8E+64	2.9E+13	1.3E+15	5.1E+73	2.2E+33	4.3E+20	1.1E+25
		-40°C	2.7E+138	7.8E+64	1.4E+16	2.3E+19	1.3E+93	7.9E+62	6.2E+15	2.4E+31
	$n_{1(85°C)}$	1σ	4.6E+06	4.6E+06	4.6E+06	4.6E+06	4.6E+06	4.6E+06	4.6E+06	4.6E+06
		2σ	1.8E+06	1.8E+06	1.8E+06	1.8E+06	1.8E+06	1.8E+06	1.8E+06	1.8E+06
		3σ	2.8E+05	2.8E+05	2.8E+05	2.8E+05	2.8E+05	2.8E+05	2.8E+05	2.8E+05
	$n_{1(20°C)}$	1σ	8.9E+05	8.9E+05	8.9E+05	8.9E+05	8.9E+05	8.9E+05	8.9E+05	8.9E+05
		2σ	3.6E+05	3.6E+05	3.6E+05	3.6E+05	3.6E+05	3.6E+05	3.6E+05	3.6E+05
		3σ	5.5E+04	5.5E+04	5.5E+04	5.5E+04	5.5E+04	5.5E+04	5.5E+04	5.5E+04
	$n_{1(-40°C)}$	1σ	2.2E+06	2.2E+06	2.2E+06	2.2E+06	2.2E+06	2.2E+06	2.2E+06	2.2E+06
		2σ	8.8E+05	8.8E+05	8.8E+05	8.8E+05	8.8E+05	8.8E+05	8.8E+05	8.8E+05
		3σ	1.3E+05	1.3E+05	1.3E+05	1.3E+05	1.3E+05	1.3E+05	1.3E+05	1.3E+05
	$R_{Z.(85°C)}$		2.2E-67	9.2E-52	2.4E-12	2.0E-14	6.2E-63	1.1E-27	6.6E-14	6.2E-16
	$R_{Z.(20°C)}$		8.2E-89	7.0E-61	2.6E-09	5.6E-11	1.1E-69	2.5E-29	1.7E-16	5.7E-21
	$R_{Z.(-40°C)}$		5.1E-134	1.7E-60	1.3E-11	7.9E-15	1.0E-88	1.7E-58	3.0E-11	6.5E-27
	$R_{Z.\Sigma}$		**2.2E-67**	**9.2E-52**	**2.6E-09**	**5.6E-11**	**6.2E-63**	**1.2E-27**	**3.0E-11**	**6.2E-16**
$R_\Sigma = R_{X.\Sigma} + R_{Y.\Sigma} + R_{Z.\Sigma}$			**2.28E-67**	**1.73E-45**	**2.44E-06**	**4.31E-08**	**3.08E-51**	**3.96E-21**	**4.29E-10**	**2.38E-15**
Remaining resource, T, h			**3.2E+70**	**4.2E+48**	**2.9E+09**	**1.7E+11**	**2.3E+54**	**1.8E+24**	**1.7E+13**	**3.0E+18**

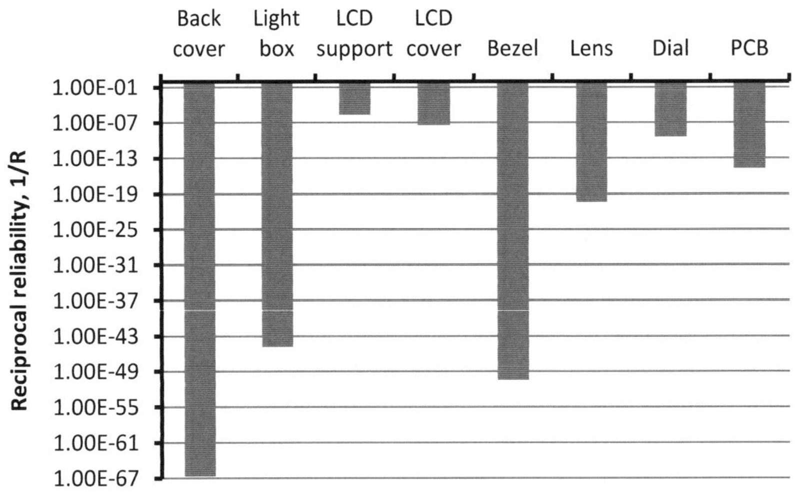

Figure 11. Comparison between components in terms of their reciprocal reliability values.

4. Conclusions

The presented methodology facilitates random vibration endurance testing using virtual prototyping and moreover – it uses conventional software tools for numerical modeling. The described methodology requires software for modal and random vibration analyses that is available in the most widely used FEA programs. It is also based on the widely used in automotive industry hypothesis of Palmgren-Miner.

This is demonstrated by a case study from the automotive industry – a cluster assembly (dashboard). Modelled assembly components are assessed as rigidity and general dynamic behavior prior to performing physical prototyping that saves time and expenses. Another important drawback is the option to investigate in detail the stress distributions inside the examined parts and – if it is needed – easily change design and test new modifications.

Generally, the study presents a methodology for a random vibration endurance test of automotive components using virtual prototyping based on common engineering analysis tools.

Acknowledgments

This work was supported by the European Regional Development Fund within the Operational Programme "Science and Education for Smart Growth 2014 - 2020" under Project CoE "National center of mechatronics and clean technologies" BG05M2OP001-1.001-0008.

References

[1] Zhao L-H, Cai H-C, Wang T and Zheng S-L 2018 Durability assessment of automotive structures under random variable amplitude loading *Advances in Mech. Eng.* **10**(4) pp 1–8

[2] Bonato M and Delaux D 2015 Synthesis and validation of accelerated vibration durability tests *Annual Realiability and Maintainability Symposium – RAMS 2015* pp 7–12

[3] Wang L, Burger R and Aloe A 2017 Considerations of vibration fatigue for automotive components *SAE Int. J. Commer. Veh.* **10**(1) pp 150–8

[4] Kulkarni A, Kapoor A, Iyer M and Kosse V 2011 Virtual prototyping used as validation tool in automotive design *19th Int. Congress on Modelling and Simulation (Perth, Australia)*

[5] Nedelchev K and Kralov I 2019 Acoustic method for identification of railway wheel disc structural vibrations using COMSOL *J. of the Balkan Tribological Assoc.* **25**(3) pp 546–57

[6] Todorov T, Nikolov N, Nikolov R, Sofronov Y and Kochev L 2019 Optimization of micro-electro-mechanical piezoelectric energy harvesters with interdigitated electrodes *J. of the Balkan Tribological Assoc.* **25**(3) pp 535–45
[7] Schoenborna S, Kaufmanna H, Sonsinoa C M and Heima R 2015 Cumulative damage of high-strength cast iron alloys for automotive applications *Procedia Engineering* **101** pp 440–9
[8] Gao H, Huang H-Zh, Zhu Sh-P, Li Y-F and Yuan R 2014 A modified nonlinear damage accumulation model for fatigue life prediction considering load interaction effects *The Scientific World Journal*
[9] Jivkov V, Zahariev E and Nikolov N 2018 Rotating cantilever beam with variable geometry – frequencies and modal vectors *Comptes Rendus de L'Academie Bulgare des Sciences* **71**(8) pp 1108–15

IOP Conf. Series: Materials Science and Engineering 1002 (2020) 012028 doi:10.1088/1757-899X/1002/1/012028

Determination the yaw moment of inertia of an SUV in laboratory conditions

Dmytro M Yashchenko[1,3], Oleksii M Timkov[1], Ihor O Samburenko[1] and Volodymyr S Hladchenko[2]

[1] Department of Motor Vehicles, National Transport University, 1 Omelianovycha-Pavlenka str., 01010 Kyiv, Ukraine
[2] Department of Scientific and Technical Expertise, State Road Transport Research Institute, 57 Peremohy Ave., 03113 Kyiv, Ukraine

[3] Email: y_d@ukr.net

Abstract. The vehicle oscillates in different planes relative to the three axes while driving. Especially important are the turns of the vehicle by dint of steered wheels. During such high-speed turns, the vehicle is subjected to additional charges of various systems, including the control system. Therefore, in the calculations of the control system of the vehicle, it is necessary to take into account the inertial mass of the vehicle. In general, the inertial characteristics of some solid body during its rotation are determined by the moments of inertia of this body. The moment of inertia of a solid body concerning the axis of rotation is simply the mass times the square of the perpendicular distance to the rotation axis. To test the results obtained by means of the analytical method, it is necessary to determine the moment of inertia of a vehicle in an experimental way. A convenient method of measuring the moment of inertia of a vehicle can be the method of torsional vibrations. The moment of inertia of a vehicle was determined using special equipment, which was constructed in the laboratory of the Department of Motor Vehicles of the National Transport University.

1. Introduction
The parameters of the motion vehicle are closely related to its construction. The unsteady turning of the vehicle is the most frequent mode of movement. During such turns, especially at high speeds, the vehicle is subjected to additional loads on many of its systems, including control systems [1-3]. Therefore, when calculating the vehicle control system it is necessary to take into account the inertial mass of the vehicle [4, 5]. It is difficult to find data on the moments of inertia of the vehicles and their systems in the literature. Researchers often determine them experimentally [6-9].

The theoretical basis for creating such an installation can be the method of torsional vibrations on a steel rod. It is necessary to create a frame for placing the vehicle and then hang it with the vehicle on a solid homogeneous metal rod, for example, to the crane beam. The fastening of a metal core both to the crane beam and to the frame has to be rigid.

2. Mathematical model
The theory of motion stability of wheeled vehicles is based on the mathematical apparatus of investigating the differential equations, developed by A. M. Liapunov [10]. He determined the properties of the perturbed system state, consisting in a tendency for the renewal of the trajectories of

unperturbed motion that took place before the appearance of perturbation. The time of the return of motion parameters to the initial ones can be accepted as the quantitative estimation of this property. With an oscillation process of the return of these parameters to the initial ones, it is possible to quantitatively estimate the stability according to the damping decrement, i.e. by the degree of reduction of vibration amplitude [11].

The mathematical model of the plane-parallel motion of a vehicle taking into account of four separate modules. Let us consider a vehicle as a system that consists the internal combustion engine, the body with a centre of mass at point C, the control wheel module with a mass centre at point A and the rear axles with a centre of mass at point B (figure 1). A vehicle has steerable wheels with rotation angles θ and θ', where $\theta \geq \theta'$, and movable details of a steering gear and a steering linkage.

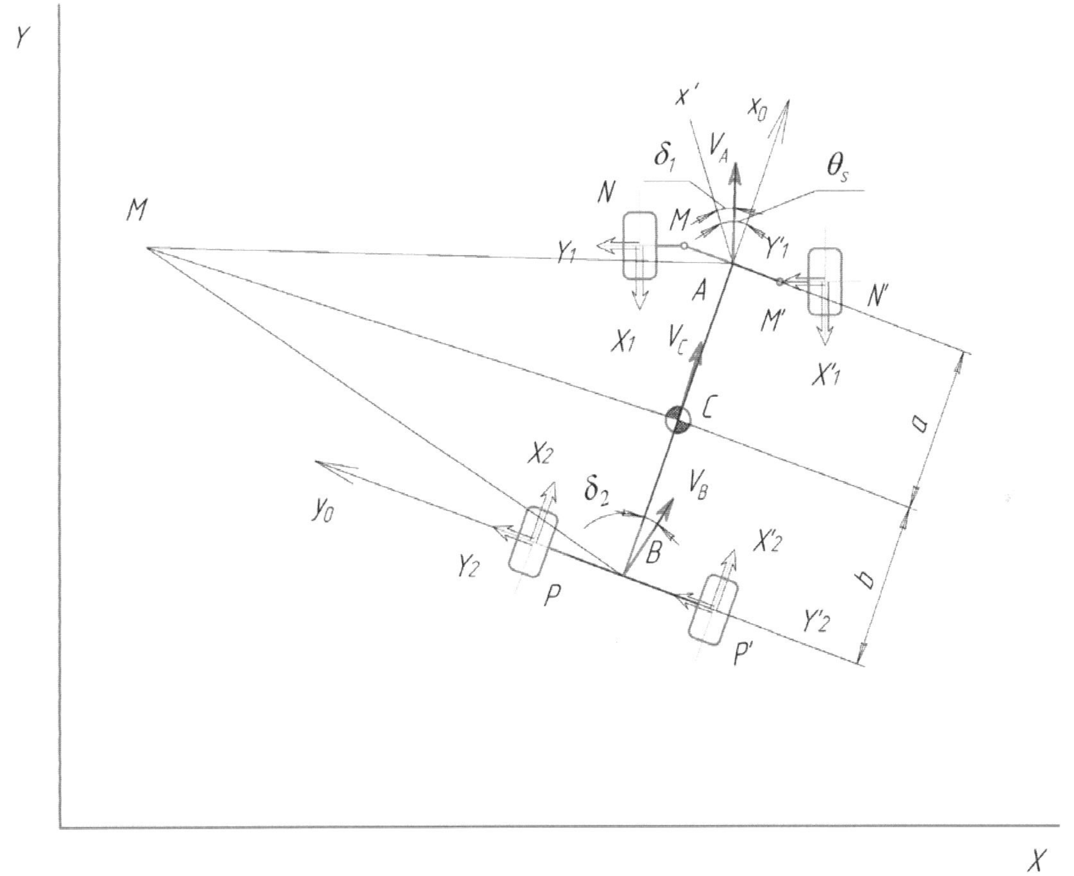

Figure 1. Design diagram of a vehicle at the unsteady turn.

The following symbols are used in the diagram and further in the text:

XOY – the two-dimensional inertial system of vehicle coordinates;

X_0BY_0 – the moving coordinates related to the vehicle;

m_i, I_i – accordingly, mass and central moment of inertia of the separate modules of the vehicle, where $i = 1 \ldots 4$ – the indexes of the related module;

v, u – longitudinal and lateral projections of the speed of point C on the axis of the moving coordinates steadily related to the modules of a vehicle;

ω_i – angular velocity of a link;

$X_1, X_1^{/}, Y_1, Y_1^{/}, X_2, X_2^{/}, Y_2, Y_2^{/}$ – longitudinal and lateral reactions of the road to the two axes;
Geometrics of a vehicle:

$$a = AC, b = CB, L = a + b, \quad h = MA, \quad c = MN .$$

When considering the stability of the rectilinear movement of the vehicle in the real constructions of automobiles the difference of angles of external and internal turns in relation to the centre of the turn of the wheel has guided the fractions of a degree on the average. Under such conditions, it is possible to conduct calculations of the middle angle of turn of the wheel $\theta_s = \left(\theta + \theta^{/}\right)2^{-1}$ with high degree of accuracy and ignore the redistribution of lateral forces [11].

After determining the longitudinal and lateral speeds and accelerations of separate links of a vehicle, the angles of a skid and lateral forces, moments of stabilizing tires, normal reactions of the bearing area on the wheels of all axes, the differential equations of the motion of a vehicle are written as follows:

$$\begin{aligned}&(m_1 + m_2 + m_3 + m_4)\dot{v} + m_1 c\dot{\omega}\sin\theta + m_1 c\ddot{\theta}\sin\theta - (m_1 + m_2 + m_3 + m_4)\omega u - \\ &-m_1 a\omega^2 + m_1 c(\omega + \dot{\theta})^2 \cos\theta = -X_1\cos\theta - Y_1\sin\theta + X_2;\end{aligned} \tag{1}$$

$$\begin{aligned}&(m_1 + m_2 + m_3 + m_4)\dot{u} + m_1(a - c\cos\theta)\dot{\omega} - m_1 c\ddot{\theta}\cos\theta + (m_1 + m_2 + m_3 + m_4)\omega v + \\ &+m_1 c(\omega + \dot{\theta})^2 \sin\theta = -X_1\sin\theta + Y_1\cos\theta + Y_2;\end{aligned} \tag{2}$$

$$\begin{aligned}&\left[I_a + m_1 a(a - c\cos\theta)\right]\dot{\omega} + m_1 a\dot{u} - m_1 ac\ddot{\theta}\cos\theta + m_1 av\omega + \\ &+m_1 ac(\omega + \dot{\theta})^2 \sin\theta = -X_1 a\sin\theta + Y_1 a\cos\theta - bY_2 - M^{//};\end{aligned} \tag{3}$$

$$\begin{aligned}&(I_a + m_1 c^2)\ddot{\theta} - [I_a + m_1 c(c - a\cos\theta)]\dot{\omega} - m_1 c\dot{u}\cos\theta + m_1 c\dot{v}\sin\theta - m_1 cv\omega\cos\theta - \\ &-m_1 c\omega u\sin\theta - m_1 ac\omega^2\sin\theta = -cY_1 - M^{/}.\end{aligned} \tag{4}$$

The acquired system of differential equations (3) describes the motion of the vehicle. The system allows for investigation of the impact of the structural parameters and operating modes of the motion of the vehicle.

Slide slip angle δ_i is the angle from the longitudinal axis of the vehicle body to the direction of the vehicle velocity. The values of the slide slip angle of wheel axles are defined by the following expressions:

$$\begin{aligned}&\delta_1 = \theta - arctg\frac{u + (a - c\cos\theta)\omega - c\dot{\theta}\cos\theta}{v + c(\omega + \dot{\theta})\sin\theta}; \\ &\delta_2 = arctg\frac{-u + b\omega}{v}.\end{aligned} \tag{5}$$

In [7], Bakker and Pacejka developed a famous lateral tire/road friction model – "magic formula". The "magic formula" assumes that the front tire force does not only depend on the front slip angle but also on the vehicle slide slip angle, steering angle and read slip angle; and so does the rear tire force.

3. Statement of the problem

In general, the inertial characteristics of any solid body during its rotation are determined by the moments of inertia of this body. The moment of inertia of vehicle I_a of a solid body relative to the axis of rotation is a physical value equal to the sum of the product of the masses of n material points on which the body is divided, on the squares of their distances to the axis, in particular in our case, to the vertical axis [12]:

$$I_a = \sum_{i=1}^{n} m_i r_i^2 = m_1 r_1^2 + m_2 r_2^2 + m_3 r_3^2 + m_4 r_4^2, \qquad (6)$$

where $m_{1,2,3,4}$ – respectively the weight of the ICE, body, front and rear axles [13]; $r_{1,2,3,4}$ – the speed and direction of rotation of the corresponding masses around the vertical axis.

From the above formula it is seen that the moment of inertia of the vehicle depends on the location of its main units relative to the vertical axis, which passes through the centre of mass [14] of the whole vehicle and is an additive value.

First, we will determine the moment of inertia of a Lada "Niva" vehicle through the graph-analytical method. To do this, we will split the vehicle into parts (figure 2) and use the formula (6) to perform calculations taking into account the mass and spatial parameters of its aggregates.

Figure 2. To determine the moment of inertia by means of the graph-analytical method.

Substituting the values in formula (6), we will get the theoretical maximum value of the moment of inertia of the Lada "Niva" vehicle:

$$I_a = 120 \cdot 0.3^2 + 240 \cdot 2.1^2 + 47 \cdot 0.7^2 + 61 \cdot 0.7^2 = 615.5 \text{ kgm}^2$$

To verify the results obtained through an analytical method, it is necessary to determine the moment of inertia of the vehicle experimentally. A convenient method of measuring the moment of inertia of the vehicle can be the method of torsional oscillations [14], namely oscillations of the vehicle on a steel rod. When the frame with the vehicle is rotated at a certain angle φ, the rod will also rotate at an angle φ, because it is rigidly connected to the vehicle frame. If you now release the frame, it will perform torsional oscillations relative to the vertical axis. Experience shows that the moment of elastic forces, which lead to torsional oscillations [12], is proportional to the angle of twist of the vehicle:

$$M_z = -C\varphi, \qquad (7)$$

where C is a constant value for a given rod, which is called its torsion modulus.

According to the basic equation of rotational motion dynamics, the moment of forces relative to the Z-axis is written as

$$I_z \frac{d^2\varphi}{dt^2} = M_z \quad \text{or} \qquad (8)$$

$$I_z \varphi'' = M_z, \qquad (9)$$

where I_z is the moment of inertia of the vehicle relative to the vertical Z – axis. Given the equation (7), we will get the differential equation of harmonic vibrations of a vehicle suspended on a rod:

$$I_z\varphi'' + C\varphi = 0. \tag{10}$$

It is not difficult to make sure by direct substitution that the solution of this equation has the form:

$$\varphi = \varphi_o \sin\left(\left(\frac{C}{I_z}\right)^{1/2} t + \alpha\right), \tag{11}$$

where the amplitude φ and phase α are determined by the initial oscillation conditions. Therefore, the period of oscillations of the "vehicle-rod" system, is equal to:

$$T = 2\pi\left(\frac{I_z}{C}\right)^{1/2}. \tag{12}$$

From the last equation you can find the torsion module. Having found I_Z and T experimentally, it is possible to find the torsion modulus C by formula (8).

When obtaining equation (7), we did not take into account the non-elastic deformations of the rod and friction forces, which are insignificant for this system and the energy losses for one period of vehicle oscillations are small compared to the energy of the system oscillations.

4. Description of the experimental stand and performance of the experiment

Determination of the moment of inertia of the vehicle was carried out using special equipment, which was manufactured in the laboratory of the Department Automobile of the National Transport University. In general, the experimental construction consists of special frames on which the vehicle is installed, after that the entire frame structure is lifted by a crane beam and suspended by a special rod. A general view of the experimental construction is shown in figure 3.

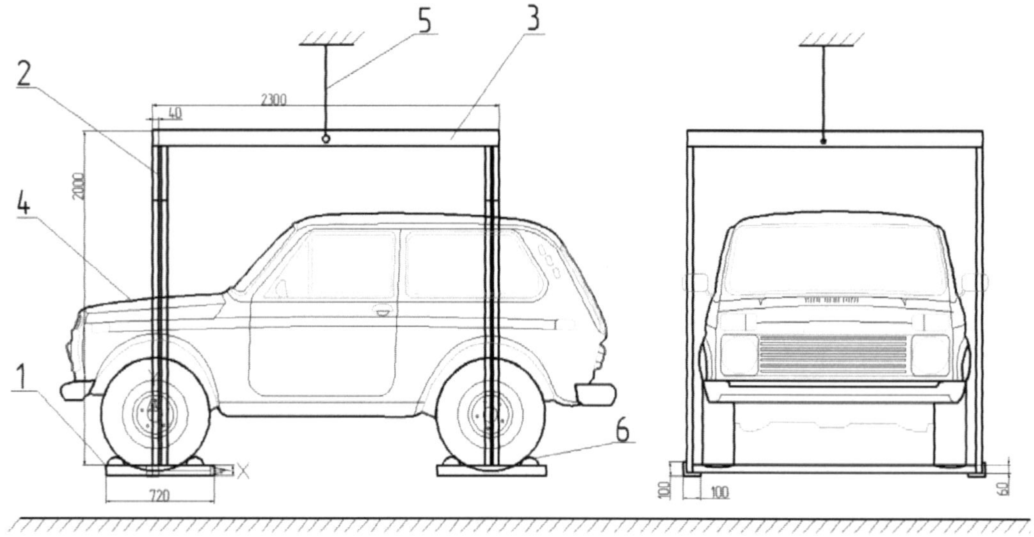

Figure 3. Constructive scheme of installation of the Lada "Niva" vehicle in the frame.

The scenario of the experiment consists of 4 stands *(1)*, two frames (*2*) are fixed, which are connected by two crossbars (*3*). Thus, the vehicle (*4*) together with the whole frame structure with the help of a two-post lift rises to a height of 0.1...0.2 m and joins the rod (*5*), which will perform the oscillating movements of the vehicle around the vertical Z-axis.

The experimental construction of the frame allows for fixing of the experimental vehicles with the help of thrust bearings (*6*), which move along the guides on the stands (*1*). The vehicle is installed so that the corresponding recesses on the surface of the stands come into contact with the wheels. After that, the guides are fixed by tightening the nuts.

After installing the Lada "Niva" vehicle in the frame and fixing it through the wheels, the entire frame structure together with the vehicle was lifted by a two-post electromechanical lift and then hung by a special rod to the crane beam. A device was pre-assembled on the crane-beam, which allowed for rigid fixing the rod. The special fastenings of the vertical support rod on the crane-beam and on the cross member of the frame did not permit it to rotate around its axis, so during torsional oscillations the whole frame structure together with the vehicle had to be twisted.

The installation of the Lada "Niva"vehicle in a frame was carried out by means of the standard electromechanical elevator (figure 4). The vehicle was fixed in a frame by means of special crossbeams in which all four wheels of the vehicle were set. In order for the vehicle to rise together with the frame, special rods were mounted on the frame, which rested on all the wheels of the vehicle from above and thus made it possible to fix the vehicle (figure 4). A special stand is located on the floor to place the measuring equipment (figure 5).

A photoelectric sensor (*1*) was installed on a special stand (*3*). The base of the sensor is a solar battery, which generated an electric current when light hit it. In turn, the oscillation indicator with laser (*2*) was fixed on the bumper of the vehicle (in its middle part). A scale (*4*) was installed on the stand (*3*), which allowed registering the deviations of the vehicle.

Also on the stand (*3*) was placed an usb-oscillograph (*5*), which recorded the oscillations of the vehicle and with the help of a special program transmitted data to a computer (*6*). On the monitor screen you could see the oscilloscope recording (figure 6) in real time, because there was registered time of recording of the oscillogram (lower part).

Figure 4. Experimental construction for determining the moment of inertia with the vehicle.

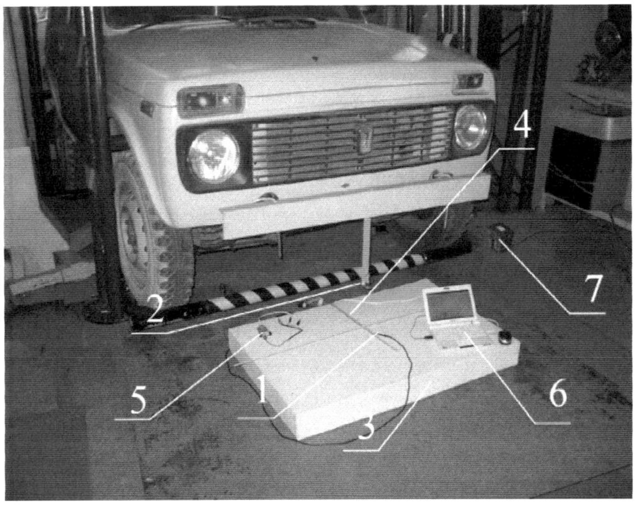

Figure 5. Devices and equipment for measuring the moment of inertia.

The experimental construction and equipment permitted the recording of the deviation of the frame before the start of each test using an electromagnet (*7*), which kept the frame in the rotated (initial) state. The operation of the electromagnet was controlled by the "Start-Stop" button.

The measurement process was as follows: before starting the measurements, the "Start-Stop" button must be free; when you press the "Start-Stop" button, the power supply of the electromagnet was turned off and the frame began to move, and the devices automatically began counting down the time and number of oscillations of the frame with the vehicle.

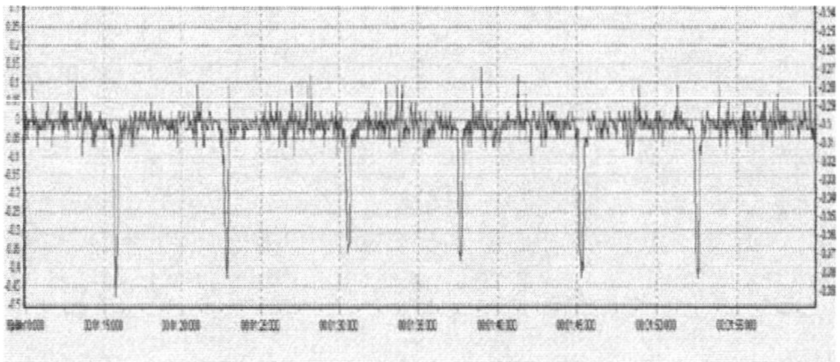

Figure 6. Oscillogram of the period of oscillation of the frame with the vehicle.

During the oscillation of the frame with the vehicle, the bracket with the laser (*2*) crosses the placement point of the photoelectric sensor (*1*), the solar battery of which generates an electric current when the light from the laser (*2*) hits it and electrical pulses are generated in the electric circuit. They are an input to the oscilloscope (*5*) and thence through special computer programs are broadcast to the computer monitor (*6*). Therefore the computer monitor is in the setup for measuring the oscillation frame and the duration of oscillations in automatic mode. By repeatedly pressing the "Start-Stop" button, you can terminate the account. Moreover, the number and duration of oscillations are fixed on the computer monitor (*6*) after the last oscillation is completed. So, if at the moment when the monitor on the oscillogram shows the 5th "peak" of oscillation (figure 4) and the "Start-Stop" button is pressed, the counting will stop after the 6th oscillation ends.

In order to facilitate the reproduction of the experimental research process, when processing the obtained data, the oscillation process of the frame with the vehicle was also recorded on a video camera in real time and could be reproduced at any time.

5. Method of determining the inertia moment of the SUV

Turning to the theoretical part of the experiment, we should note that since the moment of inertia (I_Z) and torsion modulus (C) in formula (12) are unknown to us, we will conduct the experiment in this way. First, we define the oscillation period T_1 with an empty frame, without a vehicle, and write down formula (7) for this case in this form:

$$T_1 = 2\pi\left(\frac{I_{1z}}{C}\right)^{1/2}, \tag{13}$$

where I_{1z} is the moment of inertia of the frame (without the vehicle) relative to the vertical Z-axis; C – a constant value for a given rod, which is called its torsion modulus.

Now we fix a test lever of cylindrical shape with a radius of 0.5 m and mass of 300 kg on the frame with a moment of inertia I_{2Z}, which we calculate using the well-known formula:

$$I_{2Z} = \frac{1}{2}mR^2 = \frac{1}{2}\cdot 300\cdot 0.5^2 = 37.5 \text{ kgm}^2, \tag{14}$$

where m is the mass of the test cylinder arm; R – the radius of the test cylinder arm.

After that we measure the oscillation period using the experimental construction described above T_2 frames with a test cylinder. It depends on the total moment of inertia of the frame and the test lever:

$$T_2 = 2\pi\left(\frac{I_{1z}+I_{2z}}{C}\right)^{1/2}. \tag{15}$$

According to the periods of oscillation of the empty frame determined by experiment T_1 and the frame with test cylinder T_2 and found by formula (14) I_{2Z} can be determined from the system of equations (13) and (14) we need for further calculations of I_{1Z} and C:

$$\begin{cases} T_1 = 2\pi\left(\frac{I_{1z}}{C}\right)^{1/2}, \\ T_2 = 2\pi\left(\frac{I_{1z}+I_{2z}}{C}\right)^{1/2}. \end{cases} \tag{16}$$

Once the moment of inertia of the empty frame (without the vehicle) I_{1Z} and the torsion module of C was found, it was not difficult to determine the moment of inertia of the vehicle inserted into the frame. To do this, the torsional oscillation period of the frame together with the vehicle was measured T_3:

$$T_3 = 2\pi\left(\frac{I_{1z}+I_{3z}}{C}\right)^{1/2}. \tag{17}$$

Solving the system of equations (11) and (12) together, we obtain the value of the moment of inertia of the vehicle, as well as any other unit or node placed in the experimental construction:

$$I_{3z} = \frac{T_3^2 - T_1^2}{T_2^2 - T_1^2} I_{2z}. \tag{18}$$

Therefore, by measuring the oscillation periods of the empty frame T_1, the frame with the test cylinder in the course of experimental studies T_2 and frames together with vehicle T_3 and knowing the moment of inertia of lever I_{2Z}, defined earlier by formula (14), we will find by formula (18) the moment of inertia of any I of the vehicle. Table 1 shows the results of studies to determine the oscillation periods of all the studied objects.

Table 1. Results of studies on determining the oscillation periods.

The object of research	Test number, t_{and}, s			Average value, t_s, s	The period of oscillation, T, s
	No. 1	No. 2	No. 3		
Frame (without the vehicle)	43.9	45.3	44.0	44.4	T_1 =7.4
Frame with test cylinder	49.1	48.2	48.5	48.6	T_2 =8.1
Frame with Lada "Niva" vehicle	88.7	87.6	88.3	88.2	T_{3a2121} =14.7

Accordingly, substituting the values of the oscillation periods determined experimentally T_{and} in formula (13), we will obtain the experimental values of the moment inertia I_a and enter them in table 2. In the same table there are the recorded values of the moments of inertia obtained by graph-analytical means.

Table 2. The results of studies to determine the moments of inertia.

The object of research	The moment of inertia, I_a, kgm^2		Deviation	
	theory	experiment	DI, kgm^2	ε, %
Lada "Niva" vehicle	615.5	557.6	57.9	10.4

6. Summary

In the laboratory of the Automobile Department of NTU of Kyiv, Ukraine experimental construction and equipment were made to determine the moments of inertia of a vehicle. The method of determining the moment of inertia of the vehicle around the vertical axis was used. It is demonstrated that the application of analytical methods to determine the moment of inertia leads to significant errors (up to 10 %) and this can negatively affect motion stability, especially at high speeds. For the Lada "Niva" vehicle the determined moment of inertia around the vertical Z-axis was 557.6 kgm^2.

References

[1] Li L and Wang F Y 2005 Research advances in vehicle lateral motion monitoring and control *Int. J. of Intelligent Control and Systems* **10**(1) pp 60–76
[2] Langenwalter J 2005 Embedded automotive system development process-steer-by-wire system *Proceedings of Design, Automation and Test in Europe* **1** pp 538–39
[3] Tan H-S, Bougler B, and Steinfeld A 2001 Snowplow steering guidance with gain stabilization *Vehicle System Dynamics* **36**(4–5) pp 279–305
[4] Durisek N 1993 Conceptual design for a vehicle inertia measurement facility (VIMF) *MS Thesis, The Ohio State University* p 188
[5] Lyons D 2002 Obtaining optimal results with filar pendulums for moment of inertia measurements *61st Annual Conference of the Society of Allied Weight Engineers SAWE Inc. (Virginia Beach, VA, 18-22 May)* 3237
[6] Durisek N, Heydinger G, Chrstos J and Guenther D 1997 Land vehicle roll/yaw product of inertia measurement *J. of Dynamic Systems, Measurement, and Control* **119**(2) pp 212–6
[7] Bakker E and Pacejka H 1993 Magic formula tyre model *Vehicle System Dynamics* **21** pp 1–18
[8] Genta G and Delprete C 1994 Some considerations on the experimental determination of moments of inertia *Meccanica* **29** pp 125–41
[9] Preukschat A 1985 *Fahrwerktechnik: Antriebsarten* (Vogel, Würzburg) p 226
[10] Merkin D R 1996 *Introduction to the Theory of Stability* (Springer Science & Business Media) p 356
[11] Sakhno V, Poliakov V, Timkov O and Kravchenko O 2016 Lorry convoy stability taking into account the skew of semitrailer axes *Transport Problems* **11**(3) pp 69–76
[12] Stanford A L and Tanner J M 1985 *Physics for Students of Science and Engineering* (Academic press Inc., London) p 804
[13] *Brief automobile reference book* 1982 State Scientific Research Institute of Automobile Transport 9th ed. (Transport, Moscow) p 464 (In Russian)
[14] Timkov O M, Yashchenko D M and Bosenko V M 2010 Analytical methods for determining the moments of inertia of the vehicle *Project management, systems analysis and logistics. Technical series* pp 177–81 (In Ukraine)

Service for optimization of charging stations selection for electric vehicles users during long distances drives: time-cost tradeoff

J Hassler[1,2,3], Z Dimitrova[1], M Petit[2] and Ph Dessante[2]

[1] Groupe PSA, Route de Gisy, 78140 Vélizy-Villacoublay, France
[2] GeePs | Group of electrical engineering-Paris, UMR CNRS 8507, CentraleSupélec, Université Paris-Saclay, Sorbonne Universités, 3 & 11 rue Joliot-Curie, 91192 Gif-sur-Yvette CEDEX, France

[3] E-mail: jean.hassler@geeps.centralesupelec.fr

Abstract. One of the weighty curb to electric vehicles (EV) acceptability is their constraints when performing long-distance trips. Even if improvements made on fast charging technology enable to recover up to 300 km range in 20 minutes, the low density of this kind of infrastructure and the variability in their availability can lead to significant waiting times for users. In this context, it is useful to provide users with information about the availability of charging stations to facilitate the use of EVs. In this article, we present a vehicle/infrastructure communication system. EVs ask charging stations for estimated waiting times at a specific time. Stations calculate these estimates based on notifications of intention to charge received from other vehicles. Thanks to this information, vehicles will be able to optimize their journey. This system has the advantages of being easily scalable because it is distributed, simple and only require the sharing of little personal information. However, the stations on the route may have different charging powers and prices. This raises the question of the criteria for optimizing these trips, depending on whether users favor the cost or the duration of the trip. In this article, we examine this tradeoff using multi-objective optimization and Pareto fronts, as well as the impact of such a communication system on recharge decisions.

1. Introduction

In the Intergovernmental Panel on Climate Change special report "Global warming of 1.5ºC", the first chapter [1] highlights the consequences of global warming of more than 1.5°C and links it to greenhouse gases. The second chapter [2] demonstrates the need to return to a zero-carbon footprint as soon as possible and estimates the remaining carbon credit for the temperature to exceed preindustrial temperature by no more than 1.5°C. This emission-reduction requirement is to be applied in all emitting sectors, in particular in the transport sector [3].

Within the sector, we must consider several strategies: the optimization of existing solutions as well as the use of alternative energy storage and propulsion systems. The promotion of battery electric vehicles, which do not emit any gas when driving (for a tank-to-wheel evaluation), appears promising. However, a broad approach is needed through life-cycle analysis and well-to-wheel studies. This type of study, presented in [4] and [5], clearly shows the potential reduction in greenhouse gas emissions, while pointing to the battery production phase and the electricity production mix used, as having a substantial impact on the final balance.

The advantage of electric vehicles (EV) in terms of greenhouse gas emissions demands further study of its acceptability and economic viability. First, EVs are adapted to city trips, with high efficiency, lower noise, and air pollution than thermal vehicles [6]. Major global cities are promoting its use by delineating low-emission areas. The batteries of these vehicles also permit another use with vehicle-to-grid (V2G) technologies: electric vehicle fleets can offer a frequency or voltage regulation service by extracting or supplying electrical power to the grid in a pilot-controlled way [7]. However, the primary remaining obstacle to the deployment of this technology is the ability to store energy: batteries remain expensive, heavy, and bulky (even if the mass and volume energy densities are increasing), resulting in high purchasing costs and low inherent autonomy for use over long distances. Fast charging, with powers between 50 and 150 kW, provides a solution for a growing number of vehicles.

The deployment of charging infrastructure is hence the prerequisite for the spread of electric vehicles. A well-established charging network increases effective electric vehicle range, relieving range anxiety, and reducing inconvenience in the charging process. In [8], Funke et al. compare the utility of investing in fast-charging infrastructure or longer battery ranges, pointing out the cost efficiency of an infrastructure increase.

However, heavy traffic will likely saturate the charging stations. This situations can lead to long waiting times. Also, the operators of these stations may charge different prices, which will depend in particular on the maximum power delivered.

This leads us to our research question: how can electric vehicle users optimize their travel time and cost on long-distance journeys? In this article, we will present a communication system between vehicles that permits this optimization, and then we will study its performance.

2. Methodology

2.1. Communication schema

The quality of a communication system depends not only on its performance but also on its ability to be easily deployed. Let us present all the constraints that we must consider. First, such a system must relie on wireless connections between vehicles, charging stations, and a central server, called V2X communication. These connections can be based on mobile networks or WLAN, the first technology operating for long distances, and the second one for local communication (less than a few kilometers). The risk of dead zones and the latency of these communications demands that we set apart solutions requiring continuous or rapid exchanges.

A centralized server gathering the requests of all users allows for finding optimal solutions, however, its implementation, reliability, and required computing power can be a significant hurdle, especially if the system were to be deployed on a large scale. A distributed system, without a centralized server, eliminates this problem. Users can also be reluctant to share personal data, such as destination, average speed, or state of the vehicle.

Finally, constraints of robustness and cohabitation must also be considered.

Considering all constraints mentioned in the previous paragraph, figure 1 describes the innovative communication schema that we propose. Communications between EVs and charging stations use mobile networks. This process, from the vehicle point of view, follows five steps:

1. The EV calculates several possible combinations of stops, with different stations and amounts of energy to be stored to guarantee to reach its destination. This computation is done on the vehicle CPU resources.
2. The EV requests the expected waiting time at charging stations at a specific arrival time, for all combinations.
3. Charging stations (CS) respond to this request thanks to an updated arrivals list.

4. The EV can now choose the quickest charging plan, it then sends a notification to the selected stations, including a given arrival time with an estimated charging duration.
5. The EV remove solutions that do not correspond to what are no more feasible. For example, if it already passes by a given station without stopping, it can remove solutions using this station.
6. The vehicle repeats the cycle composed of the steps 2 to 5, every period T.

The vehicle will first do this cycle when starting the trip, before entering the highway. It will then refresh according to a period T. We note $T_{adv,n}$ the time between the beginning of the trip and the moment when the vehicle enters the highway. Stations will compute expected waiting times using notifications sent by all the vehicles.

From the station point of view, the system is composed of a device that must receive requests of waiting time estimation and notification of recharge, to then compute scheduled waiting times.

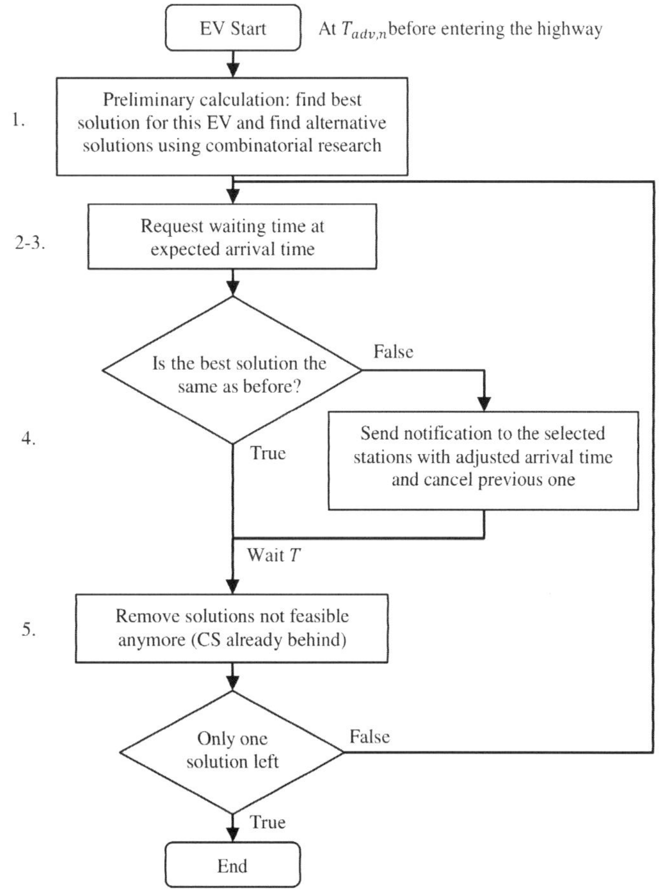

Figure 1. Communication flow chart from the vehicle point of view.

2.2. Charging price

We assume that charging sessions may have different prices depending on the charging station. The driver will, therefore, have the choice between routes with different duration but also different prices. In order to compare them, we introduce the compromise variable X, varying between 0 to 1, as well as $Sol_{perf,i}$, the performance of the solution i defined by equation (1). $Sol_{cost,i}$ (€) and $Sol_{duration,i}$ (min) respectively represent cost and duration of solution i.

$$Sol_{perf,i} = XSol_{cost,i} + (1 - X)Sol_{duration,i} \tag{1}$$

This variable X is linked to the drivers' value of time by the following relation:

$$X = \frac{1}{1 + Value\ of\ time} \tag{2}$$

There are several types of billing for recharging sessions: flat rate (€/session), price per recharging time (€/min), or price proportional to the charged energy (€/kWh). We use a price proportional to the energy, more representative of the cost borne by the recharging operator. The recharging price per kWh is assumed to be the same for all users at one charging station, even if the vehicle does not support the maximum power delivered by the terminal and charges at a lower power level. We then used Ionity network recharge prices [9] for the most powerful stations (0.8 €/kWh), and assumed that the cheapest ones were half of the most expensive (0.4 €/kWh).

2.3. Valuation method

To compare the behavior of the communication scenario, we simulate a traffic flow on a highway during one day. During these simulations, the following elements are fixed:

- Characteristics of the highway: position of the entrances and exits as well as the position, maximum power, number of charging points, and the price of the charging stations.
- A fleet of vehicles, with the characteristics of the EVs and their routes: origin and destination, departure time.

For a given situation, we vary X, in order to reproduce several value of time of the users.

2.4. Scenarios comparison

We then compared the performance of this system with two other situations. In the following, we will thus compare three scenarios:

- Without communication scenario. This is the reference case, for which there is no coordination. It is the upper bound of the problem. Users make their choices according to the positions, powers, and prices of the charging stations, but without knowing in advance the waiting times.
- Optimal solution scenario. This is an offline search for the optimal solution. The algorithm considers all the journeys over a day. It then looks for the best distribution of vehicles between the different stations. This solution constitutes a lower bound to the problem. We use a differential evolution algorithm: this kind of algorithm uses a population of solutions and repeats a process of mutation, crossover and selection. However, this scenario is difficult to achieve in practice, as it requires high computing power and knowledge of all the characteristics of the journeys in advance. An adapted NSGAII algorithm is used for the computation of the pareto front in section 3.
- Communication scenario. It is the proposal for a communication structure between vehicles and stations, described in paragraph 2.1. It is an online and deployable algorithm. Trips are known by the drivers but not by the system (charging stations).

For scenarios with and without communication, we start by simulating the behavior of vehicles for X being zero and one. Each time we obtain an average cost and an average travel time. Then we search for intermediate solutions by dichotomy. After that, we will represent all the solutions found, as well as the Pareto front.

3. Application

3.1. Case study

The generated data represents the situation of a highway in a course of a day with 100 or 150 electric vehicles. The road has one way, 11 entrances and exits, and is 553 km long. It includes six charging stations, each with three charging points of 50, 100, or 125 kW maximum available power. Figure 2 displays their implantation. There are three types of vehicles in this simulation: urban, sedan, and luxury. Table 1 gives their respective characteristics.

We based our daily highway flow modeling on the data found in [10]. Figure 3 shows the average daily flow of vehicles entering the French A6 motorway in the Ile-de-France region in France in the year 2017. These are the data of highway entrances equipped with counting loops. It indicates an almost constant flow between 8 a.m. to 9 p.m. We also notice that the traffic during the night is more than three times lower than during the day. As we want to study crowded situations, we will focus on the daytime scenario. Therefore, we modeled the distribution of departure with the previous data, between 3 a.m. to 11 p.m. We choose entry and exits number randomly for vehicles, while imposing a minimum distance of 250 km.

We also used a uniform distribution for the state of charge (SoC) at departure in the 50%-100% range. SoC at arrival (when the car exits the highway) is required to be above 30%, and SoC when entering a charging station to be above 15%. EV are charged to a maximum of 80% or below.

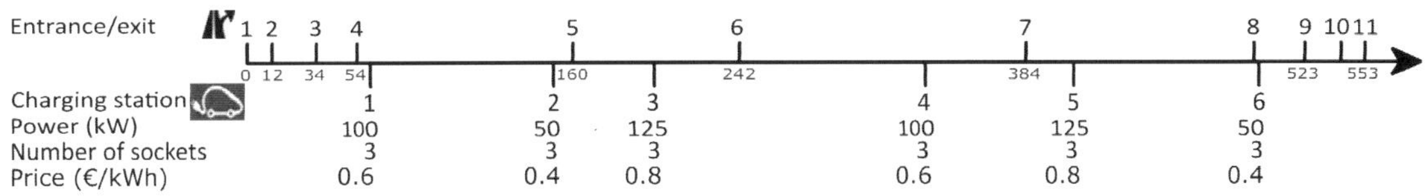

Figure 2. Highway entrances and charging stations implantation.

Table 1. Studied vehicles characteristics.

Car Type	Urban	Sedan	Luxury
Battery, kWh	50	60	100
Maximum charging power, kW	50	100	125
Consumption, kWh/km	0.15	0.18	0.18
Driving Speed, km/h	110	130	130
Generation probability	0.3	0.6	0.1

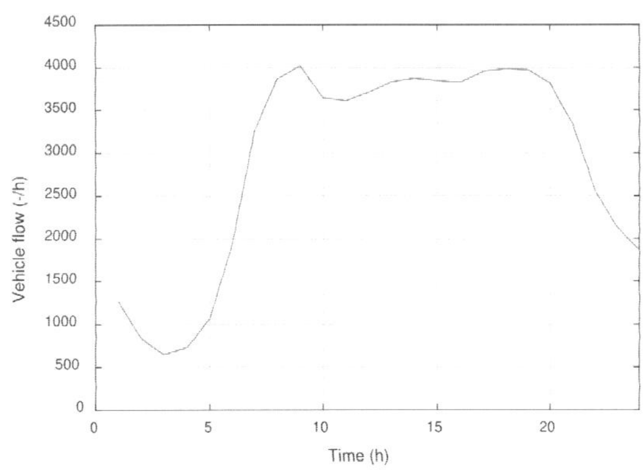

Figure 3. Average daily incoming vehicle flow.

3.2. Pareto curves comparison

Figure 4 displays the time/cost compromise for the three scenarios with 100 and 150 vehicles. For the scenarios with and without communication, it displays all the obtained solution and the Pareto front, labeled as “Best”. In the following we will focus on points labeled (1), (2) and (3).

First, we can notice that the communication allows obtaining solutions close to those obtained by the differential evolution algorithm, overall range *X*. Moreover, there are not solutions too far away.

A the point for lower cost (X=1) , the same solution is obtained with or without coordination. As waiting times are not taken into account, communication is not beneficial.

Second, without communication, a large part of the solutions is very distant from optimal solutions. As a result, communication can reduce the cost of travel by up to 10% for the same duration. This effect is even more visible in the situation with 150 EVs (figure 4(b)), which is more saturated, with a cost reduction up to 15% on equal time. We notice that the point in the pareto front for the lower time (label (1) in figure 4(a), X=0) leads to more expensive and slower solutions than if we slightly take into account the cost of recharging. Indeed, by optimizing time only, a large proportion of vehicles will want to recharge at the most powerful stations, which will saturate. Nevertheless, taking into account the cost of recharging, vehicles with only low maximum charging power will give priority to less powerful stations, which are also cheaper. In this way, there will be a better distribution between the stations. The solutions obtained minimizing the average time (label (2), X=0.954) are as good as or even better than with communication. However a priori the value of *X* that allows this is not always the same. It may depend on the traffic and the highway. Moreover, this effect does not work if there are stations on the highway that are both more powerful and cheaper than others are.

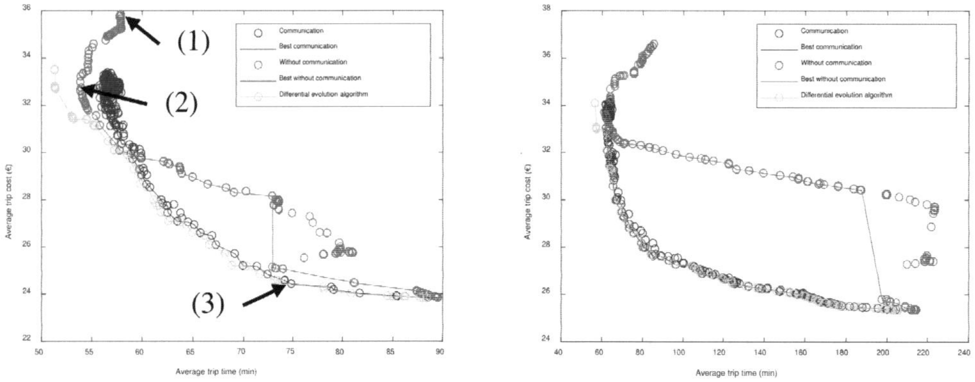

(a) (b)

Figure 4. Cost duration tradeoff for the three scenarios: a) 100 EV, b) 150 EV.

Figure 5 shows the use of charging stations as a function of vehicle type for three cases shown in figure 4. Stations numbers are defined in figure 2. The first case represents the situation obtained when vehicles do not communicate and only optimize their travel time. It can be seen that stations three and five, the most powerful, have a high demand. This situation is sub-optimal in terms of time and cost.

The second situation is the best case obtained without communication. Urban vehicles, considering prices, chose to go to cheaper stations that offer the same recharge time (as the EV side always limits the charging power). This makes it possible to rebalance the frequentation between the different stations. This situation is close to the optimum time.

The communication algorithm obtains the third situation by prioritizing cost reduction. In this case, the cheapest stations are used first. The others are used by those who have no choice (for example, if their battery level does not allow them to go to another station) or by those for whom this saves a lot of time. Indeed, moving from situation 3 to cost optimization only reduces the average cost by only 2.4% while increasing the time by 19.5%.

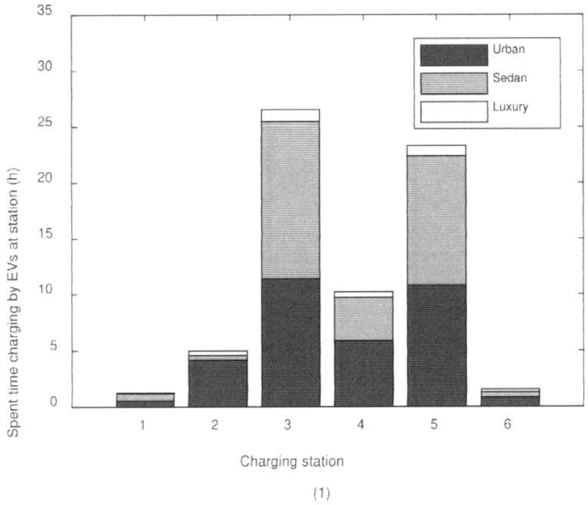

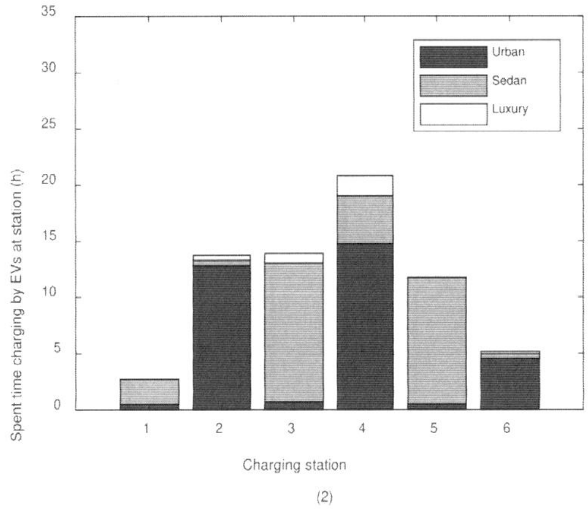

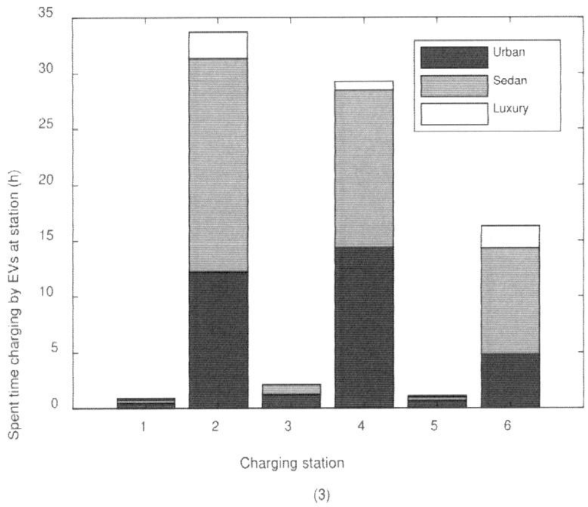

Figure 5. Utilization of the 6 charging stations: (1) No communication, time optimization only (X=0); (2) No communication, lowest average time (X=0.954); (3) With communication, mainly cost optimization (X=0.998).

The use of such a communication system can by extrapolate in more complex situations, such as situations with traffic jams, several possible routes, variable consumption and speeds. Those functions more advanced can be added by refining the calculation of charging plans, without modifying the structure of the system.

4. Conclusion

In conclusion, the research question of this article was to find a way for electric vehicle users to optimize their travel time and cost on long-distance journeys. We were therefore able to study a feasible means of communication between vehicles that allows us to find good cost/travel time compromises for long journeys in electric vehicles. This study also allowed us to show that taking into account the cost of recharging could reduce travel time and cost in situations without communication. This better distribution is only possible if the price of recharging increases with power. Later on, it will be interesting to study the robustness of such a system, for example when not all the travelers use it, and to carry out a cost-benefit analysis.

Acknowledgment
This research benefits from the support of the ANRT association and Groupe PSA.

References

[1] Allen M, Dube O P and Solecki W *Framing and Context*
[2] Rogelj J, Shindell D, Jiang K et al. *Mitigation Pathways Compatible with 1.5°C in the Context of Sustainable Development*
[3] Solaymani S 2019 CO_2 emissions patterns in 7 top carbon emitter economies: The case of

transport sector *Energy* **168** pp 989–1001
[4] Moro A and Lonza L 2018 Electricity carbon intensity in European Member States: Impacts on GHG emissions of electric vehicles *Transportation Research Part D: Transport and Environment* **64** pp 5–14
[5] Faria R, Marques P, Moura P, Freire F, Delgado J and de Almeida A T 2013 Impact of the electricity mix and use profile in the life-cycle assessment of electric vehicles *Renewable and Sustainable Energy Reviews* **24** pp 271–87
[6] Singh B and Strømman A H 2013 Environmental assessment of electrification of road transport in Norway: Scenarios and impacts *Transportation Research Part D: Transport and Environment* **25** pp 106–11
[7] Borne O, Perez Y and Petit M 2018 Market integration or bids granularity to enhance flexibility provision by batteries of electric vehicles *Energy Policy* **119** pp 140–8
[8] Funke S Á, Plötz P and Wietschel M 2019 Invest in fast-charging infrastructure or in longer battery ranges? A cost-efficiency comparison for Germany *Applied Energy* **235** pp 888–99
[9] IONITY - WHERE & HOW [Online]. Available: https://ionity.eu/en/where-and-how.html. Accessed on: 05-Feb-2020
[10] DiRIF 2018 Les comptages en Ile-de-France - Direction des routes Ile-de-France [Online]. Available: http://www.dir.ile-de-france.developpement-durable.gouv.fr/les-comptages-a174.html Accessed on: 13-Mar-2019

Experimental investigation of the motorcycle braking properties when riding on different road surfaces

Dalibor Barta, Ján Dižo[1], Miroslav Blatnický and Jozef Vaňko

University of Žilina, Department of transport and handling machines, Univerzitná 8215/1, 01026 Žilina, Slovak Republic

[1] E-mail: jan.dizo@fstroj.uniza.sk

Abstract. The presence of motorcycles on the roads significantly affects road traffic and its safety. Avoidance manoeuvres requiring braking and acceleration are closely linked to motorcycle riding. This paper points out the influence of the type of motorcycle and the type of its tyres on the braking distance and the stability of the motorcycle. Two types of motorcycles are compared - road and off-road when riding on two different surfaces and also three ways of braking - only with the front brake, only with the rear brake and with both brakes. The work analyses the problem of braking and stability of a motorcycle. When a brake lever is suddenly pushed, the stability of the motorcycle can be lost. Authors have analysed, compared and evaluated braking distances of motorcycles. Results of investigation activities have shown, that reaching of maximal safe and the shortest braking distance on the certain road surface needs to choose a proper type of motorcycle and a tyre.

1. Introduction

Today's progress places great emphasis on increasing of performance of all vehicles. It leads to increasing of speed on roads. Therefore, it is necessary to develop elements of safety. The safety of vehicles operation depends on various factors, one hand equipment of transport means, i. e. elements of active and passive safety [1, 2] and on the other hand on operation conditions, i. e. weather conditions [3, 4], quality of infrastructure etc. [5-8] The braking system is the key factor of safety in road transport as well as railway transport [9-11]. The main purpose of brakes is to slow down or to stop a vehicle at any speed. Speed plays a big role in case of traffic accidents. It often decides, whether the participants survive the accident or not. Thus, it is necessary to develop and improve brakes and braking system. The development of these key factors is significantly accelerated by new technologies, new materials and electronic systems. The main aim of braking systems, which we recently know and which are used in vehicles, is to reduce braking distance as well as to ensure steering control and stability of a vehicle.

Braking is aimed at decreasing of driving speed of a vehicle. In case of single-track vehicles, the best braking efficiency is reached on a straight road. For a motorcycle, it is redistributed in rate as following: 60% on a front wheel, 40% on a rear wheel [12-14]. It is typical for motorcycles, that the front and rear brake controls work independently of each other. Hence, distribution of braking force depends on skills of a motorcycle driver. Braking of a motorcycle causes, that the most weight is transferred to the front wheel and the rear wheel is lightened (figure 1). Parameters in this figure are: G – the gravitational force, G_P – the load of the front wheel, G_Z – the load of the rear wheel, CoG – the centre of gravity, ΔZ – change of the wheels load distribution during braking, l – the wheels, h – the height of the *CoG*, and F_a – the inertial force. Therefore, it is important to distribute properly the braking force between the front and rear wheels. The motorcycle driver motion during braking also plays the significant role [15-17]. The state, when a wheel is locked, is supposed as very dangerous [18-20]. It leads to increasing the braking distance as well as to the possible fall of the motorcycle driver. It means, adhesion of friction surfaces road/tyre is very important [12].

Vlk claims [20], braking closely relates to adhesion. Therefore, it is important to know, whether a road is dry, wet, covered by ice, snow or dirt. There must not be a situation, when the braking effect prevails over the grip of the tyre on the road. Even the strongest brakes will not help, if we are not able to transfer their effects on the road. The best adhesion is reached on a dry road. It is even increased by warming up of the tyre during driving, when a surface softens. Sometimes, the rear wheel locking is not observed until the moment, when it starts to turn on its side. If such a state happened, it is necessary a brake lever to release slowly and carefully [12, 19].

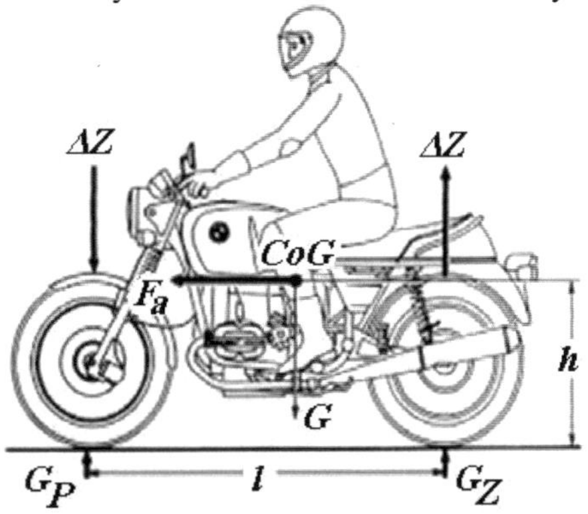

Figure 1. Wheels load distribution during braking of a motorcycle (the front wheel is loaded due to inertia effects, the rear wheel is unloaded) [12].

2. Experimental motorcycles and methods

The main goal of the experiment has been to measure and compare the braking distance of two type of motorcycles. These types of motorcycles have been compared:

- enduro motorcycle Yamaha WR450F;
- scooter Keeway ARN 125.

The enduro motorcycle Yamaha WRF450F (figure 2) is equipped with a hydraulic braking system. However, it uses a two pistons floating brake caliper on the front wheel and the disc with the diameter of 250 mm and a one piston floating brake caliper on the rear wheel. Tyres pressure has been of 1 bar.

The scooter Keeway ARN 125 (figure 3) has also a hydraulic braking system. The front brake includes a brake disc with a diameter of 220 mm and a fixed brake caliper consisted of two pistons. The rear brake is a drum brake. Tyre pressure has been of 2.3 bar and tyre tread depth of 2.5 mm.

In the preparation phase of experiment realisation, we have chosen a suitable place and conditions for measurement performance. Experiments were performed in the Trebostovo village area (Liptovský Mikuláš district, Slovak Republic). This environment has ensured suitable road and terrain conditions.

The experiment has been realised on two types of road surfaces:

- a dry asphalt road;
- a cart-road.

Motorcycles have been driving on the flat road without road incline. The experiment included three tests, during which behaviour of the motorcycles at heavy braking has been tested. The track length needed for safe acceleration up to the speed of 27 kmh^{-1} has been investigated. This speed has had to be kept until the moment of the braking initiation. After the preparation phase, the measurement of the braking distance followed. Three cases of braking have been investigated:

- braking only by the front brake;
- braking only by the rear brake;
- braking by both front and rear brakes together.

Every type of braking has been performed twelve-times. Both motorcycles has been driven by the same motorcyclist. Braking has been realised at the locking limit.

Tests have been realised afternoon about 3 p.m. Tests have been conducted without unexpected circumstances and at weather conditions as following: calm, air temperature of 22°C.

From the safety point of view, grip (adhesion) between a tyre and a road is very important. Therefore, the friction coefficient of the tyre for the given type of a road surface is calculated from measured data.

a) b)

Figure 2. The enduro Yamaha WRF450F a), detail of the tyre tread b).

a) b)

Figure 3. The scooter Keeway ARN 125 a), detail of the tyre tread b).

The aim has been to find the adhesion level of motorcycle tyre with the given road surface. Mathematic-statistic methods have been used for further analyses of reached results.

The tyre is a connecting element between a motorcycle and a road. Tyres with the friction coefficient as high as possible are the most suitable for the best transmission of braking forces to the road. The braking distance s_b is given by the known formulation as following:

$$s_b = \frac{v^2}{2g\mu}, \tag{1}$$

where v (ms^{-1}) is the motorcycle speed; g (ms^{-2}) is the gravitational acceleration and μ (-) is the friction coefficient.

Experiments have been focused on comparison of the braking distances of motorcycles. We have aimed at calculation of the friction coefficient μ. The formulation for its calculation is derived from the equation (1). Then, the friction coefficient is calculated as following:

$$\mu = \frac{v^2}{2gs_b}. \tag{2}$$

3. Measurement of motorcycles braking distances

This section contains results of experimental measurements of motorcycles braking distances. Obtained results of braking distances of the enduro motorcycle Yamaha WRF450 and for the scooter

Keeway ARN 125 driving on an asphalt road and on a cart-road are digestedly arranged in individual subsections containing graphs and tables with measured values.

3.1. Braking distances of the enduro motorcycle Yamaha WRF450

Firstly, braking distance of the enduro motorcycle has been measured. Table 1 and figure 4 show results of experiments, when it has been driving on the dry asphalt road.

Example of the calculation of the friction coefficient is given by the equation (3). It is calculated for the dry asphalt, when both brakes are activated. Thus, the friction coefficient is:

$$\mu = \frac{v^2}{2gs_b} = \frac{\left(\frac{27}{3.6}\right)^2}{2\cdot 9.81\cdot 7.23} = 0.396538, \tag{3}$$

where value of the braking distance $s_b = 7.23$ m is taken from table 1 as the average value for the case of braking by both brakes.

Table 1. Measured braking distance of the enduro motorcycle Yamaha WRF450 on the dry asphalt road.

Experiment No.	Front brake (m)	Rear brake (m)	Both brakes (m)
1st	8.17	11.88	7.50
2nd	8.84	11.20	7.20
3rd	8.07	11.30	7.15
4th	8.81	11.38	7.17
5th	8.53	11.60	7.44
6th	8.31	11.55	7.16
7th	8.20	11.33	7.20
8th	8.33	11.29	7.15
9th	8.15	11.39	7.11
10th	8.41	11.28	7.35
11th	8.22	11.43	7.20
12th	8.19	11.26	7.09
Average	8.35	11.41	7.23

Measured values of the braking distances of the enduro motorcycle on the dry asphalt road are negatively affected by the fact, that this motorcycle is designed mainly for driving in a rough terrain and cart-road. These results are likely caused due to tyre tread, which are intended to be used mostly on unpaved roads.

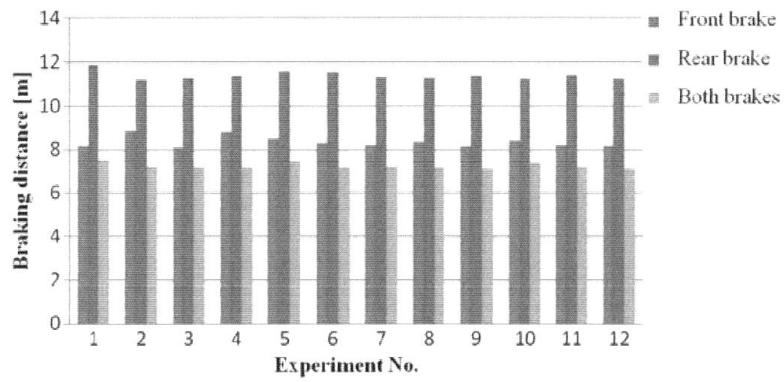

Figure 4. Braking distances of the enduro motorcycle Yamaha WRF450F on the dry asphalt road.

Only in the case of braking by the rear brake, a wheel was locked. When the endure motorcycle has been braked by the front brake or by both brakes, wheels were not locked. After the test we observed excessive wear of both tyres with naked eye.

As the second test we have performed measurement of the braking distance for the enduro motorcycle driving on the cart-road. Results of this test are listed in table 2 and graphical outputs are shown in figure 5.

Table 2. Measured braking distance of the enduro motorcycle Yamaha WRF450 on the cart-road.

Experiment No.	Front brake (m)	Rear brake (m)	Both brakes (m)
1st	5.62	17.76	5.49
2nd	5.86	17.26	5.50
3rd	6.00	17.16	5.34
4th	5.91	17.20	5.10
5th	5.95	17.33	5.25
6th	5.80	17.25	5.45
7th	5.89	17.25	5.30
8th	5.75	17.31	5.15
9th	5.86	17.50	5.27
10th	5.88	17.23	5.20
11th	5.92	17.20	5.20
12th	5.79	17.29	5.22
Average	5.85	17.31	5.29

From measured data of the braking distances we can conclude, that the off-road design and properties of the enduro motorcycle Yamaha WRF450F are positively proved during braking on the cart-road. Comparing of the braking distances we have found out this fact: when a motorcyclist has been braking this enduro motorcycle on the cart-road by the front brake or by both brake, the braking distance was shorter in comparison with braking on the dry asphalt road.

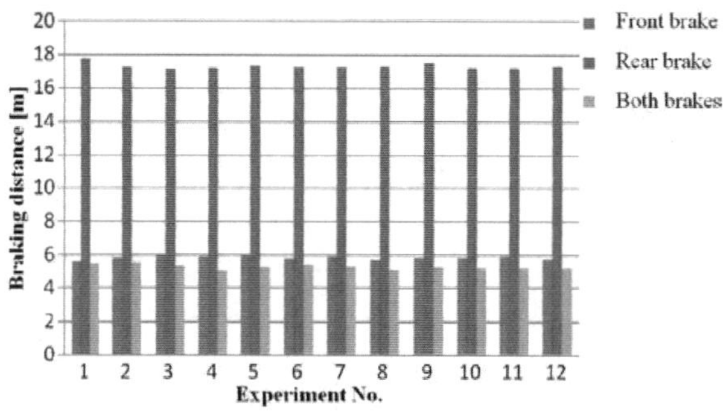

Figure 5. Braking distances of the enduro motorcycle Yamaha WRF450F on the cart-road.

Locking of a wheel has been detected just in case of braking by the rear brake and it has hold over during all time of braking. In case of braking by both brakes, the rear wheel was locked and it was detected from braking beginning till stop.

When we substitute the given driving speed (27 kmh^{-1}) and chosen value of the braking distance of 5.28 m (table 2 – average value for braking by both brakes) to the formulation (2), we get the value of the friction coefficient of 0.541961.

3.2. Braking distances of the scooter Keeway ARN 125

Experiments of braking distances measurement proceeded to investigation of braking performance of the scooter Keeway ARN 125 on the dry asphalt road and on the cart-road.

Table 3 contains values of braking distances of the scooter, when it has been driving on the dry asphalt road. Graphical comparison of measured data is shown in figure 6.

Table 3. Measured braking distance of the scooter Keeway ARN 125 on the dry asphalt road.

Experiment No.	Front brake (m)	Locking of the front wheel, sliding (m)	Rear brake (m)	Both brakes (m)
1st	4.48	-	8.82	3.48
2nd	6.40	3.00	7.43	3.23
3rd	5.84	1.00	8.08	2.75
4th	5.10	0.75	8.34	2.94
5th	4.70	-	7.93	3.39
6th	4.70	-	7.98	2.77
7th	4.99	0.75	8.16	3.09
8th	4.16	-	7.50	3.30
9th	4.00	-	8.00	3.30
10th	4.09	-	7.96	2.93
11th	4.04	-	8.03	2.69
12th	4.03	-	7.90	2.86
Average	4.71	-	8.01	3.06

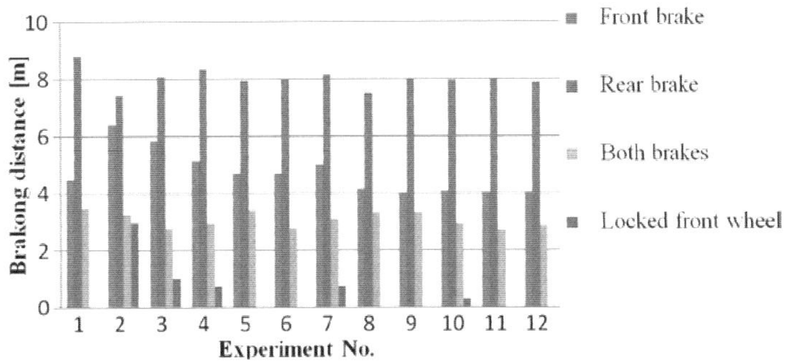

Figure 6. Braking distances of the scooter Keeway ARN 125 on the dry asphalt road.

From measured values, which are indicated in table 3 and figure 6 we can see, that on the dry asphalt road in some cases the front wheel was locked. It was in the case of braking only by the front brake. It happed, when the front brake lever was pushed stronger and more intensive. In this case, stability of the scooter was disrupted and the risk of a motorcycle driver fall was higher as well. When the scooter has been braking by the rear brake or both brakes, no wheel was locked. When we were braking by both brakes, we were sensed more stable behaviour of the scooter.

Substituting of driving speed value as well as the average braking distance for braking by both brakes, i. e. 3.06 m we can calculate the friction coefficient. Its value is relatively high, i. e. of 0.936919. It means, this value is on the great lavel.

Finally, we have performed experiments, when the scooter Keeway ARN 125 was driving on the cart-road. Obtained braking distances are written in Table 4 and comparison of these values is depicted in figure 7.

Table 4. Measured braking distance of the scooter Keeway ARN 125 on a cart-road.

Experiment No.	Front brake (m)	Rear brake (m)	Both brakes (m)
1st	5.67	7.84	3.47
2nd	5.60	6.83	3.67
3rd	5.46	7.19	3.47
4th	5.47	6.50	2.93
5th	5.92	6.54	3.47
6th	5.61	6.80	3.50
7th	5.80	6.61	3.45
8th	5.71	6.94	3.48
9th	5.49	6.84	3.70
10th	5.46	6.55	3.52
11th	5.50	6.58	3.48
12th	5.46	6.55	3.45
Average	5.60	6.81	3.47

From detected data of braking distances for the cart-road we can see, that wheels were locked during the whole braking process. Locking occurred in case of using both the front brake and the rear brake.

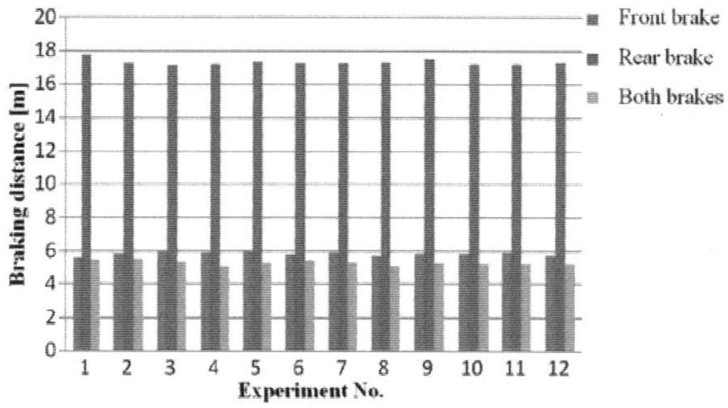

Figure 7. Braking distances of the scooter Keeway ARN 125 on the cart-road.

When the scooter was braked by both brakes, only the rear wheel was locked.

The scooter behaved stabile on the dry asphalt road as well as on the cart-road in case of braking by both brakes.

The friction coefficient calculated using equation (2) for the driving speed of 27 kmh^{-1} and for the average braking distance of 3.47 m (both brakes) is of 0.826217. Difference of the friction coefficient value for the braking on the cart-road and on the dry asphalt road is minimal.

3.3. Summary of findings

Calculated values of average data of braking distances of the enduro motorcycle Yamaha WRF450F have shown, that differences have been wide. When motorcycle driver has been braked by the front brake, average values have been higher of 2.5 m and in case of using both brakes, the difference has been of 1.94 m. Both values were on the behoof of the braking on the cart-road. Using of the rear brake was critical, because the difference of average values of braking distances was of 5.9 m on the behoof of braking on the dry asphalt road.

The scooter Keeway ARN 125 has not shown such a wide difference of braking distance depending on road surface quality. Nevertheless, braking on the dry asphalt road is safer than braking on the cart-road. Based on obtained data we can see, that the difference of average values of braking distances for braking on the dry asphalt road for the front brake is of 0.89 m and for the both brakes of 0.41 m lower than braking on the cart-road. Wider difference is found out in case of braking just by the rear brake, which is of 1.2 m lower for the cart-road.

From average values listed in table 5 and in figure 8 we can see small difference for the scooter Keeway depending on the road surface quality. The enduro motorcycle Yamaha shows wider differences. We can notice a significantly longer braking distance, when it is braked just by the rear brake on the cart-road than on the dry asphalt road.

Table 5. Comparison of average values of braking distances of the tested motorcycles.

	The dry asphalt road			The cart-road		
Motorcylce	Front brake (m)	Rear brake (m)	Both brakes (m)	Front brake (m)	Rear brake (m)	Both brakes (m)
Yamaha (enduro)	8.35	11.41	7.23	5.85	17.31	5.29
Keeway (scooter)	4.71	8.01	3.06	5.60	6.81	3.47

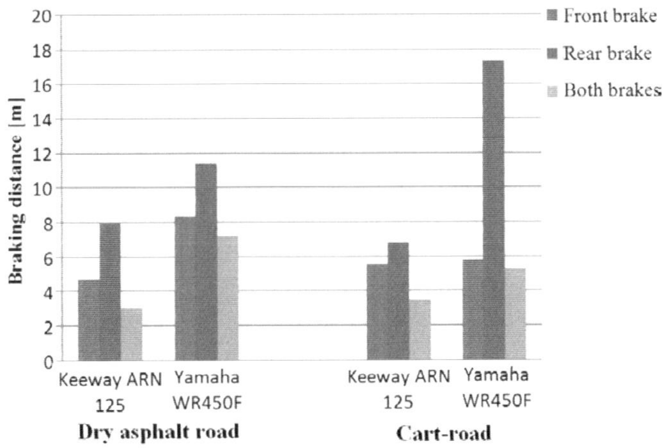

Figure 8. Average values of the braking distances of the tested motorcycles.

Comparing the values of the friction coefficients (table 6) we have observed, that the level of tyre friction is greater for the scooter Keeway on the dry asphalt road. After all, this tested scooter Keeway has good friction level on both road surfaces. The tested enduro motorcycle Yamaha has better tyre friction on the cart-road and the tyre friction is significantly lower on the dry asphalt road. We suppose, that such a low value of friction coefficient could be caused because of the different type of the tire tread, which is primary given for heavy off-road conditions. Nipples of tires reduce friction surface and it can lead to lower friction coefficient on the rigid road surface.

Table 6. Comparison of the calculated values of the friction coefficient μ depending of the type of a road surface for tested motorcycles.

	Type of the road surface	
Motorcycle	**The dry asphalt road (μ)**	**The cart-road (μ)**
Yamaha (enduro)	0.396538	0.541961
Keeway (scooter)	0.936919	0.826217

Generally, we can state, that the best braking performance has been reached, when motorcycles have been braking by both brakes. On the contrary, the worst braking performance has been achieved in the case of braking only using the rear brake.

4. Conclusion

The main goal of the experimental measurement was to determine and to compare the braking distances of two types of motorcycles, when they are driving on two different types of road surface qualities as well as to find out the tyre/road friction level. Moreover, we have wanted to justify the fact, that the proper tyre type is very important element for the safe driving on a motorcycle.

From measured data and calculated average values of braking distances of motorcycles we have determine, that the scooter Keeway ARN 125 has similar braking performance on both types of road surfaces. For the dry asphalt road, the braking distance has been shorter for the front brake and for both brakes in comparison with braking only by the rear brake. On the contrary, using of the rear brake has been more efficient on the cart-road. Achieved differences of braking distances on both road surfaces for the enduro motorcycle Yamaha WRF450F have been very wide. The braking distance on the cart-road has been shorter for the front brake and for both brakes. The rear brake has been more efficient on the dry asphalt road. From investigated data results, that the tyre friction depends not only on the tyre tread and on climatic conditions, but also on the road surface quality.

Acknowledgments
This work was supported by the Cultural and Educational Grant Agency of the Ministry of Education of the Slovak Republic in the project No. KEGA 044ŽU-4/2019: Implementation of innovative elements in the education process within the study program Maintenance of Means of Transport.

References

[1] Savino G, Lot R, Massare M, Rizzi M, Symeonidis I, Will S and Brown J 2020 Active safety systems for powered two-wheelers: A systematic review *Traffic Injury Prevention* **21**(1) pp 78–86
[2] Chen C P and Chiang M H 2018 Mathematical simulation and analyses of proportional electro-hydraulic brakes and anti-lock braking systems in motorcycles *Actuators* **7**(3)
[3] Luskova M, Dvorak Z and Leitner B 2015 Impact of extreme weather events on land transport infrastructure *19th Int. Sc. Conf. on Transport Means (Kaunas, Lithuania)* pp 306–9
[4] Luskova M and Leitner B 2018 Extreme weather impact on transportation and energy infrastructure *22nd Int. Sc. Conf. on Transport Means (Trakai, Lithuania)* pp 569–73
[5] Buczaj A, Krzysiak Z, Pecyna A, Caban J and Brumercik F Safety during chemical transport of dangerous goods, *Przemysl Chemiczny* **98**(8) 1276–80
[6] Rybicka I, Caban J, Vrábel J, Šarkan B, Stopka O and Misztal W 2018 Analysis of the safety systems damage on the example of a suburban transport enterprise *11th Int. Science and Technical Conference Automotive Safety (Casta Papiernicka, Slovakia)* pp 1–5
[7] Droździel P and Wrona R 2018 Legal and utility problems of accidents on express roads and motorways *11th Int. Sc. and Technical Conference Automotive Safety (Casta Papiernicka, Slovakia)* pp 1–5
[8] Madlenak R, Hostakova D, Madlenakova L, Drozdziel P and Torok A 2018 The analysis of the traffic signs visibility during night driving *Advances in Science and Technology – Research Journal* **12**(2) pp 71–6
[9] Kurčík P, Gerlici J, Lack T, Suchánek A and Harušinec J 2019 Innovative solution for test equipment for the experimental investigation of friction properties of brake components of brake systems *Transportation Research Procedia* **40** pp 759–66
[10] Gerlici J, Gorbunov M, Kravchenko K, Prosvirova O, Lack T and Hauser V 2018 Assessment of innovative methods of the rolling stock brake system efficiency increasing *Manufacturing Technology* **18**(1) pp 35–8
[11] Gerlici J, Gorbunov M, Kravchenko K, Prosvirova O and Lack T 2017 The innovative design of rolling stock brake elements *Communications – Scientific Letters of the University of Zilina* **19**(2) pp 23–8
[12] Vlk F 2003 *Dynamics of Engine Vehicles* (František Vlk, Brno) p 431 (In Czech)
[13] Kovácsová N, Grottoli M, Caliberti F, Lemmens Y, Happee R, Hagenzieker P M and de Winter J C F 2020 Emergency braking at intersections: A motion-base motorcycle simulator study *Applied Ergonomics* **82**
[14] Ariffin A H, Hamzah A, Solah M S, Paiman N F, Mohd Jawi Z and Md Isa M H 2017 Comparative analysis of motorcycle braking performance in emergency situation *J. of the Society of Automotive Engineers Malaysia* **1**(2) pp 137–45
[15] Lack T and Gerlici J 2017 Integration methods for rail vehicle ride dynamic solution assessment *23rd Conf.: Current problems in Rail Vehicles (Ceska Trebova, Czech Republic)* pp 217–34
[16] Gerlici J, Sakhno V, Yefymenko A, Verbitskii V, Kravchenko A and Kravchenko K 2018 The stability analysis of two-wheeled vehicle model *MATEC Web of Conferences* **157**
[17] Rabinovich E et al. 2018 Evaluation of the powertrain condition based on the car acceleration and coasting data *SAE Technical Papers* **2018**

[18] Koetniyom S, Dangchat M, Mongkonlerdmanee S, Carmai J and Chanta S 2018 Identification of handbrake patterns of young motorcycle riders in Thailand using a newly invented force measuring device *Engineering Journal – Thailand* **22**(5) pp 185–97
[19] Huertas-Leyva P, Savino G, Baldanzini N and Pierini M 2020 Loss of control prediction for motorcycles during emergency braking maneuvers using a supervised learning algorithm *Applied Science* **10**(5)
[20] Chindamo D, Gadola M, Armellin D and Marchesin F 2017 Design of a road simulator for motorcycle applications *Applied Science* **7**(2)

Investigation of power split schemes for modern hybrid cars transmissions

Boyko Gigov[1] **and Evgeni Dimitrov**

Department of Combustion Engines, Automobile Engineering and Transport, Technical University – Sofia, 8 Kliment Ohridski Blvd., 1756 Sofia, Bulgaria

[1] E-mail: bgigov@tu-sofia.bg

Abstract. The modes of operation of the transmissions for hybrid cars are studied in this work, using planetary gears for connecting the internal combustion engine (ICE) and the electric machines (motor-generators). The peculiarities of the two-stream variants with an electric branch – a closing block consisting of two electric machines is analysed. This achieves a stepless gear change and an increase in overall efficiency with respect to the efficiency of electrical machines. The theoretical formulation is based on the dependences between the losses of a two-stream circuit, which proves that this increase is a function of the relative share of power transmitted through the electrical branch. The characteristics of typical two-stream circuits with input differential and output differential with one mechanical and one electrical branch are determined. The aim is to evaluate not only the change in the total efficiency, but also the kinematic gear ratio and the ratio of the torques and their ranges for the whole circuit, as a function of the parameters of the electrical branch. The development of the schemes in order to improve their performance in different operating conditions is considered, which allows for combining the variants and modes of operation – serial hybrid and parallel hybrid in two variants, at low speed and high speed. Analytical and graphical methods are used to analyse and illustrate the results, which are more visual and suitable for composite planetary gears with more units. It is expected that cars with such transmissions will have better dynamics compared to conventional automatic transmissions and lower energy consumption or do less damage to the atmosphere, as a result of the optimal combination of the operating modes and dimensions of ICE and electric machines.

1. Introduction

Electric machines – motors and generators in combination with planetary gears are known in transport engineering as electromechanical transmissions and in the past were used mainly in tractors and locomotives [1]. When all the power for driving the transport or traction machine is transmitted from the internal combustion engine (ICE) through the electric machines, a stepless change of the gear ratio is achieved in a sufficiently wide range, but undesirable losses occur due to the double conversion of the energy – from mechanical to electrical and vice versa. With planetary gears the total power flow is separated into two or smaller parallel flow – mechanical and electrical. According to the power ratio, an increase in efficiency and preservation of the stepless change in the gear ratio of the transmission as a whole is achieved. In addition, it is possible to use smaller electrical machines, due to the lower power transmitted through the electrical branch. Generally it is considered that this solution has the disadvantage of greater complexity and smaller gear range. However, since electric machines are reversible (rotating in different directions) and multi-mode (the same machine works both as a motor and as a generator) it is possible to achieve a larger range.

Recently, such schemes have also gained popularity in hybrid drives for passenger cars [2]. In connection with the trends for the transition of the car fleet to electric propulsion, modern high-speed electric motor-generator units (MGU) with very small relative mass and dimensions and advanced control and regulation systems are being developed. As a result, higher values of the efficiency and a more appropriate mechanical characteristics required for a transport machine engine are achieved, close to the ideal hyperbola. However, the use of further extending the range of gear ratio changes by the mechanical part is not superfluous and has a beneficial effect on the entire drive system [3].

The aim of the present study is to show the kinematic and force dependences of two-stream electromechanical transmissions and to evaluate the impact of the parameters of the mechanical and electrical branch on the performance of the whole system, to illustrate the connections between the individual units of the system and to investigate the possibilities of given scheme for realization of optimal operating modes. This will make it possible to increase the efficiency of energy conversions and use the results to create an adequate strategy for the management of internal combustion engine and electrical machinery.

2. Determining the features of electromechanical power split transmissions

In hybrid cars, there is a wide variety of ways to connect the internal combustion engine, the electric machines and drive wheels [4]. They are classified and analyzed in the specialized literature according to the number of electrical machines, the number of planetary rows, the number of used clutches and brakes, with which achieve to different modes of operation and more [5]. Power split systems can be basically of three types – single split, double split and a combination of the first two options.

2.1. Analytical method for determining the parameters of power split transmissions with electric branch

The idea of the two-stream transmissions is to increase their total efficiency (η) compared to the efficiency of the continuously variable part. In this case, that is possible if the efficiency of the mechanical branch (η_M) is higher than the efficiency of the electric branch (η_E) and the relative share of the power diverted through the electric branch (ε) is a positive number smaller than one (efficiency theorem: if $0 < \varepsilon < 1$ and $\eta_M > \eta_E$, to $\eta > \eta_E$). This is proven by the dependences shown below.

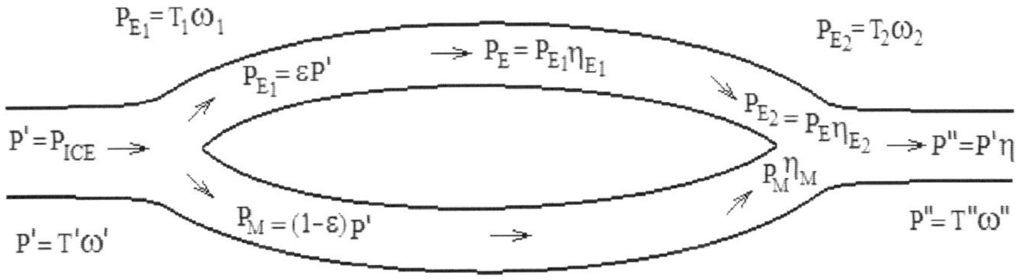

Figure 1. Flow scheme for single power split transmission.

Figure 1 shows the power transmitted through the mechanical branch (P_M) and through the electrical branch and the corresponding losses, taken into account by the efficiency coefficients of the mechanical and the electrical part. The mechanical power at the input of the electrical branch is marked with P_{E1}, and at the output of the electrical branch – with P_{E2}. The electric power that is exchanged between the two electric machines is denoted by P_E. The relationship between them is expressed by the efficiency of the electric machine E_1 (η_{E1}), the electric machine E_2 (η_{E2}) and the whole electrical branch ($\eta_E = \eta_{E1}\eta_{E2}$). The relative share of mechanical power diverted to electrical machines is denoted by $\varepsilon = P_{E1}/P'$ and the relative share of electric power – by $\varepsilon_E = P_E/P'$.

$$\eta = \varepsilon_{\mathrm{E}}\eta_{\mathrm{E}_2} + \left(1 - \varepsilon_{\mathrm{E}}/\eta_{\mathrm{E}_1}\right)\eta_{\mathrm{M}}. \tag{1}$$

Equation (1) is derived by expressing the output power (P'') through the input power (P') as the sum of the input powers of the two branches. To prove the theorem, the simplified variant of the equation (1') is used [6].

$$\eta = \eta_{\mathrm{M}} - (\eta_{\mathrm{M}} - \eta_{\mathrm{E}})\varepsilon. \tag{1'}$$

It shows that the smaller the relative share ε, the greater the increase in total efficiency. For example, if we accept the efficiency of the mechanical branch to be of the order of 0.98 and of the electrical branch to be of the order of 0.72 (for one machine 0.85 together with the control), then the total efficiency will be above 0.9 if the relative share ε is below 30%. It should also be borne in mind that the efficiency of electrical machines depends on their operating mode.

The basic equation of the kinematics of the planetary mechanisms and the relations between the torques are used to determine the transforming properties of the two-flow scheme. Figure 2 shows the principal kinematic schemes of the two types of power split gearboxes with simple division.

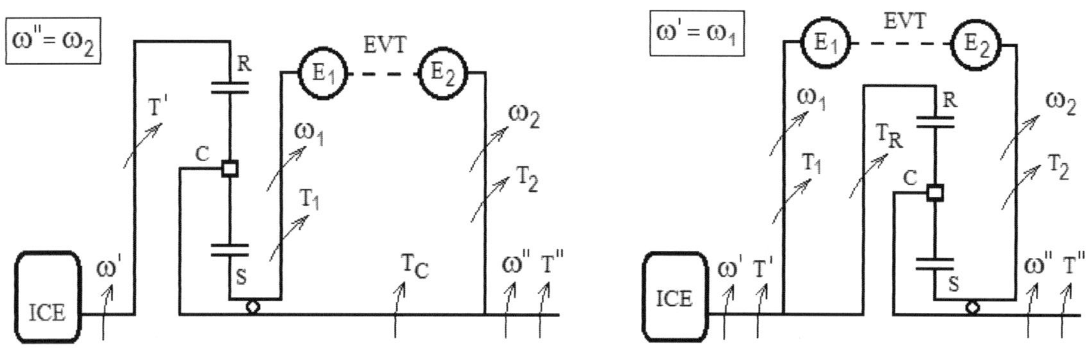

a) Input split b) Output split

Figure 2. Power split gears with single separation.

The task is to determine the speed ratio (i) and the torque ratio (K) of the respective scheme, expressed as functions of the same indicators (i_{E}) and (K_{E}) of the electrical stepless part – EVT (Electric Variable Transmission):

- For the whole scheme:

$$i = \omega''/\omega' \quad , \quad K = T''/T'; \tag{2}$$

- For the electrical part:

$$i_E = \omega_2/\omega_1 \quad , \quad K_E = T_2/T_1. \tag{3}$$

The parameters of the planetary mechanisms are also required, for which the designations are accepted: "R" – for the ring gear-wheel, "S" – for the sun gear-wheel and "C" – for the planetary gears carrier and for the torques respectively – T_{R}, T_{S} and T_{C}. The ratio of the number of teeth of the ring wheel (Z_{R}) to the number of teeth of the sun wheel (Z_{S}), which is a positive number greater than one, is a characteristic design parameter for each planetary set and is denoted by $\alpha = Z_{\mathrm{R}}/Z_{\mathrm{S}}$. Friction losses in gear pairs are reported approximately with an efficiency of gear pair with external gearing (η_{EX}) and an efficiency of gear pair with internal gearing (η_{IN}), which can be represented by a total mechanical efficiency ($\eta_0 = \eta_{\mathrm{EX}}\eta_{\mathrm{IN}}$), internal to the planetary set. It is usually accepted for external gearing $\eta_{\mathrm{EX}} =$

0.97 and for internal – η_{IN} = 0.985. However, the manufacture quality of the gear wheels and the size of torque transmitted must be taken into account, which significantly affects these values.

For the scheme in figure 2 (a) the following dependencies are valid:

- By using the basic equation of kinematics for a simple planetary set

$$\omega_S + \alpha\omega_R = (1+\alpha)\omega_C \quad \longrightarrow \quad \omega_1 + \alpha\omega' = (1+\alpha)\omega'', \tag{4}$$

and by dividing each member in it by ω'' or ω_2 (since $\omega'' = \omega_2$) using equation (2) and equation (3) the required relationship between the kinematic ratios and parameter alpha is obtained:

$$i = \frac{\alpha}{1+\alpha-1/i_E}; \tag{5}$$

- By using the equilibrium condition of the torques of the three external units of the planetary set

$$T_C = T_R + T_S = T' + T_1\,, \tag{6}$$

the internal relationship between the torques of the sun wheel and ring wheel

$$T_S = \frac{1}{\alpha}\eta_0 T_R, \tag{7}$$

and equations (3), the torques of the electrical machines and the carrier by the torque of the input can be expressed as follows:

$$T_1 = T'\eta_0/\alpha\,; \quad T_2 = T'\eta_0 K_E/\alpha\,; \quad T_C = T'(1+\eta_0/\alpha). \tag{8}$$

On the other hand, for the output torque is valid

$$T'' = T_C + T_2. \tag{9}$$

The dependence for the torque ratio (2) is obtained by substituting (8) in the expression (9) for T'':

$$K = 1 + \eta_0(1+K_E)/\alpha. \tag{10}$$

For the internal connection between the torques of the sun wheel and of the ring wheel of the planetary mechanism another case is also possible

$$T_R = \alpha\eta_0 T_S \quad \longrightarrow \quad T' = \alpha\eta_0 T_1, \tag{7'}$$

when in the relative movement (at conditionally stationary carrier) the driving unit is the sun wheel, and the driven one is the ring wheel – in case of reversing or forwarding with high speed. This is determined by the directions of forces and velocities in the units by constructing plans of forces and velocities [7]. In this case, instead of equation (10), equation (10') is obtained through similar calculations:

$$K = 1 + (1+K_E)/\alpha\eta_0. \tag{10'}$$

For the scheme in figure 2 (b) similar dependences are used, taking into account the differences of the connections between the electric machines and the external units of the planetary mechanism and the characteristic feature $\omega' = \omega_1$:

- By dividing each member in equation (4) by ω' or ω_1 in this case for the kinematic gear ratio (2) is valid:

$$i = (i_E + \alpha)/(1+\alpha); \tag{11}$$

- And to determine the dependence of the torque ratios (2), the torques of the input and output by the torque of the electric machine E_2 are expressed by using equations (3), (6) and (7):

$$T_C = T_S + T_R \quad \longrightarrow \quad T'' = T_2 + T_2\alpha/\eta_0 = T_2(1+\alpha/\eta_0); \tag{12}$$

$$T' = T_1 + T_R = T_2/K_E + T_2\alpha/\eta_0 = T_2(1/K_E + \alpha/\eta_0); \quad (13)$$

$$K = \frac{1+\alpha/\eta_0}{1/K_E + \alpha/\eta_0}. \quad (14)$$

For modes, where the sun wheel is driving and the ring wheel is driven – by equation (3), (6) and (7'):

$$K = \frac{1+\alpha\eta_0}{1/K_E + \alpha\eta_0}. \quad (14')$$

Equations (5), (10), (10') and (11), (14), (14') can be used to estimate the change in the converting properties and efficiency ($\eta = Ki$) of the power split transmission as a whole as a function of the change in the electrical part (EVT) parameters and the planetary mechanism parameter for different operating modes.

2.2. Graphical method for visual representation of the connections between the elements of the power split transmission with double power separation

In order to improve the use of electric machines at high speeds of the car, the division of the input power flow into two mechanical and one electric branch is applied [8]. Two planetary sets are used, connected according to the kinematic schemes in figure 3 and figure 4, which show one possible way for the flow of the streams.

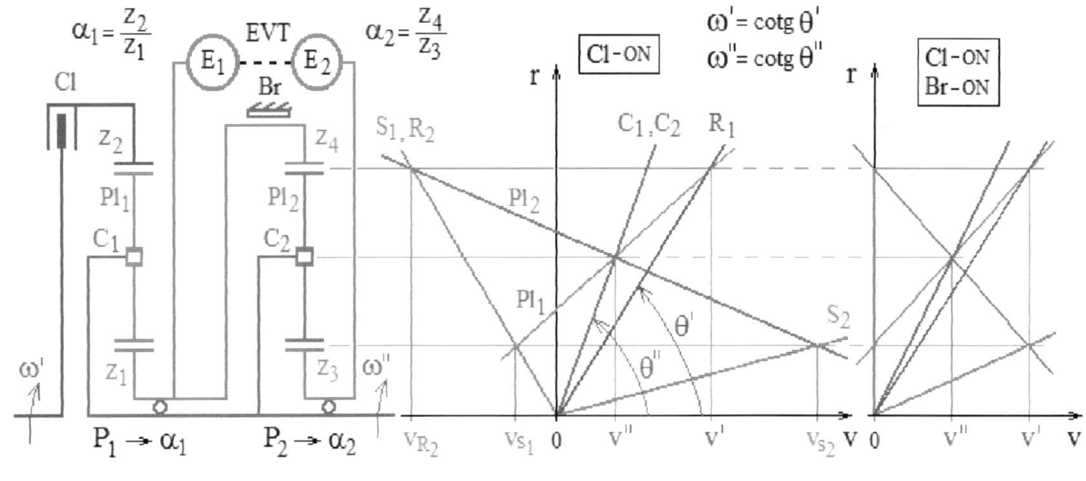

Figure 3. Scheme of a hybrid transmission with double power split and speed plan at two operating modes.

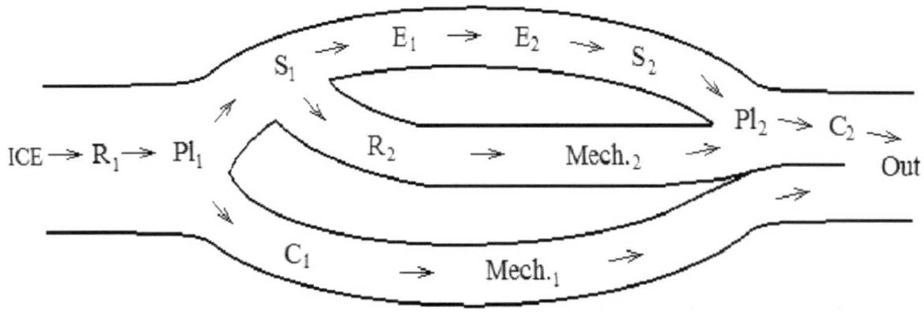

Figure 4. Schematic diagram of the flows in double separation.

Figure 3 shows also a graphical representation of the input (ICE) and output speeds of the entire drive system and of the individual units of the mechanism (electrical machines) as straight lines. On each unit (gear wheel or carrier) corresponds a straight line with a certain slope. Since the linear velocities of the units (v) are plotted on a certain scale along the x-axis of the velocity plan, and the radii (r) are plotted on the y-axis, the slope of the straight lines (θ) relative to the x-axis shows visually the angular velocity of their respective units: $\omega = v/r = \text{cotg}\,\theta$. This presentation is more convenient when the mechanisms have more units and makes it possible to estimate the change in the gear ratio $i_g = \omega'/\omega''$ by comparing the angular velocities (slope of the lines) of the input and output unit. Different operating modes achieved as a result of changes in the speed of rotation of the internal combustion engine or the electric machines can also be compared. Two modes with the same ICE speed are given for illustration – in the first case it refers to the state of the system when all units rotate at angular speeds other than zero (most common case), and in the second – when the brake is applied and the speed of the electric machine E_1 is zero (θ=90°). When starting, the speed of the exit is initially zero and this forces the two electric machines to rotate in different directions (θ_{E1}>90° and θ_{E2}<90°) because they are connected respectively to the ring wheel and the sun wheel on planetary set P_2, and its carrier is stationary. Then the power through the electrical branch is the greatest. When accelerating, the output speed gradually increases and this leads to a decrease in absolute value of the angular velocities of electric machines at the same mode of ICE. This also implies a gradual reduction of the power, transmitted through the electrical branch, at acceleration of the car and an increase in the efficiency of the transmission as a whole – equation (1').

When the electric machine E_1 reduces its speed to zero, the brake Br is applied automatically, which allows the movement to continue either only with the ICE (Conventional with gear ratio $i_g = 1 + 1/\alpha_1$), when the electric machine E_2 is switched off, or only with electric machine E_2 (Electric 1 with gear ratio $i_g = 1 + \alpha_2$) – with disconnected clutch Cl and ICE. When the brake and clutch are on, it is also possible to charge the battery from an electric machine E_2, when it works as a generator and is driven by the ICE, or by the kinetic energy of the car when braking (Regeneration with the clutch disengaged).

Starting can be done either in two-stream mode (clutch on) with the largest share of electrical power, or only with an electric machine E_2 (brake included), if there is sufficient battery power – for example in urban traffic conditions. Several more modes are possible:

- Parallel hybrid 1, the electric machine E_2 works as a motor together with the ICE;
- Parallel hybrid 2, both electric machines operate in motor mode with mechanical differential connection together with the ICE, with brake off;
- Electric 2, with two electric machines, connected differentially by planetary set P_2 at brake off.

The described operating modes are systematized in table 1.

When the brake is applied, the mechanical constant gear ratio is determined by applying equation (4) for the respective planetary set (P_1 and/or P_2), which is active, or for both and taking into account which unit is input, which is output and which is stationary.

Table 1. Modes of operation of the transmission with double separation of the mechanical flow.

Mode	Cl	Br	ICE	E_1	E_2	Batt	Gear ratio
Conventional	On	On	On	Off	Off	Off	$1 + 1/\alpha_1$
Charging by ICE	On	On	On	Off	G	On +	$(1 + 1/\alpha_1)/(1 + \alpha_2)$
Parallel hybrid 1	On	On	On	Off	M	On -	ICE: $1 + 1/\alpha_1$; E_2: $1 + \alpha_2$
Parallel hybrid 2	On	Off	On	M	M	On -	variable
Regeneration	Off	On	Off	Off	G	On +	$1/(1 + \alpha_2)$
Electrical 1	Off	On	Off	Off	M	On -	$1 + \alpha_2$
Electrical 2	Off	Off	Off	M	M	On -	variable
Double power split	On	Off	On	G/M	M/G	Off	variable

Battery status: On + charge; On – discharge

3. Characteristics of combined power split transmissions

The dual power split mode allows for optimal use of electric machines with almost twice smaller power compared to the single power split mode but in a narrower speed range shifted to high speeds. For example, when using internal combustion power in the range of 50 to 250 kW, the required power of electrical machines in single separation is in the range of 40 to 160 kW, and in double separation – from 50 to 80 kW, but at higher internal combustion power – from 150 to 250 kW. Therefore, it is appropriate to combine the two variants of power split schemes. Such hybrid drive systems that allow for operation on both schemes are called Two-Mode Hybrid [9]. Figure 5 shows an example of such a transmission with three planetary sets, two electric machines, two brakes and two clutches.

Table 2 shows the possible modes that are provided by the scheme for different combinations of turn on clutches and brakes.

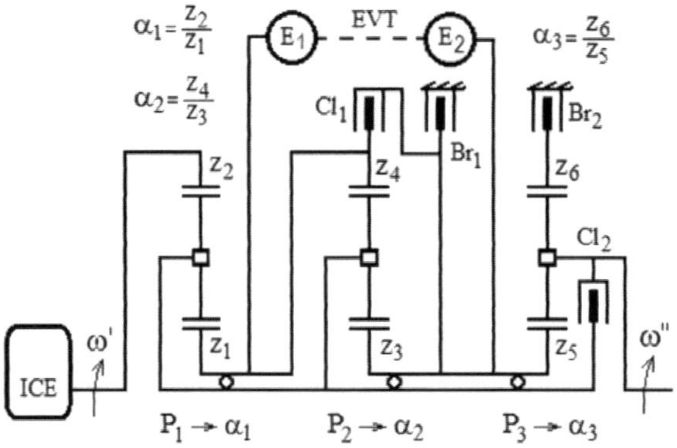

Figure 5. Kinematic scheme of Two-Mode Hybrid.

With this scheme, two stepless ranges are achieved, which cover the entire required working range of gear and in principle correspond to the single and double power separation. Stepless ranges are suitable for low and high speed starting and acceleration modes, and fixed gear ratios (Conventional mode with four mechanical gears) can be used when driving in established mode to avoid losses in electrical machines.

Table 2. Operating modes of Two-Mode Hybrid power split transmission.

Mode	Cl_1	Cl_2	Br_1	Br_2	ICE	E_1	E_2	Gear ratio	Value
Simple power split (1)	Off	Off	Off	On	On	G/M	M/G	variable	$\infty \div 1.7$
Fixed gear 1	On	Off	Off	On	On	Off	Off	$1+\alpha_3$	3.69
Fixed gear 2	Off	On	Off	On	On	Off	Off	$1+\alpha_3/\alpha_1\alpha_2$	1.7
Double power split (2)	Off	On	Off	Off	On	G/M	M/G	variable	1.7÷0.5
Fixed gear 3	On	On	Off	Off	On	Off	Off	direct	1
Fixed gear 4	Off	On	On	Off	On	Off	Off	$1-1/\alpha_1\alpha_2$	0.74
Electrical	Off	Off	Off	On	Off	Off	M	$1+\alpha_3$	3.69
Regenerative braking	Off	Off	Off	On	Off	Off	G	$1/(1+\alpha_3)$	0.27

The gear ratios are expressed as functions of the parameters (α) of the active planetary sets by applying equation (4). For example for gear ratio 2:

$$P_3 \longrightarrow \omega_S + \alpha_3 0 = (1+\alpha_3)\omega'' \longrightarrow \omega_S = (1+\alpha_3)\omega''; \quad (15)$$

$$P_2 \longrightarrow (1+\alpha_3)\omega'' + \alpha_2\omega_R = (1+\alpha_2)\omega'' \longrightarrow \omega_R = \omega''(\alpha_2-\alpha_3)/\alpha_2; \quad (16)$$

$$P_1 \longrightarrow \omega''(\alpha_2-\alpha_3)/\alpha_2 + \alpha_1\omega' = (1+\alpha_1)\omega'' \longrightarrow i_2 = \omega'/\omega'' = 1+\alpha_3/\alpha_1\alpha_2. \quad (17)$$

These capabilities of the scheme, as well as the reduced size of electric machines, make it suitable not only for passenger cars but also for trucks [10].

A similar solution, which uses only a single power split, but with two stepless ranges, is shown in figure 6 (a). These ranges are realized by an additional Ravigneaux mechanism, connected in series in the electric stream, and two brakes, with which the ranges are switched [11]. The aim is to expand the range of the whole system, while reducing the size of the electric machines to be built into the transmission, without significantly increasing its mass and dimensions.

The Ravigneaux mechanism is often used in conventional automatic transmissions and is a unified unit. It allows for the use of three different planetary sets combined in one block. The first is ordinary (R, Pl_2, S_2), the second consists of two sun wheels and two sets of successively engaged planetary wheels on a common carrier (S_2, Pl_2, Pl_1, S_1) and the third – of a ring wheel, two sets of successively engaged planetary wheels of common carrier and small sun wheel (R, Pl_2, Pl_1, S_1). In this case, only the first two planetary rows are used, for which also applies equation (4) when determining the gear ratios A peculiarity of the second planetary set is that in the calculations the large sun wheel can be regarded as a fictitious ring wheel. Then the parameter α_2 equals Z_{S_2}/Z_{S_1}, and in the equation instead of ω_R, ω_{S_2} is used and instead of ω_S, ω_{S_1} is used. Nevertheless, the signs in the equation are preserved, because instead of one external and one internal gearing, there are three external gearing.

When the brake Br_1 is applied, the following is obtained:

$$\omega_{S_1} + \alpha_2\omega_{S_2} = (1+\alpha_2)\omega_C \longrightarrow 0 + \alpha_2\omega_2 = (1+\alpha_2)\omega'' \longrightarrow \omega_2/\omega'' = 1 + 1/\alpha_2. \quad (18)$$

The other range is achieved by applying brake Br_2 and disabling brake Br_1, when the first planetary set is active. The gear ratio in the electrical branch is obtained by analogous calculations:

$$\omega_2/\omega'' = 1 + \alpha_1. \quad (19)$$

In this case, the following operating modes are possible, systematized in table 3.

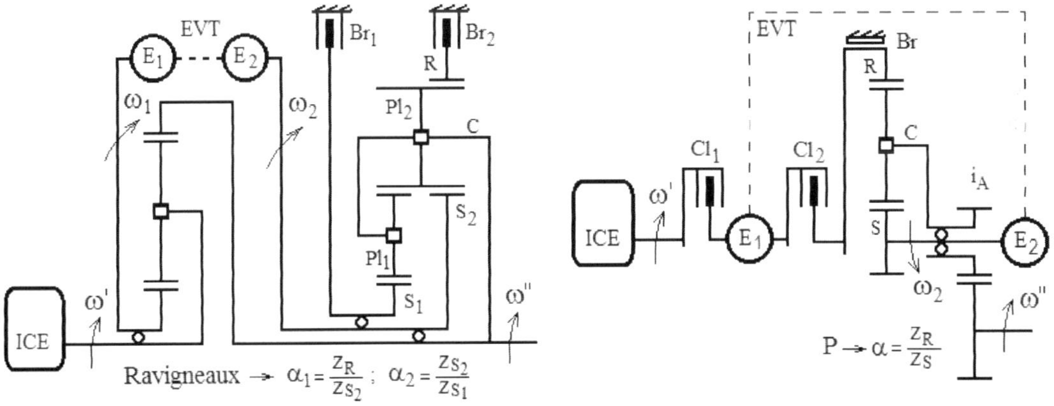

a) Power split transmission with two EVT ranges b) Power split transmission with E-REV

Figure 6. Kinematic schemes of power split with single division and additional possibilities.

Table 3. Modes of operation of the scheme of figure 6 (a).

Mode	Br_1	Br_2	ICE	E_1	E_2	Gear ratio
Power split 1 (EVT at low driving speed)	Off	On	On	G/M	M/G	variable
Power split 2 (EVT at high driving speed)	On	Off	On	G/M	M/G	variable
Electrical 1	Off	On	Off	Off	M	$1+\alpha_1$
Electrical 2	On	Off	Off	Off	M	$1+1/\alpha_2$
Regenerative braking 1	Off	On	Off	Off	G	$1/(1+\alpha_1)$
Regenerative braking 2	On	Off	Off	Off	G	$1/(1+1/\alpha_2)$

The scheme in figure 6 (b) provides further expansion of the range of autonomous vehicle movement by ICE, when the battery charge is exhausted (E-REV, Extended-Range Electric Vehicle). This corresponds in principle to a serial hybrid, but smaller electric machines are used, as they are mainly designed to work in the electrical branch of the power split transmission on a simple split scheme (Output split) [12]. To increase the acceleration when driving only with electric machines, the power of both machines (E_1 and E_2), differentially connected by the planetary set, can be used, while in economical movement only E_2 is used. These modes are achieved by two clutches and one brake (table 4).

Table 4. Operating modes of the scheme of figure 6 (b).

Mode	Cl_1	Cl_2	Br	ICE	E_1	E_2	Gear ratio
Electrical 1	Off	Off	On	Off	Off	M	$\omega_2/\omega'' = -(1+\alpha)i_A$
Electrical 2	Off	On	Off	Off	M	M	variable
Serial hybrid	On	Off	On	On	G	M	variable
Power split	On	On	Off	On	G/M	M/G	variable
Regenerative braking	Off	Off	On	Off	Off	G	$\omega''/\omega_2 = -1/(1+\alpha)i_A$

4. Conclusion

The proposed methodology makes it possible to obtain the characteristics of power-split schemes (i, K) as functions of the characteristics of the electric branch (i_E, K_E) and the planetary mechanism (α, η_0). They can easily be used to calculate the degree of impact of each parameter on the total efficiency, on the range of gear ratio change and on the operating modes of the ICE and the electrical machines. The proposed graphical method can be used to study and visualize the changes in the angular velocities of the individual elements of the system – input (ICE), output, E_1 and E_2 and the resulting ratios between them (gear ratios) in different operating modes.

In this way, the possibilities of power split schemes for realization of different stepped and stepless ranges of gear ratio change and for expansion of the range of operation with high efficiency can be explored. The methodology allows for experimentation with different combinations of circuits, which provide not only an optimal combination of single and double separation, but also additional simpler drive options – only with internal combustion engines, only with electric machines and regenerative braking.

The research shows that with power split transmissions a combination of the optimal modes of the machines and the use of the mechanical characteristics in the necessary range can be achieved through appropriate control. Depending on the operating modes, it is expedient to reduce the relative share of the electric power at lower values of efficiency in the electric branch and vice versa.

The present work can also be used to develop simulation models corresponding to different control strategies for hybrid drives in order to optimize them according to different criteria.

References

[1] Ivanchenko P N, Savelev N M, Shapiro B Z and Vovk V G 1962 *Electromechanical transmissions (theory and calculations)* (Moscow, Mashgiz) p 432 (In Russian)

[2] Yang Y, Hu X, Pei H and Peng Z 2016 Comparison of power-split and parallel hybrid powertrain architectures with a single electric machine: Dynamic programming approach *Appl. Energy* **168** pp 683–90

[3] Yao M, Qin D, Zhou X, Zhan S and Zeng Y 2019 Integrated optimal control of transmission ratio and power split ratio for a CVT-based plug-in hybrid electric vehicle *Mech. Mach. Theory* **136** pp 52–71

[4] Bertoluzzo M, Bolognesi P, Buja G and P Thakura 2007 Role and technology of the power split apparatus in hybrid electric vehicles *IECON – 33rd Annual Conf. of the IEEE Industrial Electronics Society (Taipei)* pp 256–61

[5] Wang L, Cui Y, Zhang F and Li G 2020 Architectures of planetary hybrid powertrain system: Review, classification and comparison *Energies* **13**(2) pp 329–53

[6] Morchev E, Gigov B and Dimitrov E 2011 Fahrverhalten und kraftstoffverbrauch eines hubstaplers mit leistungsverzweigtem wandlergetriebe *Proc. Int. Sci. Conf. BulTrans-2011 (Sozopol, Bulgaria)* **3** pp 155–60
[7] Gigov B 2011 Untersuchung der mehrsteg-planetensätze für PKW-wandlerautomaten *Proc. Wissenschaftliche Konferenz Innovationen und Wettbewerbsfähigkeit (24-25 November FDIBA, Sofia, Bulgaria)* pp 149–58
[8] Conlon B 2005 Comparative analysis of single and combined hybrid electrically variable transmission operating modes *SAE Transactions* **114** pp 1265–75
[9] Nitz L,Truckenbrodt A and Epple W 2006 The new two mode hybrid system from the global hybrid cooperation *27th International Vienna Motor Symposium, Reihe* **12**(622)
[10] Tamai G, Reeves S and Grewe T 2010 Truck utility & functionality in the GM 2-mode hybrid *SAE Int. J. of Engines* **3**(1) pp 620–30
[11] Hermance D and Abe S 2006 Hibrid vehicles lessns learned and future prospects *SAE Technical Papers* 2006-21-0027
[12] Grebe U and Nitz L 2011 Voltec – das antriebssystem für Chevrolet Volt und Opel Ampera *MTZ Motortech Z* **72**(5) pp 342–51

IOP Conf. Series: Materials Science and Engineering 1002 (2020) 012032 doi:10.1088/1757-899X/1002/1/012032

Efficiency of automotive electric supercharging compressors

Pierre Podevin[1,6], Christelle Périlhon[1], Michael Deligant[2], Plamen Punov[3], Adrian Clenci[4] and Guillaume De La Bourdonnaye[5]

[1] Laboratoire LIFSE, Le Cnam, Conservatoire national des arts et métiers, 75003 Paris, France
[2] Laboratoire LIFSE, Arts et Métiers ParisTech, 75013 Paris, France
[3] Department of Combustion Engines, Automobile Engineering and Transport, Technical University – Sofia, 1756 Sofia, Bulgaria
[4] Department Automobiles and Transport, University of Pitesti, 110040 Pitesti, Romania
[5] Air charging system, Valeo, 95892 Cergy Pontoise, France

[6] Email: pierre.podevin@lecnam.net

Abstract. The automotive electric supercharger provides high torque at low engine speeds. During accelerations, it fills the power gap due to the turbocharger's inertia and avoids turbo lag. Its operation is of short duration because once this deficit has been juggled, it must be stopped in order not to generate unnecessary electricity consumption and to avoid overheating of the electric motor which would compromise its lifetime. Usually the power of automotive turbocharger is calculated assuming an adiabatic air compression and the adiabatic efficiency is assessed. At low speed or for low compression ratio, which is the case of electric supercharger, this hypothesis is no more valid and special experiments have to be done. On our test bench the compressor is driven by a turbine supplied with cold compressed air. Between the compressor and the turbine, a torque meter is inserted to measure the power given to the compressor shaft. The tests were conducted with a non-insulated compressor and an insulated compressor for eight iso-speeds. Additional tests were carried out to assess the actual power supplied to the fluid including bearing losses measurements and to assess the influence of different preload springs used to counter the axial thrust of the compressor. This article presents and analyses the results of this work.

1. Introduction

The automobile transport is responsible to a significant part of greenhouse gases due to burning the fossil fuels in internal combustion engines. In order to reduce global warming, European commission imposed a new CO_2 emissions target for passenger cars that will be fully applied in 2021. The new reference target value was set to 95 g/km which means reduction by 27% compared to previous legislation [1]. Moreover, more restrictive targets were defined for 2025 and 2030 that account to 15% and 37.5% CO_2 reduction compared to 2021. Based on the real vehicle CO_2 emissions, it is obviously that in long term a significant number of zero and low emission vehicle need to be in the fleet. However, in short term the main energy source will still remain the internal combustion engine with rising level of electrification and hybridization [2]. In order to increase the engine thermal efficiency turbocharging technology has been applied for many years both in diesel and gasoline engines. However, turbochargers cannot meet all the engine requirements, especially the engine performance at low speed and transient operating model. That could be overcome by applying modern technics such as: variable geometry turbine, dual stage turbocharging (widely used in diesel engines), twin scroll turbines or using additional compressor [3-5]. This additional compressor could be mechanically driven or electrically assisted. Electrically assisted supercharger could be adapted to nominal voltage within the range of 12 V to 48 V [6], thus it can be applied to mild-hybrid passenger cars as well. Usually, it operates for limited time (less than 20 s) during acceleration phase at low engine speed and full load. Therefore, it offers better engine acceleration at low speed in combination with high torque and lower fuel consumption up to 25% [7].

In order to evaluate overall engine efficiency, the electrical consumption of the additional supercharger has to be taken into accounts. This consumption is related to operation strategy, mechanical losses, motor efficiency and compressor efficiency. The most significant losses are aerodynamic, thermal and mechanical losses within the compressor. A literature review shows that the performance of this type of electric supercharger is not covered in publications. For that reason, this article aims to study the isentropic efficiency on an electrically assisted supercharger based on experimental test. The tests were conducted at different operating conditions as the effect of insulation and bearings loads was evaluated. The efficiency was evaluated based on thermodynamic parameters and compressor power balance.

2. Mathematical background

The power and efficiency of a turbomachine can be expressed according to the first law of thermodynamics. Applied to a compressor between inlet point 1 and outlet point 2, for a steady-state flow and neglecting the potential energy, the internal work W_i is [8]:

$$W_i = (h_{i2} - h_{i1}) - Q, \tag{1}$$

where W_i is internal work on the compressor wheel; h_i – stagnation enthalpy; Q – heat exchanged with the external environment; $Q < 0$ if the heat is transferred to the outside; $Q > 0$ if the heat is received from the outside; $Q = 0$ for an adiabatic evolution.

It should be noted that the internal work W_i on the transmission shaft is the work received by the compressor wheel in contact with the fluid. It is therefore equal to the work of the mobile channels added to that absorbed by disk friction [8]:

$$W_i = W_{ch} + W_{frd}, \tag{2}$$

where W_{ch} is impeller channel work; W_{frd} – disk friction work.

This internal work is also equal to the work on the shaft measured by a torque meter, reduced by the work absorbed by the bearings:

$$W_i = W_{shaft} - W_{frb}, \tag{3}$$

where W_{shaft} is work on the shaft measured by the torquemeter; W_{frb} – work of the bearing system.

In the case of an adiabatic machine, the work for a perfect ideal gas is thus simply written:

$$W_{adia} = h_{i2} - h_{i1} = c_p(T_{i2} - T_{i1}), \tag{4}$$

where T_i is stagnation temperature; c_p – specific heat at constant pressure.

This work is then "easily" determined by measuring the two inlet and outlet temperatures and isentropic efficiency is:

$$\eta_{is} = \frac{W_{is}}{W_{adia}} = \frac{T_{i2is} - T_{i1}}{T_{i2} - T_{i1}}, \tag{5}$$

where T_{i2is} is stagnation temperature of the isentropic theoretical evolution obtained by:

$$T_{i2is} = T_{i1}\left[\frac{p_{i2}}{p_{i1}}\right]^{\frac{\gamma-1}{\gamma}}, \tag{6}$$

where p_i is stagnation pressure.

In fact, while the adiabatic assumption can be used for large turbomachines, the operation of small turbomachines diverges from it at low speeds and low loads. Particularly on test benches with a turbine supplied with hot gas, heat transfers are established from the turbine to the colder compressor through mechanical components and lubricating oil. Many studies have been done to estimate heat exchange [9-11]. If Q is known, calculation of W_i according to equation (1) does not present major difficulties.

Another way to determine the internal work, is to measure W_{shaft} the mechanical work on the shaft using a torque meter and to determine the mechanical losses W_{frb} of the bearings. Then W_i is calculated from equation (3). Calculations of mechanical losses are presented in various papers [12-15].

In our laboratory, we have used a special torque meter to measure directly the mechanical work and done some experiments to determine the mechanical losses presented in this paper.

3. Test bench overview

The compressor is a prototype electric compressor developed by Valeo. It is driven by an 48V electric motor. It reduces turbocharger turbo lag and provides high torque at low engine speeds. Its design and purpose make it necessary for its operation to be intermittent. Also, the electric motor could not be used for our experiments.

To assess the torque transmitted to the compressor, we have used a special torquemeter able to run at 120000 rpm, the maximum speed of the compressor being 80000 rpm. The torque is measured with an accuracy of ±0.00021 Nm. Compressor speed and torque values are given by the torquemeter. This torquemeter is placed between the compressor and a turbine. The turbine acts as the electric motor for the compressor with the benefit of being able to operate continuously at a given operating point.

The turbine is fed by dry compressed air close to ambient temperature, which avoids hot gas supply problems but requires a large air compression and drying plant. A schematic of the test bench is given in figure 1. A specific assembly was set up in order to mount it on our bench, figure 2. The bearings and the preload spring were kept.

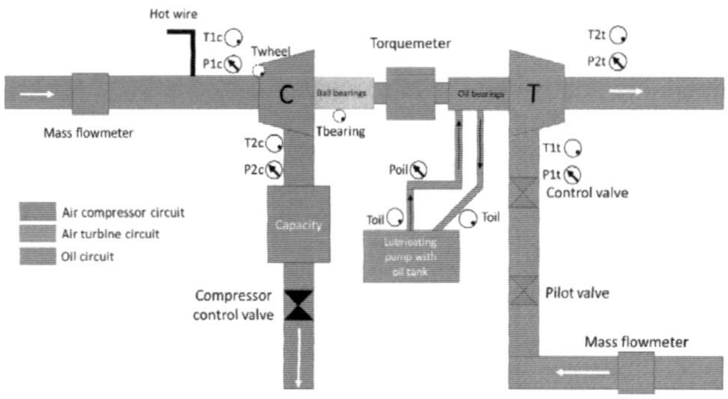

Figure 1. Schematic of the test bench.

Figure 2. Compressor Assembly.

The usual measurements were made, i.e. temperatures and pressures at the compressor inlet and outlet; the air flow rate was measured by a thermal mass flow meter. The temperatures were measured with platinum resistance thermometers PT100 and pressure with piezoresistive sensors. Additional measurements were made on the turbine side.

A capacity was fitted on the circuit to increase the surge phenomenon. The definition of the surge line and appearance of surge was presented in [16-17].

All the signals are converted into voltage in the range 0-10 V and sent to a data acquisition card. The data are visualized and acquired using a LabVIEW program. All the data are recorded in a file and processed in an Excel file.

4. Experimental results

The compressor characteristic curves are expressed as iso-speed values.

Since the measured temperature is not really the stagnation temperature, a α correction is applied to the measurement to calculate it in accordance with usual practice in our laboratory.

$$T_m = T_s + \alpha \frac{v^2}{2Cp}, \tag{7}$$

$$T_i = T_s + \frac{v^2}{2Cp}, \tag{8}$$

where T_m is measured temperature; T_s – static temperature; v – air flow speed.

The stagnation pressures are calculated from the static pressures measured assuming a reversible adiabatic evolution.

$$\frac{p_i}{p_s} = \left(\frac{T_i}{T_s}\right)^{\frac{\gamma}{\gamma-1}}. \tag{9}$$

The pressure ratio is:

$$Pr = \frac{p_{i2}}{p_{i1}}. \tag{10}$$

Characteristic curves represent stagnation Pressure ratio versus Air flow rate. Air flow and speed are corrected according to the following reference values: T_{ref} = 298 K, p_{ref} = 100 kPa.

According to equation (1), the internal compressor power, P_i is:

$$P_i = \dot{m}_a Cp(T_{i2} - T_{i1}) - \dot{Q}, \tag{11}$$

where $\dot{m}_a$ is air flow rate, which can also be written:

$$P_i = P_{enth} - \dot{Q}, \tag{12}$$

where P_{enth} being the enthalpic power:

$$P_{enth} = \dot{m}_a \, Cp(T_{i2} - T_{i1}). \tag{13}$$

This enthalpic power is equal to adiabatic power P_{adiab} for an adiabatic compression or if heat exchange can be neglected. Isentropic efficiency is calculated according to equation (5).

4.1. *Isentropic efficiency evaluation with and without insulation*

Measurements are made by steps of 5000 rpm from 30000 to 75000 rpm. Particular attention has been paid to the temperature stabilization. The isentropic efficiency is displayed directly on the computer control screen. When it is stabilized to within 1%, the point is recorded. This takes approximately 20 min for the first point (compressor outlet vane fully open) and then about 5 min for the following points.

First measurements were done without insulation, see figure 2. Then to reach adiabatic conditions mineral wool was wrapped all around the compressor figure 3.

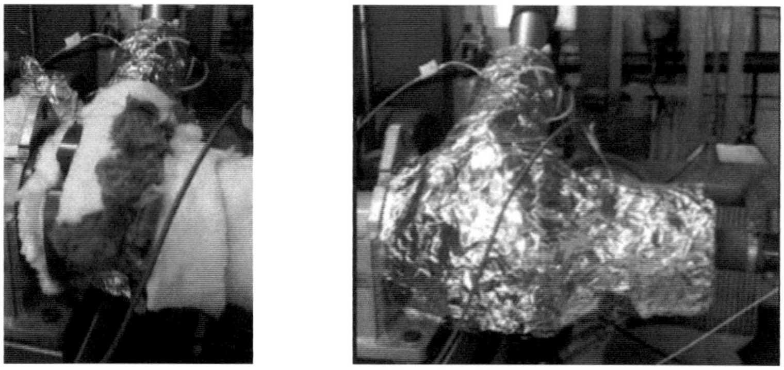

Figure 3. Insulation of the compressor.

Figure 4 shows the results of pressure ratio versus Air flow rate and figure 5 Isentropic efficiency versus air flowrate with and without insulation. For reasons of clarity, 3 speed lines are shown.

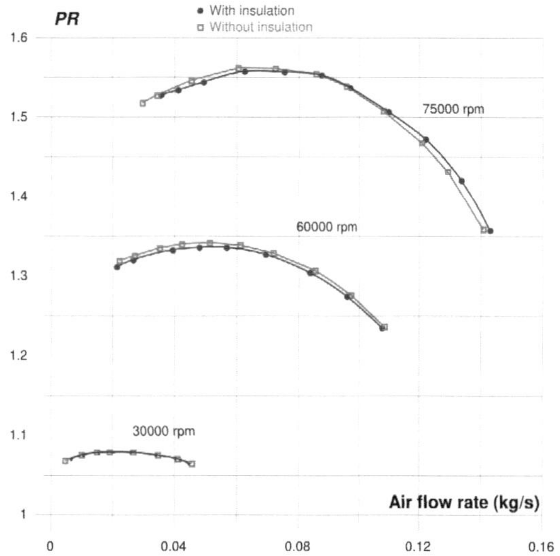

Figure 4. Comparison of pressure ratio vs air flow rate with and without insulation.

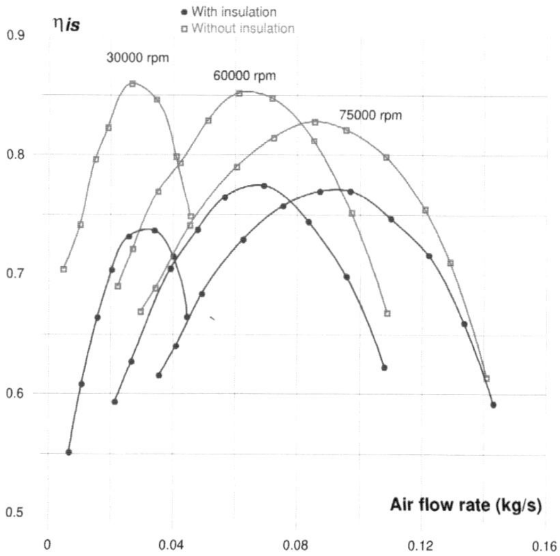

Figure 5. Comparison of isentropic efficiency vs air flow rate with and without insulation.

Insulation has no effect on the characteristic curve of pressure ratio vs air flow rate. The efficiency values with insulation are now in the range of 75 to 77%, maximum expected values from Valeo calculations.

By subtracting the enthalpic power before and after insulation, the heat losses during the initial setup can be estimated, figure 6.

Heat losses at low speeds are obviously lower than at high speeds, but due to a lower power delivery they have a much greater impact on the calculation of isentropic efficiency.

During the 75000 rpm test with insulation, a temperature reading of the surfaces was carried out with an infrared thermometer. It appears that the compressor is not completely isolated and that a small heat exchange still takes place with the frame of the torque meter. During this reading, the compressor outlet temperature was T_{2m} = 78°C and the temperature of the torquemeter bearings varied from 38 to 40°C, figure 9.

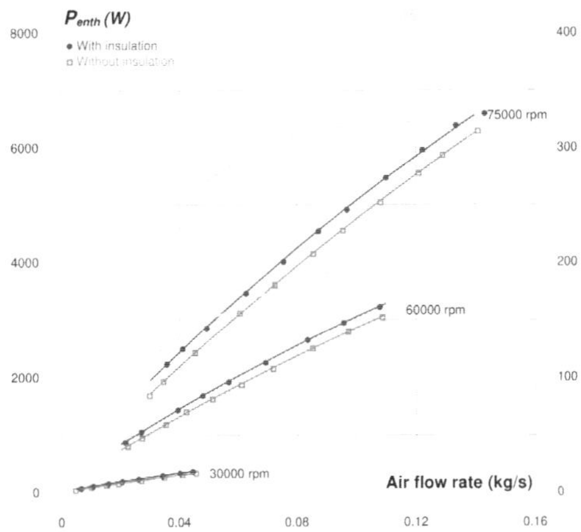

Figure 6. Comparison of heat losses vs air flow rate with and without insulation.

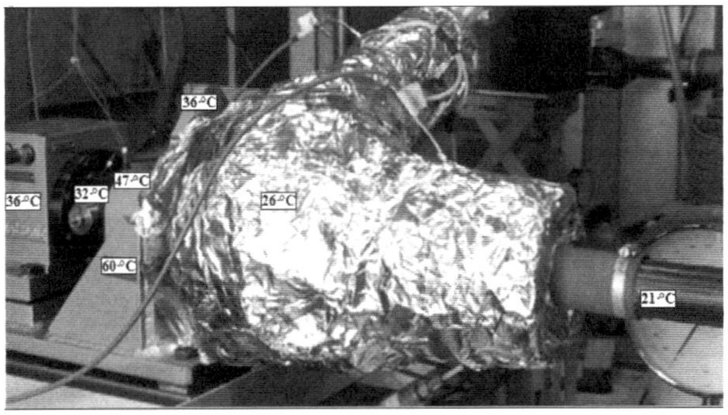

Figure 7. Measurements of the surface temperatures.

As it seemed difficult to provide better insulation, we conducted a test to observe the influence of insulation around the bearings by removing this insulation.

Temperature around the bearings reaches 80°C, figure 8, and the effect of this insulation on isentropic efficiency can be clearly seen, figure 9. Less insulation causes an artificial increase in isentropic efficiency.

Figure 8. Measurements of the surface temperatures without insulation of the bearings.

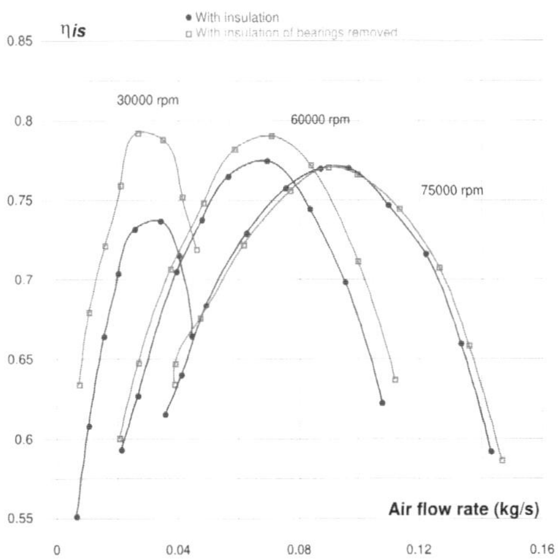

Figure 9. Isentropic efficiency vs air flow rate with and without insulation of the bearings.

From these experiments it appears that heat exchange cannot be totally avoided. We therefore tried to calculate the real power given to the air flow, considering the torque measured by the torquemeter.

4.2. Evaluation of shaft power with full insulation

The use of a torquemeter makes it possible to measure the real power given to the compressor shaft.

Here we considered the tests with full insulation. A comparison of the mechanical power measured by the torquemeter P_{shaft} and the compressor power P_{enth} is plotted in figure 10. The shaft power, P_{shaft} is obviously greater than the adiabatic compressor power P_{enth}. At high speed, i.e. 75000 rpm, there is a maximum gap of around 400 W. At medium speed, 60000 rpm, the medium difference is about 200

W. At 30000 rpm, the difference cannot be clearly seen in this figure. A zoom in at 30000 rpm figure 11 shows that P_{shaft}, is always greater than P_{enth}.

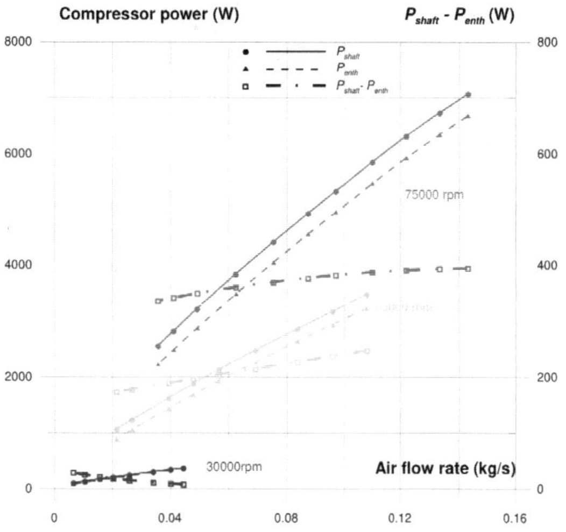

Figure 10. Comparison of insulated compressor enthalpic power and shaft power vs air flow rate.

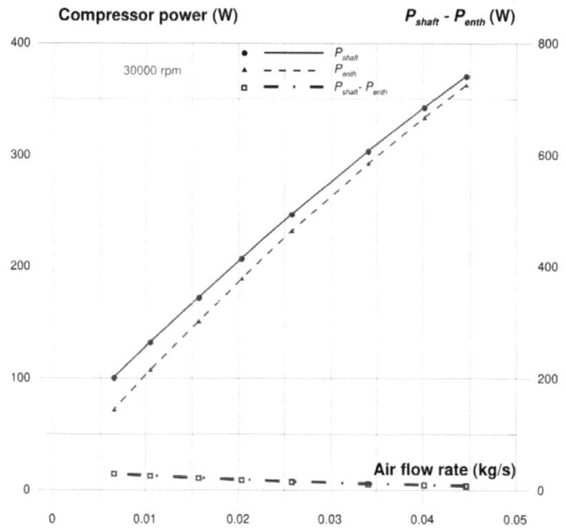

Figure 11. Comparison of insulated compressor enthalpic power and shaft power vs air flow rate (zoom).

However, it seems very difficult to measure the adiabatic power of the compressor P_{enth} accurately in this operating area due to a very low temperature rise between the compressor inlet and outlet. This temperature rise at the maximum flow rate is 8.1°C and 10.6°C at the minimum flow rate.

To determine the real internal power given to the fluid P_i, it is necessary to know the bearing friction losses P_{frb}. The internal power being equal to the power given to the shaft P_{shaft} minus the power due to bearing friction losses P_{frb}, equation (3) becomes:

$$P_i = P_{shaft} - P_{frb}. \tag{14}$$

To estimate these bearing losses, we performed the experiments detailed below.

4.3. Bearing friction losses

A new rotating assembly with a spacer instead of the wheel was supplied by Valeo figure 12.

The bearings were new. After running for 15 minutes at 30000 rpm, the torque decreased from 0.045 Nm (140 W) to 0.015 Nm (47 W). Nevertheless, this running-in period was insufficient, and a minimum friction power of 9 W was later obtained for 30000 rpm.

As bearing friction may be impacted by axial loads, measurements were performed varying axial load. The first experiment was conducted with the original spring and then a washer was added to increase the preload. Then Valeo provided us with several springs of different stiffnesses.

In real operating conditions, the aerodynamic gas load is estimated to be between 10 to 60 N by Valeo. Spring stiffnesses were measured on the pulling machine in Valeo's Cergy metrology laboratory.

At low load, the effect of spring stiffness can be clearly observed on the torque curves in figure 13. We measured the torque at 30000 rpm then up to 80000 rpm by steps of 10000 rpm. The speed was then decreased with the same step. Some hysteresis can be seen.

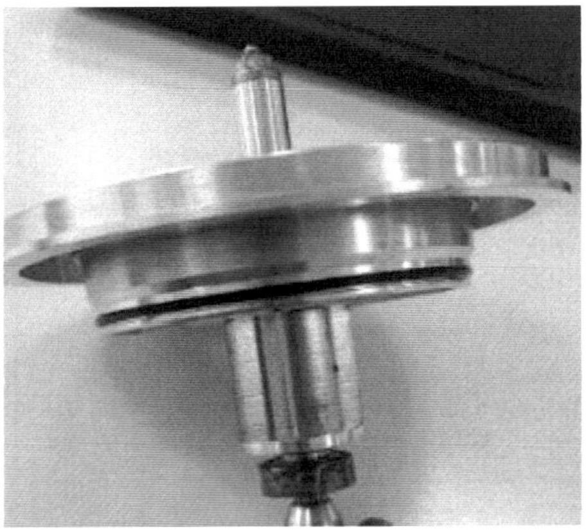

Figure 12. Rotor assembly with the wheel replaced by a spacer.

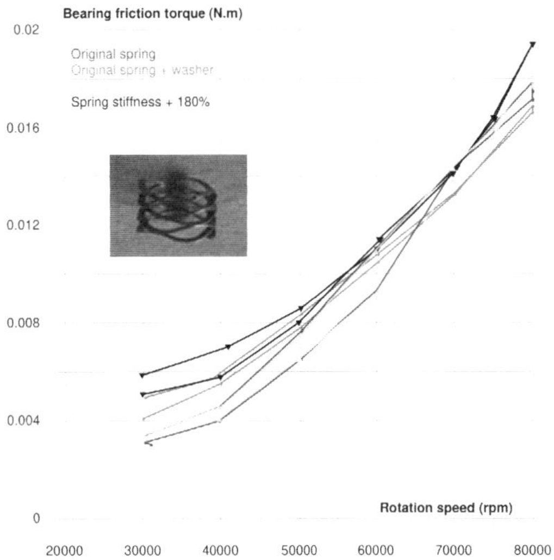

Figure 13. Bearing friction for different spring loads and rotation speeds.

As the axial thrust at low speed is weak, it appears necessary to choose the spring suited to the maximum load and not to oversize it in order to reduce friction losses at low speed.

Figure 14 plots friction power versus rotation speed for the original spring and for the spring with maximum stiffness. The friction of the original spring can be estimated by a power law. This law was later applied to calculate internal power.

The last experiments on bearing friction aimed to see the influence of temperature on bearing losses. Hot air was introduced by a heat gun at the outlet of the volute, to increase the temperature of the body which maintains the bearings from 30 to 80°C. The reduction in friction losses was relatively small, about 10 W at 75000 rpm, figure 15.

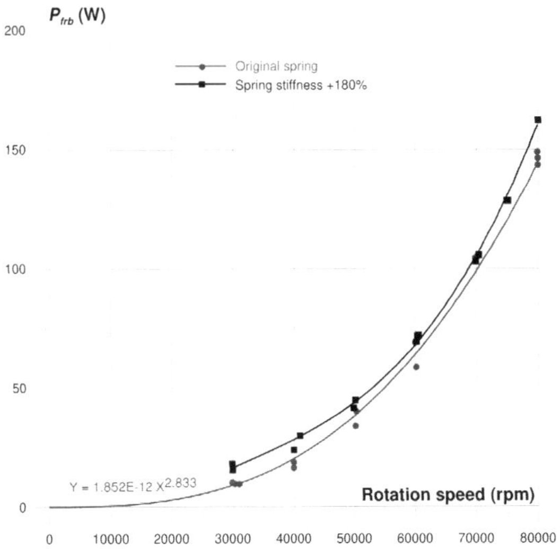

Figure 14. Bearing friction for different spring loads vs rotation speeds.

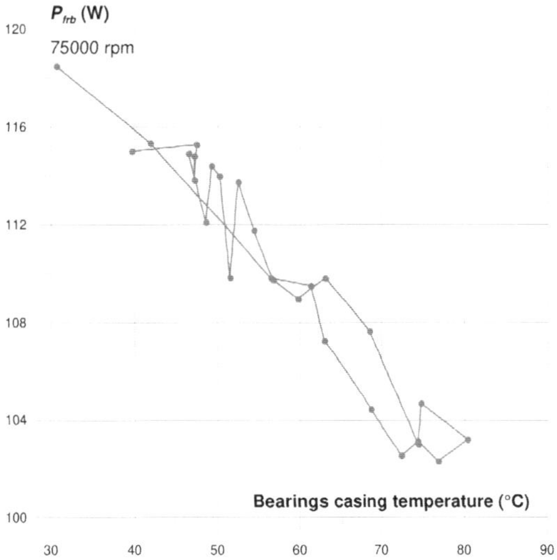

Figure 15. Variation of bearing friction power in the case of heated body at 75000 rpm.

4.4. Analysis of the results

From the previous tests, it is possible to calculate the internal power P_i supplied to the fluid according to equation (14). The mechanical power is measured by the torquemeter and the bearing friction loss power P_{frb} calculated from the proposed power law figure 14. According to equations (12) and (14) thermal losses $\dot{Q}$ in this insulated case are given by:

$$\dot{Q} = P_{enth} - P_{shaft} + P_{frb}. \tag{15}$$

In figure 16, we added the speed 50000 rpm to observe the evolution of these thermal losses more closely. As Q is negative, which means that heat is transferred from the fluid to the ambient, on the figure $-\dot{Q}$ is represented.

For 60000 and 75000 rpm, the thermal losses increase with the flow rate, are almost constant at 50000 rpm and decrease at 30000 rpm. It should be noted that at 30000 rpm there are negative losses ($-Q<0$), which means that the fluid receives heat from bearing power losses.

As a first approximation by averaging the thermal losses $-\dot{Q}$, it is possible to express this as a function of the compressor rotation speed and compare these against the friction losses figure 17.

As a result, despite the care taken with the insulation, the thermal losses are significant.

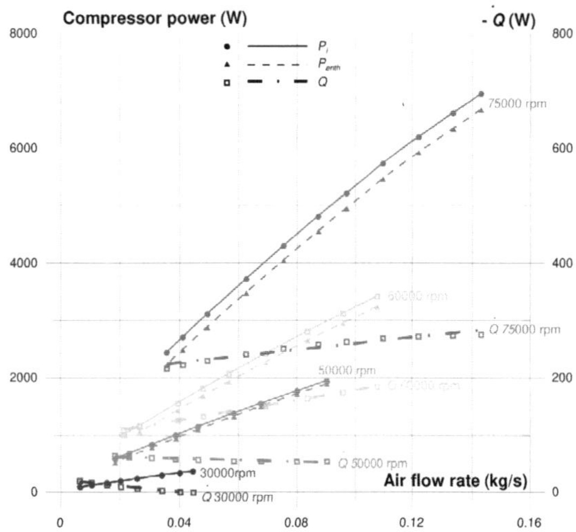

Figure 16. Insulated compressor power and thermal losses vs air flow rate.

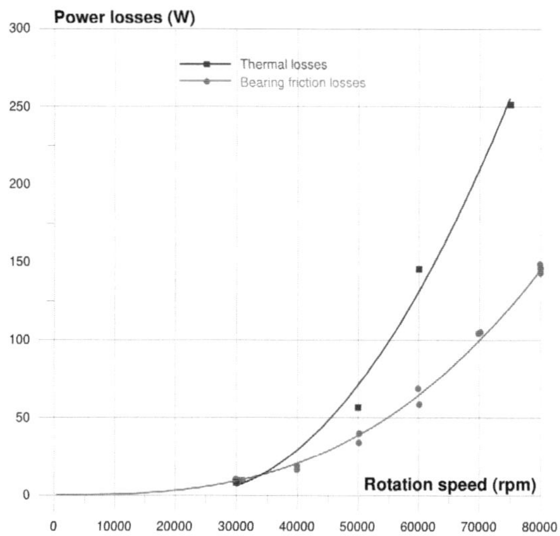

Figure 17. Friction and thermal losses for the insulated compressor vs rotation speed.

From equation (14), the power measured by the torquemeter and the proposed bearing friction power law figure 16 it is possible to calculate the internal power P_i supplied to the fluid and thus compare the isentropic efficiencies calculated from this power (16) with results obtained directly by measuring temperatures with the insulated compressor (5), see figure 18.

$$\eta_{is} = \frac{P_{is}}{P_i} = \frac{P_{is}}{P_{shaft} - P_{frb}} = \frac{\dot{m}_a c_p (T_{i2is} - T_{i1})}{P_i}. \tag{16}$$

As the calculated internal power is higher than the adiabatic power, the result is a decrease in efficiency for results obtained from the torquemeter rather than temperatures. This is expected as, even when insulated, the compressor is not adiabatic and because thermal losses are not taken into account in the efficiency calculated by equation (5).

The maximum isentropic efficiency of the compressor decreases from 77.5 to 72.5%. The maximum efficiency is practically constant for each iso-speed.

From the test results, the mechanical efficiency η_m can be plotted, figure 19, P_i being calculated with equation (14).

$$\eta_m = \frac{Pi}{Pshaft}. \tag{17}$$

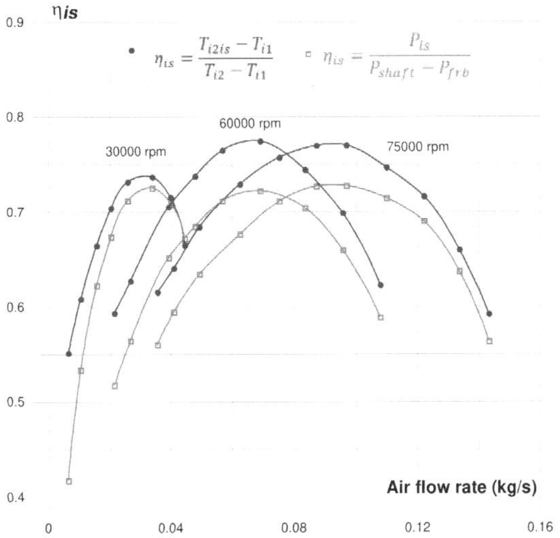

Figure 18. Isentropic efficiency vs air flow rate.

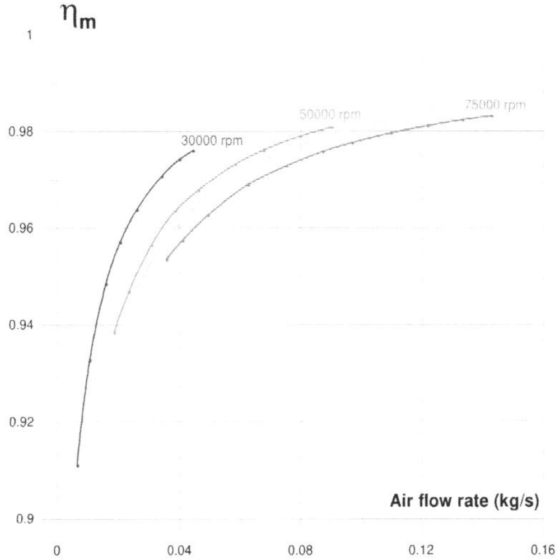

Figure 19. Mechanical efficiency vs air flow rate.

5. Conclusion

The tests revealed the influence of insulation on the performance measurement of the compressor. While the insulation has no influence on the Pressure ratio-Air flow rate characteristic curves, it does have an influence on the power transmitted from the compressor shaft.

If during compression a quantity of heat is exchanged with the outside environment, the power required for compression decreases. As a result, the compressor efficiency is improved.

However, if the measurement of compressor power is deduced from the first law of thermodynamics by considering the flow as adiabatic, the increase in isentropic efficiency calculated in this way is unrealistic.

By using ball bearings instead of oil bearings, it was hoped to encounter fewer difficulties in determining the actual performance of the compressor.

This is true for the determination of bearing losses. It has been shown that these are not very sensitive to axial thrust and therefore it is possible to calculate the internal power supplied to the fluid relatively easily.

As far as heat loss is concerned, simple insulation with mineral wool does not eliminate heat exchange. Some of the compression heat is transmitted by conduction to the compressor body and then to the torquemeter frame. At low engine speeds it is likely that it is the heat from friction losses of the bearings that is transmitted to the compressor.

It would therefore be advisable to ensure the connection of the compressor torquemeter with an insulating material in order to reduce thermal losses and possibly control the temperature of the bearings by a hydraulic circuit.

When considering the motorization of the compressor, what is important to know is the power that must be supplied to the compressor (not insulated). This is what the direct measurement with the torquemeter allows. However, this information is not sufficient. It is also necessary to ascertain that the compressor is adapted to its task by the adequacy of its operation with the demands required by the powertrain.

To be able to assess the efficacy of this machine, its efficiency is a perfectly adequate variable. Isentropic efficiency, why not? Polytropic or isothermal efficiency why not? All the powers of these theoretical evolutions are easy to calculate. However, the main problem remains to easily determine (without a torquemeter) the internal power when heat exchanges with the external environment cannot be neglected.

Acknowledgements
We would like to thank Valeo for funding this study and allowing us to use this compressor for further surge studies. We also thank the BTS Arras and BTS La Joliverie trainees as well as the engineering students of Cnam who made it possible to carry out these experiments successfully. The study was conducted within the frame of project "Amélioration de l'efficacité énergétique des véhicules routiers par optimisation du couplage moteur-turbocompresseur" funded by joint research program AUF-FRS 2018 (КП-06-Франкофония/3).

References
[1] European Commission. CO2 emission performance standards for cars and vans (2020 onwards) https://ec.europa.eu/clima/policies/transport/vehicles/regulation_en
[2] Ajanovic A and Haas R 2017 The impact of energy policies in scenarios on GHG emission reduction in passenger car mobility in the EU-15 *Renewable and Sustainable Energy Reviews* **68**(2) pp 1088–96
[3] Feneley A J, Pesiridis A and Andwari A M 2017 Variable Geometry Turbocharger Technologies for Exhaust Energy Recovery and Boosting-A Review *Renewable and Sustainable Energy Reviews* **71** pp 959–75
[4] Romagnoli A, Vorraro G, Rajoo S, Copeland C and Martinez-Botas R 2017 Characterization of a supercharger as boosting & turbo-expansion device in sequential multi-stage systems *Energy Conversion and Management* **136** pp 127–41
[5] Zi D, Zhang L, Chen B and Zhang Q 2019 Study of the electric-booster and turbo-generator system and its influence on a 1.5 L gasoline engine *Applied Thermal Engineering* **162** 114236
[6] Tran H H, Richard B, Gray K and Hall J M 2016 Developing a performance specification for an electric supercharger to satisfy a range of downsized gasoline engine applications *SAE Technical Paper*
[7] Baek S, Woo S, Kim Y and Lee K 2019 Prediction of turbocharged diesel engine performance equipped with an electric supercharger using 1D simulation *Energy* **185** pp 213–28
[8] Pluviose M and Périlhon C 2017 Turbomachines - Description et principes de base *Techniques de l'ingénieur* BM4280(2017)**v2** pp 1–17
[9] Bohn D, Heuer T and Kusterer K 2003 Conjugate flow and heat transfer investigation of a turbocharger, part 1: numerical results *Proc. of Turbo Expo 2003 (Atlanta, Georgia, USA)* Paper GT2003-38445 pp 715–22
[10] Baines N, Wygant K D and Dris A 2010 The Analysis of Heat Transfer in Automotive Turbochargers *Journal of Engineering for Gas Turbines & Power* **132**(4) 042301
[11] Chesse P, Chalet D and Tauxia X 2011 Impact of the heat transfer on the performance calculations of automotive turbocharger compressor *Oil & Gas Science and Technology - Revue d'IFP Energies nouvelles* **66**(5) pp 791–800
[12] Brouwer M D, Sadeghi F, Lancaster C, Archer J and Donaldson J 2013 Whirl and Friction Characteristics of High Speed Floating Ring and Ball Bearing Turbochargers *J. Tribol.* **135**(4) 041102
[13] Ashtekar A and Sadeghi F 2011 Experimental and Analytical Investigation of High Speed Turbocharger Ball Bearings *J. Eng. Gas Turbines Power* **133**(12) 122501
[14] Lamquin T and Gjika K 2009 Power losses identification on turbocharger hydrodynamic bearing systems: test and prediction *Proc. of ASME Turbo Expo (Orlando, Florida, USA)* GT2009-59599 pp 153–62
[15] Vanhaelst R, Kheir A and Czajka J 2016 A systematic analysis of the friction losses on bearings of modern turbocharger *Combustion Engines* **164**(1) pp 22–31
[16] Deligant M, Danlos A, Podevin P, Clenci A and Guilain S 2017 Surge detection on an automotive turbocharger during transient phases *IOP Conf. Ser.: Mater. Sci. Eng.* **52** 012082
[17] Podevin P, Danlos A, Deligant M, Punov P, Clenci A and De La Bourdonnaye G 2018

Automotive compressor: effect of an electric throttle in the upstream circuit on the surge limit *MATEC Web Conf.* **234** 03006

A study on characteristics of vacuum solenoid valves

Stoyan Stoyanov[1] and Veselin Mihaylov[1,2]

[1] Technical University of Varna, Department "Transport Engineering and Technologies", 9010 Varna, Bulgaria

[2] E-mail: v_mihaylov@tu-varna.bg

Abstract. Vacuum solenoid valves have been widely used in internal combustion engine engines. They are used to regulate various processes in the car's systems by controlling actuators – for example, variable geometry of the turbocharger, exhaust gas recirculation (EGR) valves, various bypass or throttle valves. The vacuum solenoid valves are controlled by the electronic control unit (ECU) by means of a constant frequency and pulse width modulation of the controlling signal. In operation, these valves change their characteristics over time, which leads to a change in the control of the actuators, and hence to the overall operation of the engine and systems related to the environmental performance of the vehicle. Often there is no feedback on the actual value compared to the demand and their correct diagnosis is difficult. In this work an experimental study was performed, which shows dependence of pressure characteristics from possible deviations of some parameters (vacuum system, power supply) in operation of the vehicle.

1. Introduction

Internal combustion engines are one of the main air pollutants in big cities [1]. Today, strict standards have been introduced for harmful components in exhaust gases, and significant progress has been made in reducing them [2]. The emission control technologies developed in recent years are very effective in reducing the emissions of most cars, but if that all elements of the engine management system work according to the characteristics set by the manufacturer [3-8]. In modern internal combustion engines additional control systems are used, such as exhaust gas recirculation (EGR), variable geometry turbine control, wastegate valves, throttle valves, charge geometry control etc. In many cases, these systems are controlled by the electronic control unit (ECU) through PWM solenoid valves, connected to pneumatic actuators using vacuum from the vehicle's vacuum circuit, figure 1.

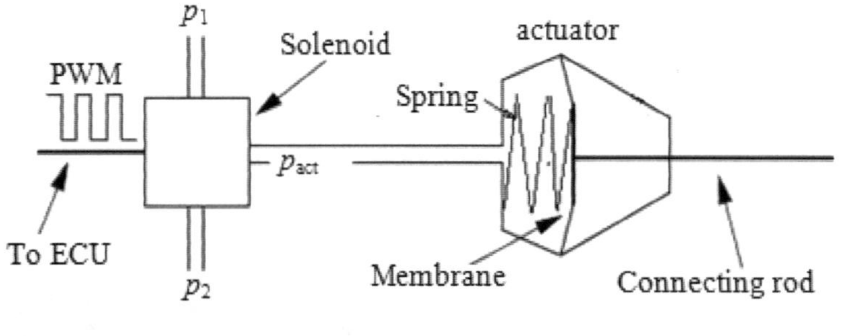

Figure 1. Schematics of pneumatic actuator system for control of various engine systems (adapted from source [10]).

The pneumatic actuator is connected to the controlled device by a mechanical connection and its chamber is divided into two chambers by an elastic membrane. The vacuum solenoid valve maintains the pressure in one of the actuator chambers, which generates a force on diaphragm due to pressure difference. A spring action is used to bring the diaphragm to its initial position. The pressure in the actuator results in a force on the membrane that transfers to the connecting rod that is attached to the controlled system. The pressure in the actuator chamber p_{act} is maintained by a vacuum solenoid valve between atmospheric pressure p_2 and the vacuum p_1 in the car installation and the operation of the controlled system depends on its characteristics [9].

Number of studies related to the pneumatic control of various devices in automobile engines can be found in scientific literature. In his study Mehmood [9] examined various sources that describe physical modelling of the solenoid valves. He found that in most cases its magnetic and pneumatic models are discussed separately. In his work, he presented both the magnetic and pneumatic model of the vacuum solenoid valve, defining detailed mathematical model for a variable geometry turbocharger pneumatic actuator controlled by a vacuum solenoid valve. In his work, he does not present the dependences between the fill factor of PWM and the pressure supplied by the valve, as well as the dependence of the pressure on the supply voltage. In [10] the author examined many works in which the modelling and control of pneumatic actuators are considered and found that very few authors have treated the modelling of vacuum solenoid valves. In his work, the solenoid valves were studied by conducting an experiment in which he used a frequency of 300 Hz for the control signal and the supply voltage from 10.3 to 11.9 volts.

The purpose of this study is to determine dependence of pressure characteristics from possible deviations of some parameters (vacuum system, power supply) during normal usage of the vehicle, which can be useful in diagnosis. For this purpose, a stand was made for testing the parameters of valves. To determine the operating conditions of the solenoid valves, measurements were made on some vehicles at typical driving conditions. It has been found that for this type of solenoid valves, made by Pierburg, control signal with constant frequency of 250 Hz and variable fill factor of 10 to 99% is used. The pressure maintained in the vehicle system provided by the vacuum pump of the brake system is around 50 mbar(a), and in the case of intensive frequent use of the brakes can reach up to 400 mbar(a) for a brief period of time.

2. Construction and principle of operation

The principle of operation is based on the balance of forces acting on the armature of the electromagnet – figure 2 and figure 3. The pressure supplied by the electro valve is included between the atmospheric pressure and the vacuum from the vehicle's vacuum circuit. The force which acts under the plunger is proportional to the atmospheric pressure. Two forces act in the opposite direction – the magnetic force created by the electromagnet and the force proportional to the pressure in the actuator chamber. If the coil is not energized (figure 3a), the pressure on both sides of the diaphragm is equal to atmospheric pressure and plunger is in equilibrium. Valves 1 and 2 are closed. If there is little leakage from valve 1 or if it is not completely closed, the pressure in the actuator chamber decreases slowly. This causes the piston to move in the direction of the actuator chamber and valve 1 closes. When the coil is supplied with PWM (Pulse Width Modulation), it causes a force to be applied to the armature, proportional to the fill factor of PWM. This force causes the plunger to move in the direction of the coil and the valve 1 to open (figure 3b). At this time, the valve 2 is still closed. The air from actuator chamber passes through the open valve 1 into vacuum chamber and the pressure decreases to a new equilibrium (green line). When the PWM fill factor decreases or completely shuts off, the electromagnetic force decreases or disappears causing the piston to move in the direction of the actuator chamber and the valve 2 to open (figure 3c). The pressure in the actuator chamber increases in proportion to the fill factor of the PWM [10].

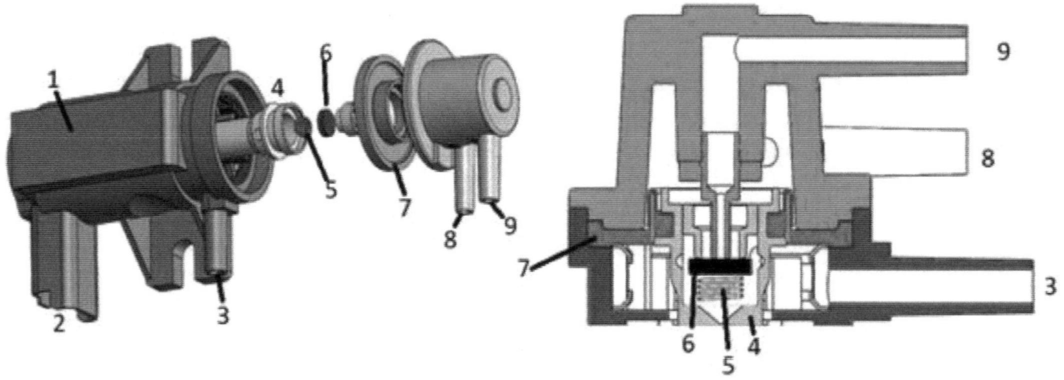

Figure 2. Construction of vacuum solenoid valve:
1 – body with electromagnet; 2 – electrical connector; 3 – connection to the atmosphere; 4 – armature; 5 – spring; 6 – valve plate; 7 – membrane; 8 – to actuator; 9 – to vacuum source.

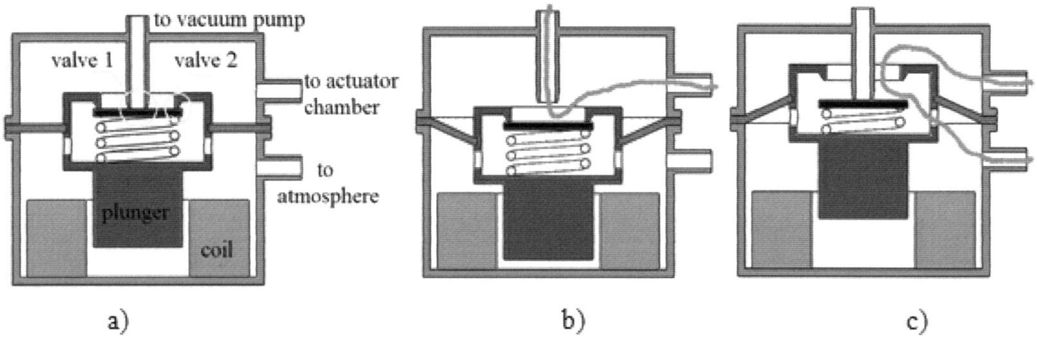

Figure 3. Principle of operation: a) idle; b) decrease of absolute pressure; c) increase of absolute pressure.

3. Experimental set-up

To determine the static and dynamic characteristics, the following test stand has been built and used. The stand enables to be explored the set pressure at different duty cycle of PWM signal, at different voltages and at various system pressures. The stand is shown in figure 4.

Constant adjustable pressure can be maintained at the inlet of the tested valve by means of the pressure regulator, the tank and the vacuum pump. If a mechanical pressure regulator is not available, another vacuum solenoid valve can be used for this purpose. The valve outlet is connected to an actuator chamber, in our case an EGR valve. For monitoring and recording of pressures, the inlet and outlet are connected to a 2-chanel absolute pressure sensor (MAP sensor). The rate at which the pressure in the actuator chamber changes depends also on its characteristics and is not investigated in this study. A stabilized power source with the possibility of changing the output voltage is used to supply the valve. The control signal with variable pulse width (PWM) can be supplied by a commercial PWM generator for manually controlling the duty cycle. Another option is with a DAC device which is capable of PWM generation. The second choice is more complicated in case of programming but provides more flexibility in case some automation or dynamic characteristics – time for reaching the set value, are needed. The control signal is fed to the tested valve through an amplifier designed to control electromagnetic devices with a built-in freewheeling diode. The values from the pressure sensors and the control signal are recorded by a digital oscilloscope.

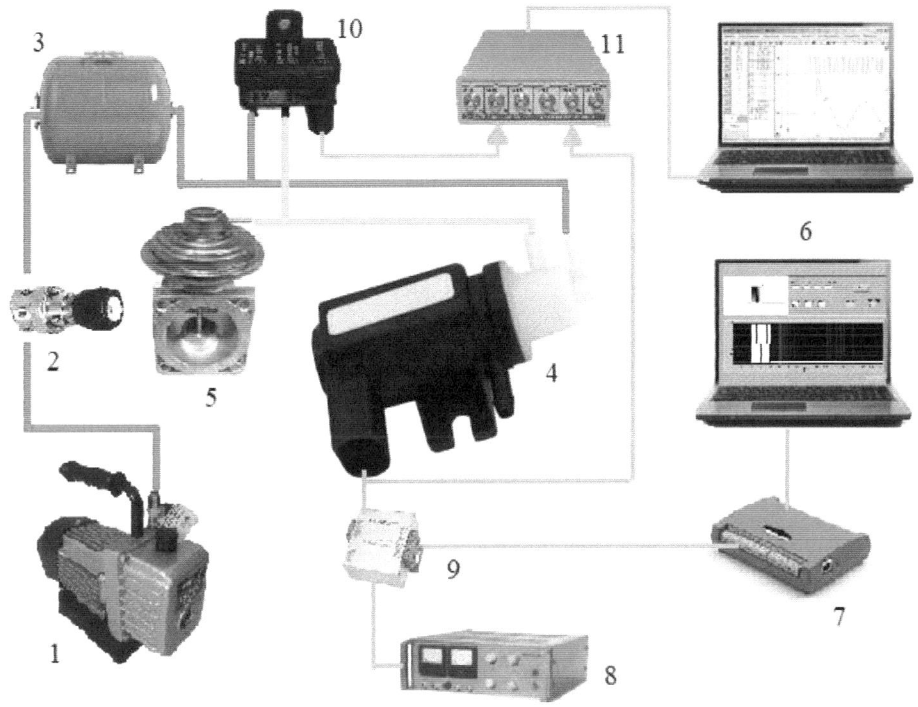

Figure 4. Experimental test stand:
1 – vacuum pump; 2 – pressure regulator; 3 – vacuum tank; 4 – vacuum solenoid valve; 5 – actuator; 6 – laptops; 7 – DAC for control signal generation; 8 – power supply; 9 – amplifier; 10 – 2-channel absolute pressure sensor; 11 – DAQ device.

4. Results and discussions

For current study were used electro pneumatic valves, made by Pierburg (figure 4, pos.4), which are commonly used by Europeans car manufacturers. The valve is controlled by signal with constant frequency of 250 Hz and varying duty cycle. The valve starts opening at 12÷13% duty. At idle (closed) state the system is feeding the valve with 10% duty. The usable range of duty cycle is 10 to 90%.

The characteristics pressure vs duty are determined by increasing the duty cycle to 100 % and back at steps of 10%. Figure 5 and figure 6 show the change of regulated pressure p_{act} at different values of the system pressure p_I, that's the pressure of the cars vacuum system (in the article "mbar(a)" stands for unit of absolute pressure). This change can happen in case of malfunction of the car's vacuum system – worn vacuum pump or leak in the system. For comparison between system pressure p_I=50 mbar(a) and p_I=250 mbar(a) there is very little change in the characteristics – less than 1.8%. In the case of p_I=450 mbar(a) there is the same small change in the characteristics until 70% duty (p_{act}=500 mbar(a)), that's 50 mbar above the system pressure. After that point the regulated pressure p_{act} reaches the system pressure, value beyond which it cannot fall.

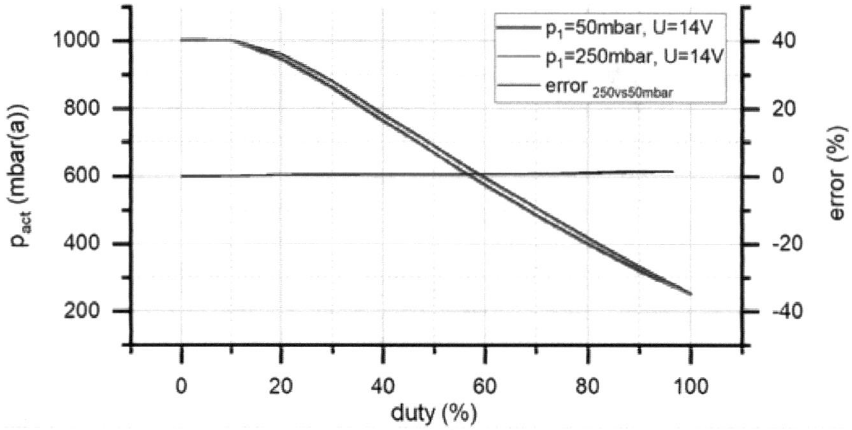

Figure 5. Regulated pressure vs duty cycle at two different system pressures: 50 mbar(a) and 250 mbar(a).

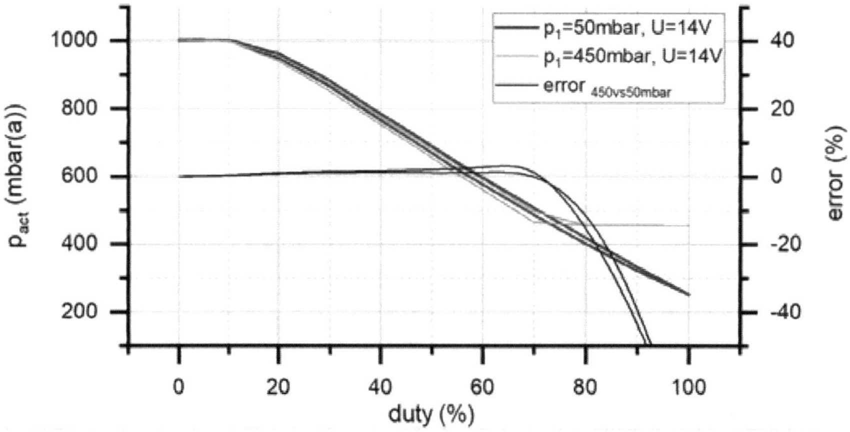

Figure 6. Regulated pressure vs duty cycle at two different system pressures: 50 mbar(a) and 450 mbar(a).

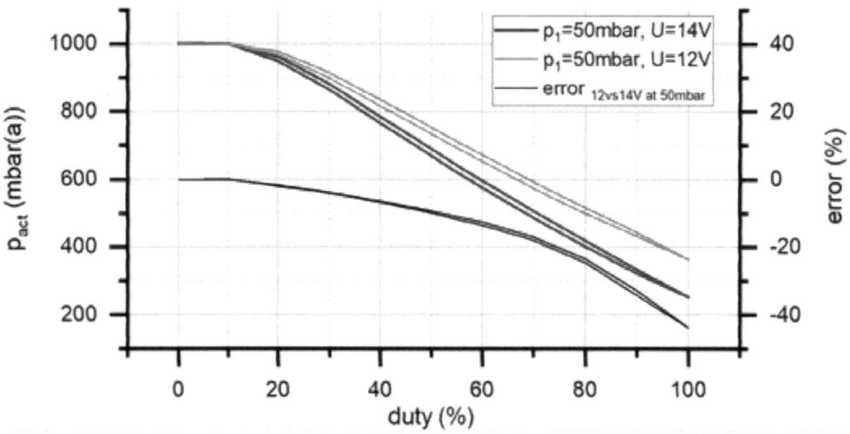

Figure 7. Regulated pressure vs duty cycle at two different supply voltages: 14 V and 12 V at system pressure 50 mbar(a).

The dependence of regulated pressure p_{act} from the supply voltage to the valve is shown on figure 7 and figure 8. It can be seen a gradual, but significant decrease of actual value of pressure against demand one, the error is around –33% at 90% duty for both cases.

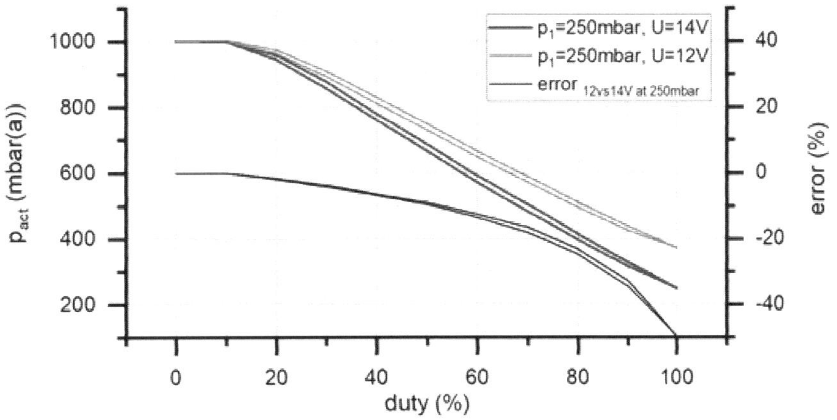

Figure 8. Regulated pressure vs duty cycle at two different supply voltages: 14 V and 12 V at system pressure 250 mbar(a).

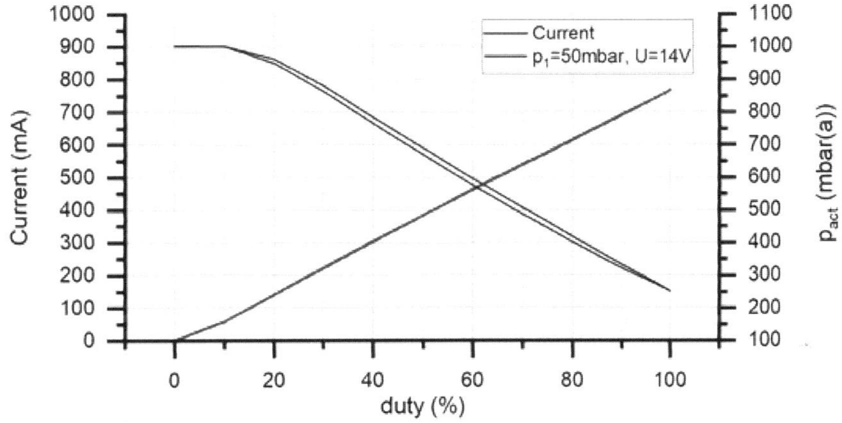

Figure 9. Current through the valve vs duty cycle.

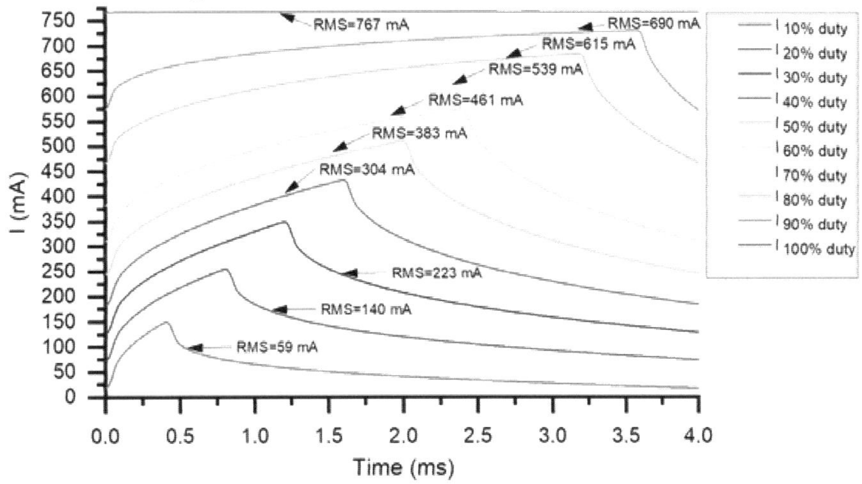

Figure 10. Current through the valve for one cycle of the control signal.

Difference of current through the coil at different system pressures is negligible. On figure 9 is shown current (rms value) for the case with system pressure p_1 = 50 mbar(a) and supply voltage U=14 V. There is linear trend of the current at the range 10÷100% duty cycle. More detailed picture of the current for one cycle of the control signal is shown on figure 10. The current in all cases is not falling to zero, instead is gradually decreasing.

5. Conclusion

Experimental research has been done and multi-parameter characteristics of a type of electromagnetic pneumatic valves are determined. These characteristics can help greatly in the diagnostics process of valves, due to the lack of feedback signal. It can be concluded that the system pressure doesn't pay an important role on the regulated pressure. In case the demand value is 50 mbar(a) or more above the vacuum pump pressure, the valve will be capable to maintain the regulated pressure with error less than 2 percent. The valve is much more sensitive to the supply voltage. A difference of 2V at the valve will lead to significant error in maintaining the set value, which can reach up to –32% at 90% duty cycle of the control signal.

Acknowledgment

This research was supported by the Bulgarian Ministry of Education and Science through National Programme "Young scientists and postdoctoral students".

References

[1] Weiss M et al. 2011 On-road emissions of light-duty vehicles in Europe *Environ. Sci. Technol* **45**
[2] Commission Regulation UE 2017/1151 of 1 June 2017 r. supplementing Regulation (EC) No 715/2007 of the European Parliament and of the Council on type-approval of motor vehicles with respect to emissions from light passenger and commercial vehicles (Euro 5 and Euro 6)
[3] Ivanov Z, Mihaylov V and Chekelov G 2017 Exhaust gas recirculation parameters in a diesel engine *Proc. of University of Ruse* **56** 52-57 (In Bulgarian)
[4] Kazmierczak A, Miksiewicz K and Wrobel R 2010 The diagnosis of EGR system's failuers depended on detection and measure of vibroacustic vibrations *Diagnostyka* **4** pp. 27-35
[5] Cui H, Zhao Z, Geng Z and Liu Y 2019 Study on Diesel Engine Characteristics under Large Transient EGR *IOP Conf. Ser.: Earth Environ. Sci.* **300** 032042
[6] De Serio D, De Oliveira A and Sodré J 2016 Application of an EGR system in a direct injection diesel engine to reduce NOx emissions *J. Phys.: Conf. Ser.* **745** 032001
[7] Krakowian K 2018 Influence of the single EGR valve on the concentration level of selected toxic compounds in exhausts of a multi-cylinder compression ignition engine *IOP Conf. Ser.: Mater. Sci. Eng.* **421** 042039
[8] Jaworski A, Lejda K, Madziel M, and Ustrzycki A 2018 Assessment of the emission of harmful car exhaust components in real traffic conditions *IOP Conf. Ser.: Mater. Sci. Eng.* **421** 042031
[9] Mehmood A 2012 Modeling, simulation and robust control of an electro-pneumatic actuator for a variable geometry turbocharger *Computers and Society [cs.CY] Universite de Technologie de Belfort-Montbeliard NNT BELF 0188*
[10] Thomasson A 2014 Modeling and control of actuators and co-surge in turbocharged engines *Linköping studies in science and technology* Dissertations No 1590

Comparison between Model Predictive (MPC) and Model Reference Adaptive Controllers (MRAC) for Electrohydraulic Steering System Implemented as Real-Time Simulink® Program

Alexander Mitov[1,3], Jordan Kralev[2], Tsonyo Slavov[2] and Ilcho Angelov[1]

[1] Dept. of Hydroaerodynamics and Hydraulic Machines, Technical University of Sofia, Bulgaria
[2] Dept. of Systems and Control, Technical University of Sofia, Bulgaria

[3] E-mail: a_mitov@tu-sofia.bg

Abstract. The purpose of the article is to investigate tracking performance of steering cylinder when controlled with two nonlinear controllers. The system under study is a laboratory electrohydraulic steering test rig. Two controls strategies are examined - model reference adaptive control (MRAC) and model predictive control (MPC). The MPC controller design is based on an identified model from experimental data in open-loop. The model is represented in state-space form and the future values of the state trajectory and output variables are used as optimization objective with respect to the future values of the control signal. Since this optimization procedure have to be performed every sampling interval and involve iterative calculation the implementation of MPC requires considerable computational resources. Hence, we have decided to develop a distributed computational platform, where the controller is executed in real-time as a Simulink model. The control and measurement signals are processed by the PLC and are synchronized with the real-time model over high-speed CAN channel. The process model for the MRAC is assume to be a simple integrator with unknown gain, possibly slowly varying with time due to physical parameter drift. The MRAC is not so computationally intensive as MPC, however for the sake of comparison it is implemented in the same architecture. The structure of the MRAC is selected with respect of the *Lyapunov* stability criteria to guarantee asymptotic convergence of the cylinder position error.

1. Introduction

The customization of the modern mobile machinery steering ratio provides additional possibilities to achieve better system performance. This functionality is guaranteed by digital electro-hydraulic actuator and steering wheel rotational sensor which allow to precise tune steering performance. The digital actuator has an embedded steering software that has been advanced for machines with various application, till steering sensor is added to measure the real position and rotational speed of the steering wheel. With help of the embedded control system operator can set the number of times which the steering wheel is turned to adapt steering performance and to make it appropriate to specific working prerequisite or driving environment [1]. All of this is a motivation for research and development of embedded control software to implement advanced control laws [2] that meet these requirements and safety function [3].

The purpose of the article is to investigate tracking performance of steering servo cylinder when controlled with two nonlinear embedded controllers in order to evaluate their applicability to power steering of low speed mobile machines [4]. The system under study is a laboratory electrohydraulic steering test bench [5] composed of hydraulic steering cylinder, proportional spool valve, digital switching valve actuator for pilot control of the spool valve and programmable logic controller (PLC) for vehicle applications. Two controls strategies are examined - model reference adaptive control (MRAC) and model predictive control (MPC). MPC controller design is based on an identified model from experimental data in open-loop. The model is represented in state-space form and the future values of the state trajectory and output variables are used as optimization objective with respect to the future values of the control signal. Since this optimization procedure have to be performed every sampling interval and involve iterative calculation the implementation of MPC requires considerable computational resources. Hence, we have decided to develop a distributed computational platform, where the controller is executed in real-time as a Simulink® model running on a host workstation. The control and measurement signals are processed by the PLC and are synchronized with the real-time Simulink® model over high-speed CAN channel. The advantage of such hardware-in-the-loop (HIL) architecture is that we are able to employ a large variety of functional blocks available in Simulink® libraries in the real-time model. Also the Simulink® model of the controller can be used for direct code generation and targeting to a particular computation platform. However, the MPC is not suitable for implementation of conventional PLC for vehicle applications and requires more powerful platforms.
The process model for the MRAC is assume to be a simple integrator with unknown gain, possibly slowly varying with time due to physical parameter drift. The MRAC is not so computationally intensive as MPC, however for the sake of comparison it is implemented in the same architecture. The structure of the MRAC is selected with respect of the *Lyapunov* stability criteria to guarantee asymptotic convergence of the steering cylinder position error. The MPC objective is finite time quadratic cost function accounting for output error and control amplitude. The achieved closed-loop performance with both controllers is similar as steady state errors and transients.

2. Multivariable system identified model

The main purpose is to find a parametric stochastic *"black box"* model [6, 7], which describes in detail the plant and noises dynamics for whole working frequencies. To measure input-output data an experiment in accordance to structure presented in figure 1 is performed. The sample rate is 40 *Hz*.

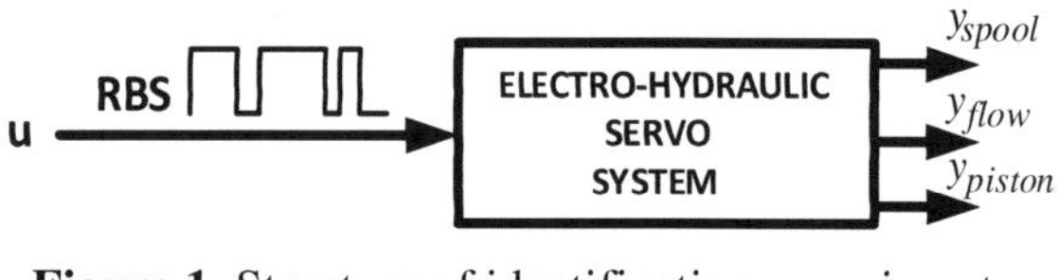

Figure 1. Structure of identification experiment.

The absolute value of consumed flow rate $|y_{flow}|$ is measured by gear flow meter, which is connected before EHSU. The movement direction of steering cylinder piston depends on supplying flow rate in accordance of chambers. To use this signal in model estimation and in feedback its sign should be known. The information of the flow rate sign is obtained from opening of proportional spool throttling area. The flow rate is equal to zero caused by overlapping in positive direction of the proportional valve construction. Such dead band is imposed to comply with the safety requirements of the EHSU devices by preventing unacceptable random displacements in the steering axle. Hence in this study we assume that

$$sign(y_{flow}) = sign(y_{spool}). \tag{1}$$

Thus, the actual value of y_{spool} is obtained as $sign(y_{flow})|y_{spool}|$.

The system used in identification procedure is composed of steering servo cylinder, hydraulic steering valve, piston position transducer, and microprocessor. The obtained input output data are centered. The centered measurements is divided into several data arrays, which are used to estimate the parameters of the model or to check model validity. Estimation data set is presented in figure 2 where the manipulated variable is depicted in mV for increased quantization. First, we estimate an autoregressive multivariable model with exogenous excitation (ARX) [7] with the following formal description

$$\begin{pmatrix} A_{11}(q) & A_{12}(q) & A_{13}(q) \\ A_{21}(q) & A_{22}(q) & A_{23}(q) \\ A_{31}(q) & A_{32}(q) & A_{33}(q) \end{pmatrix} \begin{pmatrix} y_1(q) \\ y_2(q) \\ y_3(q) \end{pmatrix} = \begin{pmatrix} B_1(q) \\ B_2(q) \\ B_3(q) \end{pmatrix} u(q) + \begin{pmatrix} e_1(q) \\ e_2(q) \\ e_3(q) \end{pmatrix}, \tag{2}$$

where $y_1(q)$ is a spool position, $y_2(q)$ is a flow rate , $y_3(q)$ is a cylinder piston position and q is the shift operator. The polynomials $A_{ii}(q) = 1 + \sum_l a_{ii,l} q^{-l}$ are chosen from 4th order The polynomials $A_{ij}(q) = \sum_l a_{ij,l} q^{-l}, i \neq j$ are from 2nd order and the polynomials $B_i(q) = \sum_l b_{i,l} q^{-l-\tau}$ are from 8th order with single input delay element. The residual vector $e(q) = \left[e_1(q) \quad e_2(q) \quad e_3(q)\right]^T$ represents the difference between measured output and model output.

The obtained multivariable ARX structure has three 4th order polynomials A_{ii}, six 2nd order polynomials A_{ij} and three 8th order polynomials B_i. Thus, the minimal state space realization of estimated ARX structure is 24. In the control applications, such model order is not quite acceptable. In figure 3, the *Hankel singular values* of estimated model are shown. It is seen that a lower order model as high as 3rd can be selected to describe steering system.

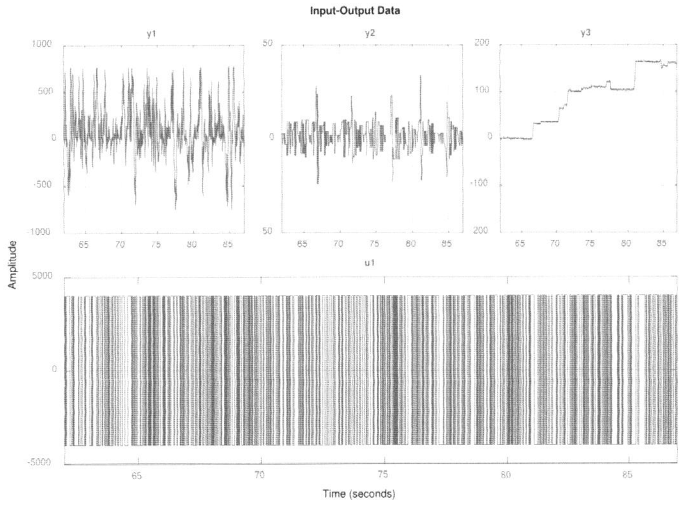

Figure 2. Experimental data used for parameter estimation. Below – input voltage. Above – spool position, flow rate, piston position.

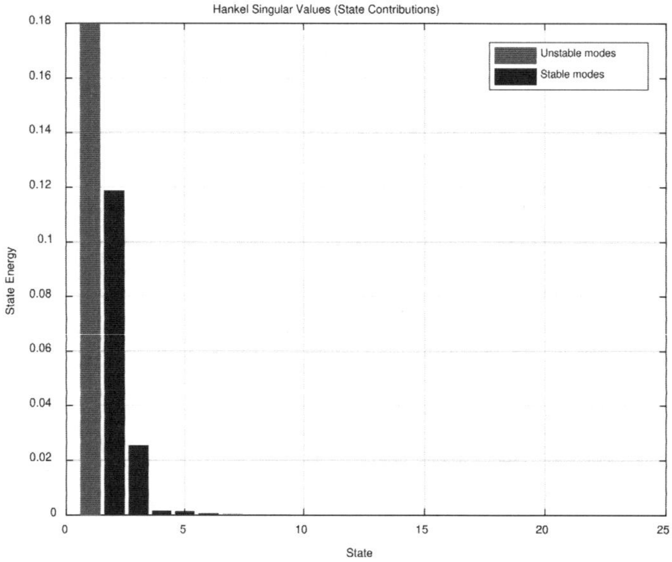

Figure 3. Hankel singular values for model reduction of the ARX structure.

The instantons values of the parameters are used to make a prediction of the output data and prediction error is calculated [8]. In case of state-space structure estimation the predictor has a representation of *Kalman* filter. The model parameters are adjusted by iterative algorithm, which minimizes a cost function based on difference between actual and predicted outputs [9, 10]. The matrices of the identified state-space model are

$$A=\begin{pmatrix} -1.05 & 0.19 & 6.27 \\ -0.04 & 1 & 0.198 \\ -0.371 & 0.033 & 2.044 \end{pmatrix}, B=10^{-3}\begin{pmatrix} -0.37 \\ -0.034 \\ -0.23 \end{pmatrix},$$
$$C=\begin{pmatrix} -31 & 0.67 & -6.58 \\ -0.584 & 0.007 & -0.25 \\ 0.059 & -2.019 & 0.125 \end{pmatrix}, D=10^{-3}\begin{pmatrix} 0.117 \\ 0.014 \\ 0.0021 \end{pmatrix}, \quad (3)$$
$$K=\begin{pmatrix} -0.025 & -0.023 & -0.022 \\ -0.002 & 0.0066 & -0.429 \\ -0.007 & -0.01 & 0.0033 \end{pmatrix}$$

and covariance matrix

$$M(ee^T)=10^3\begin{pmatrix} 7.92 & 0.034 & 0.011 \\ 0.034 & 0.03 & 0.001 \\ 0.011 & 0.001 & 0.003 \end{pmatrix}. \quad (4)$$

3. MPC and MRAC Controllers Design

3.1. MPC Design

Model predictive control problem [5] is defined by the objective function

$$\vec{u}_t = Arg\min_{\vec{u}_t} \sum_{k=0}^{N-1} q(x_{t+k}, u_{t+k}), \tag{5}$$

which have to be minimized with respect to the control signal for the prediction horizon of N sampling intervals from the system future subject to initial state $x_t = x(t)$, system state equation (3), fixed control horizon $M < N$, state and control amplitude constraints and rate constraints. The idea is to use internal system model to predict system behaviour [11]. The main benefit is explicit account for constraints in the controller design.

$$\vec{u}_t = (u_t, u_{t+1}, \ldots, u_{t+N-1}) \tag{6}$$

with $u(t+i-1) = u(t+m-1)$ for $i > m$. When the optimal sequence if determined only the first control action u_t is implemented. In the next sample interval the optimization procedure is repeated again and new input sequence is generated. In most applications, the controller must keep selected plant outputs at or near specified reference values so the sample cost becomes

$$q(x_i, u_i) = w_1 y_{1,i}^2 + w_2 y_{2,i}^2 + w_3 (r - y_{3,i})^2, \tag{7}$$

where the LVDT, flow rate and cylinder piston position error are taken into account. The MPC solves a quadratic program at each sampling interval and the solution determines the value of manipulated variable to be held until the next sampling event.

3.2. MRAC design

The design of a MRAC controller is done based on development closed-loop system scheme presented in figure 4.

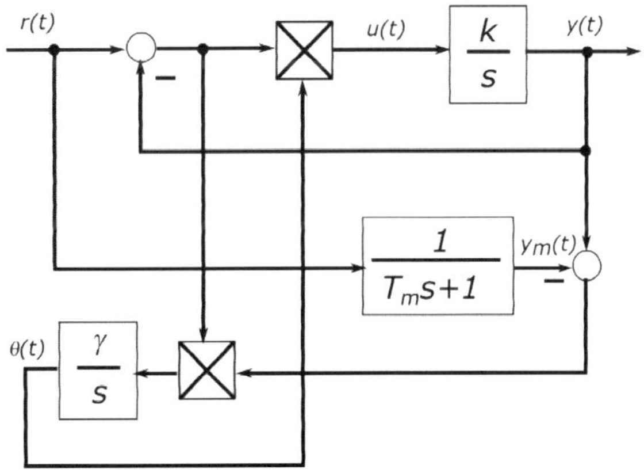

Figure 4. Closed-loop system structure for MRAC controller design.

Let assume that the steering cylinder and the hydraulic systems are modeled by a simple integrator equation as

$$\dot{y}(t) = ku(t), \tag{8}$$

where $y(t)$ is the position of the steering cylinder, $u(t)$ is the voltage PWM signal applied to the PVE transducer and k is an unknown system gain, possibly slowly varying with time due to physical parameter drift. The proposed controller is selected as an adaptive P-based output feedback as

$$u(t) = \theta(t)\left(r(t) - y(t)\right), \tag{9}$$

where $r(t)$ is the desired steering angle and $\theta(t)$ is the adaptive gain. The closed-loop system then is described by the following expression

$$\dot{y}(t)=k\theta(t)(r(t)-y(t)). \tag{10}$$

The adaptation of the parameter $\theta(t)$ will be according to the output reference model

$$\dot{y}_m(t)=-\frac{1}{T_m}y_m(t)+\frac{1}{T_m}r(t), \tag{11}$$

This represents an aperiodic model with a time-constant T_m and unit gain. So after as much as $5T_m$ the output $y_m(t)\to r(t)$. The goal is to select the parameter $\theta(t)$ such that reference model error $e_m(t)=y(t)-y_m(t)$ to be minimized. In order to investigate the convergence of the $e_m(t)$ we find its first derivative

$$\dot{e}_m(t)=\dot{y}(t)-\dot{y}_m(t). \tag{12}$$

After substituting the corresponding terms with the expressions define above we get

$$\dot{e}_m(t)=k\theta(t)r(t)-k\theta(t)y(t)+\frac{1}{T_m}y_m(t)-\frac{1}{T_m}r(t). \tag{13}$$

After grouping the term according to their places in the control loop the derivative of the model reference error becomes

$$\dot{e}_m(t)=-\frac{1}{T_m}e_m(t)+\left(\frac{1}{T_m}-k\theta(t)\right)(r(t)-y(t)). \tag{14}$$

From this form it is evident that we can select the following *Lyapunov* function

$$V(e_m(t),\theta(t))=e_m^2(t)+\left(\frac{1}{T_m}-k\theta(t)\right)^2. \tag{15}$$

To achieve stability of the closed-loop according to the second *Lyapunov* method [12] the total derivative of the *Lyapunov* function with respect to time must be non-increasing

$$\frac{dV(e_m(t),\theta(t))}{dt}<0. \tag{16}$$

Hence by differentiating the expression for V we get

$$\dot{V}=2e_m\dot{e}_m-2k\left(\frac{1}{T_m}-k\theta(t)\right)\dot{\theta}(t). \tag{17}$$

Now substituting the expression for the derivative of the model error

$$\dot{V}=2e_m\left(-\frac{1}{T_m}e_m(t)+\left(\frac{1}{T_m}-k\theta(t)\right)(r(t)-y(t))\right)-2k\left(\frac{1}{T_m}-k\theta(t)\right)\dot{\theta}(t). \tag{18}$$

By grouping the terms the resultant expression becomes

$$\dot{V} = -\frac{2}{T_m} e_m^2(t) + \left(\frac{1}{T_m} - k\theta(t) \right) \left(2e_m e(t) - 2k\dot{\theta}(t) \right), \tag{19}$$

where $e(t) = r(t) - y(t)$ is the feedback error. Now the sufficient condition for the stability condition $\dot{V}(t) < 0$ to be satisfied is

$$2e_m e(t) - 2k\dot{\theta}(t) = 0 \tag{20}$$

or

$$\dot{\theta}(t) = \gamma e_m(t) \left(r(t) - y(t) \right), \tag{21}$$

where $\gamma = \frac{1}{k}$ is a tunable parameter.

3.3. Frequency response comparison

In order to compare the performance of the closed-loop system with both regulators we decided to investigate the tracking performance in simulation and in frequency domain. The problem here is that both controllers – the MPC and the MRAC are nonlinear so even though the system model obtained from identification is linear, the formed closed-loop system is again nonlinear.

The MPC nonlinearity comes from selecting the current value of control signal as a minimizer of quadratic cost function of predicted model output considering signal amplitude and rate constraints. The optimal solution for quadratic cost and infinite time horizon corresponds to linear-quadratic *Gaussian* (LQG) control so we may assume that MPC output is

$$u_{MPC}(t) = -Kx(t) + \eta(t), \tag{22}$$

where $x(t)$ is current state of the model and $\eta(t)$ is nonlinear distortion due to finite time horizon, signal noise and iterative procedure of calculation. So here we can see that nonlinearity introduced by MPC is not considerable in terms of equivalent signal-to-noise ratio. And from here we can represent the closed-loop system with linear approximation which will help us to investigate frequency domain properties of the model.

The MRAC nonlinearity can be thought as linear-parameter varying representation of the control algorithm where the controller gain $\theta(t)$ is calculated in real-time to minimize the difference between reference model and system output. However from experimental observation of the MRAC performance we know that the controller gain though variable quickly enough settles around its steady state value with some variation around it, or

$$u_{MRAC}(t) = \left(\theta_0 + d(t) \right) e(t), \tag{23}$$

where θ_0 is the constant or slowly varying steady state of the controller gain and $d(t)$ is its variation due to adaptation process. However if the amplitude of the variation is small enough compared to the amplitude of the steady state gain we may assume that controller behavior will be equivalent to linear controller response with some random distortions. Hence again we can represent the closed-loop system with the MRAC with its linearized approximation.

Keeping this analysis in mind now we need to decide how to obtain the linear approximation of both closed-loop systems. Due to statistical properties of the commented nonlinear distortions the best approach for linearization is through system identification. For this purpose we use a frequency sweep signal with amplitude ± 10 mm, additively introduced into generated reference signal so we get

$$r(t) = r_{prog}(t) + s(t, \omega_0, \omega_1, T), \tag{24}$$

where $r_{prog}(t)$ is the programmed reference trajectory [13] (steps between different cylinder piston positions in our case) and $s(t, \omega_0, \omega_1, T)$ is a frequency range between $\omega_0 = 0$ rad/s and $\omega_1 = 3.14$ rad/s with a period of $T = 1000$ s. The response of the system is presented in figure 5.

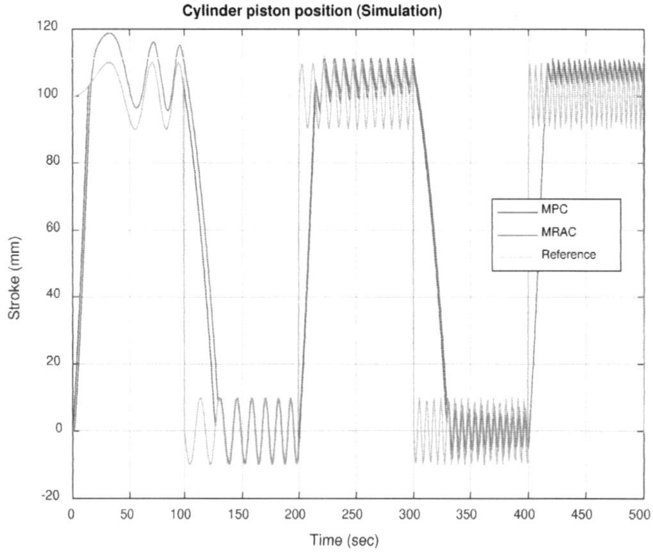

Figure 5. Identification experiment for simulated closed-loop steering system with nonlinear controllers.

Based on the collected data two regression *Box-Jenkins* models with 42 parameters of the closed-loop system are identified – one for each controller MPC or MRAC.

$$y_{cyl}(z^{-1}) = \frac{B(z^{-1})}{F(z^{-1})} r(z^{-1}) + \frac{C(z^{-1})}{D(z^{-1})} \epsilon(z^{-1}), \tag{25}$$

where B, C, D, F are polynomials and ϵ is model latent variable. The model input is the reference signal and the model output is the cylinder position. Both models are with structural parameters $nb = 6$, $nc = 12$, $nd = 12$, $nf = 12$ and $nk = 0$. The increased order of the noise model is required to account for nonlinear distortions introduced by the nonlinear controllers. For the closed-loop system with the MPC controller the achieved fit to training data is 94.8%, the total final predicted error is 70.58 mm and the total mean square error is 64.9 mm. For the closed-loop system with the MRAC the achieved fit to training data is 95.08%, total final predicted error is 65.5 mm and total mean square error is 60.23 mm.

On the next figure 6 and figure 7 are compared both closed-loop systems with MPC and MRAC with respect to their bandwidth and output sensitivity function. As can be seen the frequency response of the both controllers is very similar. They share common bandwidth of 0.1 rad/s and common damping frequency of 0.6 rad/s. This can be explained partly due to the fact that a same model structure is used for both dataset even though the parameter space dimensionality of 42 is large enough. The more important contribution here is that both controllers share common adaptive behavior where their steady state performance is characterized as stable static gain and small enough nonlinear distortions around it. Of course such observations are representative only for the particular system and cannot be generalized for all cases.

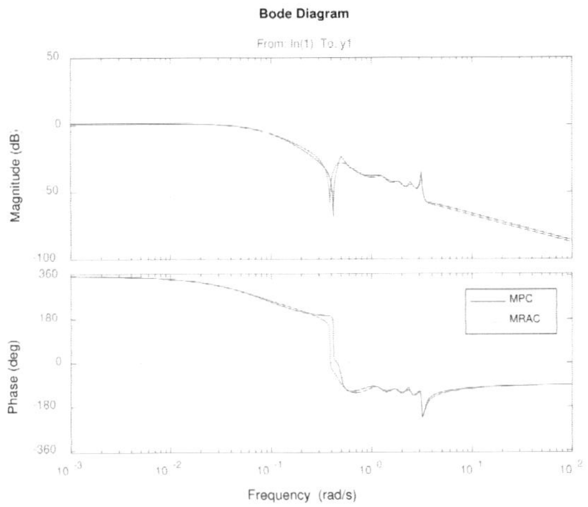

Figure 6. Complementary sensitivity of the closed- loop system with MPC and MRAC.

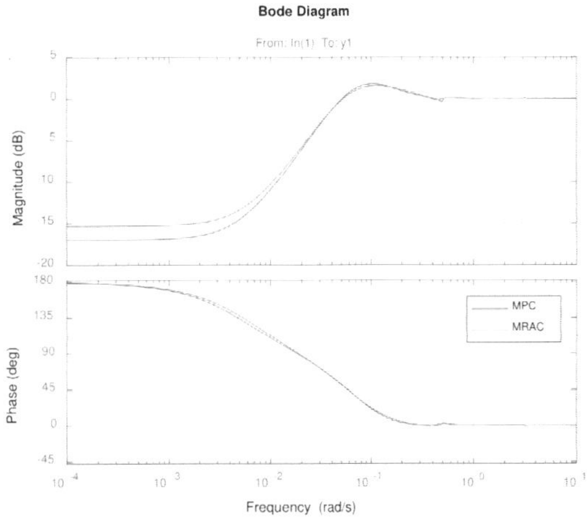

Figure 7. Output sensitivity function of closed- loop system with MPC and MRAC.

Another interesting observation is that both adaptive controllers reach a comparable levels of closed-loop performance which is indicative for their optimality. Moreover the MPC is based on a high order regression model and the MRAC is based on a simple low-order order model. This elevated order for the MPC comes from introduction of additional output signals (flow rate and spool position) which carry information about internal system processes. So generally we could expect a better performance of the MPC with respect to those additional output channels in comparison with MRAC where the information from these variables is not fed into controller.

The output sensitivity function of the closed-loop system with both controllers reveal a common disadvantage of the adaptive systems, i.e. elevated sensitivity to low frequency load disturbances in comparison to other systems like LQG. However we still have around 10 times attenuation of output disturbance in the closed-loop bandwidth which is enough for some applications. The sensitivity is a bit lower for the MPC than MRAC. The sensitivity can be further improved by taking additional measures – like explicit introduction of disturbance into cost function, adding feedforward controller term for disturbance compensation, and more.

4. Experimental results

4.1. Implementation of designed controllers

The Simulink® diagram used for execution of the experiments with MPC and MRAC is presented on the figure 8. The central block is MATLAB® function block which supports the communication on the CAN channel between MC012-022 and the workstation. Also this block guarantees that the model will run in a blocking mode. The designed controller is implemented with the *Controller* block. The MPC is realized via *MPC Controller* block. In figure 9 the implementation of MRAC in Simulink® is shown on the base of the structural scheme for synthesis presented in figure 4.

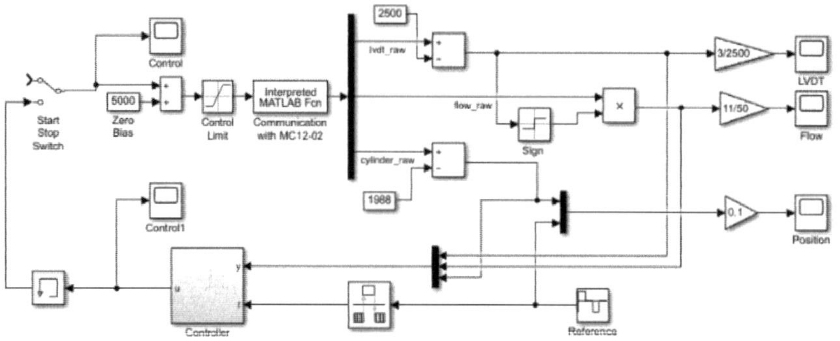

Figure 8. Simulink® diagram for implementation of designed controller.

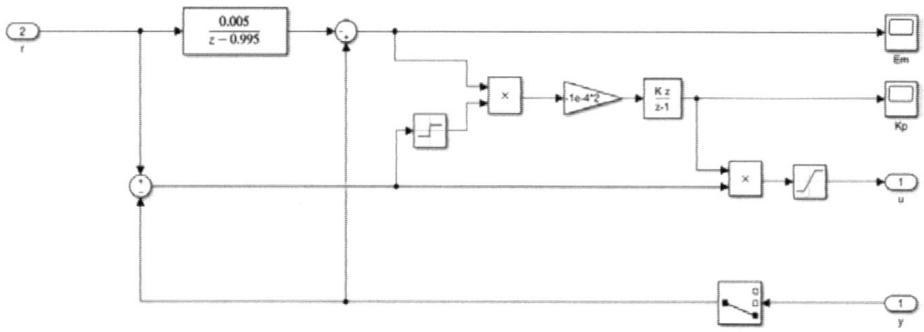

Figure 9. Simulink® diagram for implementation of MRAC.

4.2. Comparison of measured data

Figure 10 compares the measured cylinder position for the two tunings of the MPC and the MRAC. The output weights of MPC cost function are set for $w_1 = 0$, $w_2 = 0.8$ and $w_3 = 1$ for first variant of controller (MPC1) and $w_1 = 0$, $w_2 = 0.9$ and $w_3 = 1$ for the second variant of the controller (MPC2). These experiments confirm the observation from simulation that both controllers share similar performance in terms of settling time and steady state accuracy. It can be seen that the MRAC need some periods to settle its gain in the beginning.

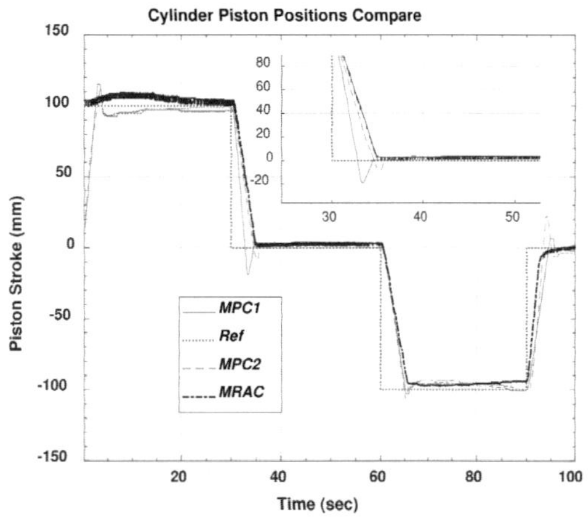

Figure 10. Measured step response of the closed-loop system.

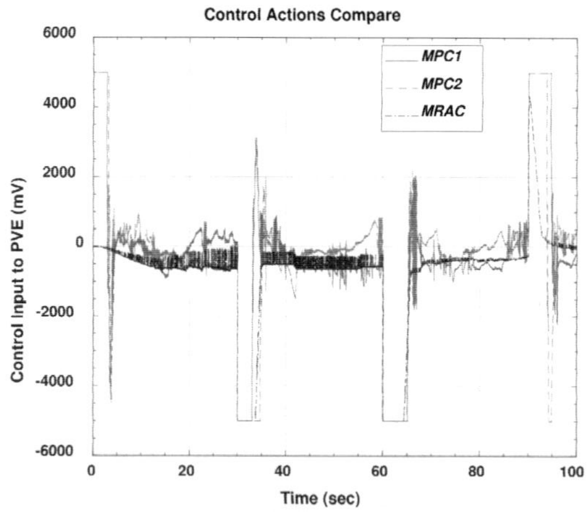

Figure 11. Measured control action of the closed-loop system.

Figure 11 compares the respective control signals of MPC and MRAC systems. Their amplitudes are matching during the transient between two cylinder positions. In steady state the MRAC is less oscillatory than MPC which can be explained with its lower sensitivity to output disturbances.
Figure 12 and figure 13 compares the internal system signals – spool position and flow rate. Only MPC system use these signals in control action calculation the MRAC does not. So we can observe that these signals are with less variation for the case of MRAC and more affected by MPC.

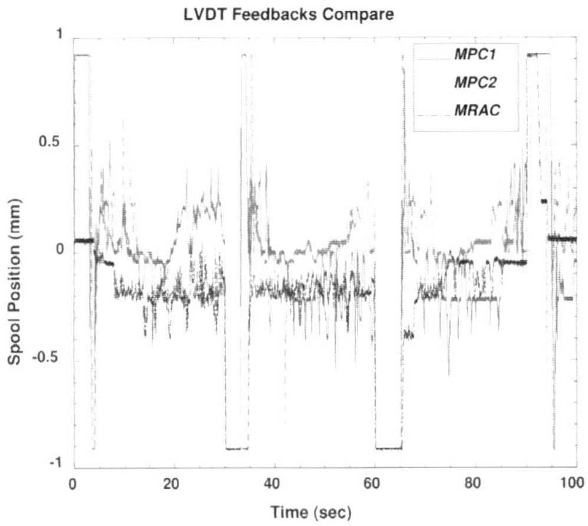

Figure 12. Measured proportional spool valve position.

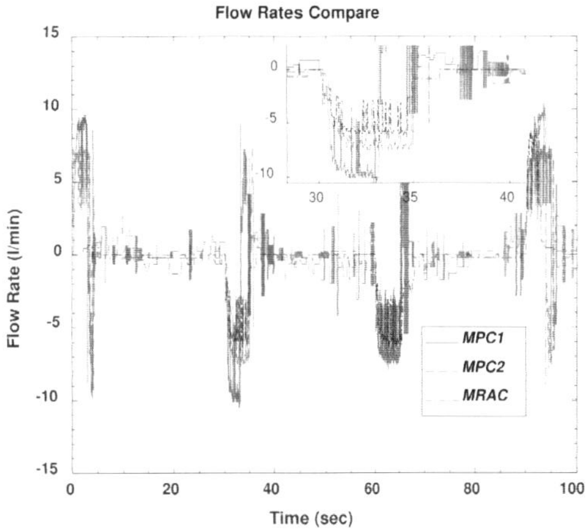

Figure 13. Measured flow rate.

5. Conclusion

The article presents design and implementation of embedded system for control of electrohydraulic power steering base on two different advanced control strategies – model predictive control (MPC) and model reference adaptive control (MRAC). The MPC controller is synthesis via multivariable model obtained by identification approach. The MRAC controller is designed on the basis of simple integrator with time varying gain. The designed controllers are implemented by the PLC and are synchronized with the real-time Simulink® model over high-speed CAN channel. The Simulink® model of the controller is used for direct code generation and targeting to a particular computation platform of the electrohydraulic steering test rig. The comparative analysis in time and frequency domain of designed nonlinear MPC and MRAC controllers in simulation conditions are performed. The obtained results present the similar performance of the both closed-loop system. A number of experiments have been executed, which confirm the results from simulation analysis of the electrohydraulic power steering control system with the MPC and MRAC strategies.

Acknowledgments

The authors would like to thank the Research and Development Sector at the Technical University of Sofia for the financial support.

References

[1] Danfoss 2018 *Steering. EHi Steering Valve* (Technical Information BC00000379en-US0202)
[2] Goodwin G C, Graebe S F and Salgado M E 2001 *Control System Design* (Prentice-Hall Inc.)
[3] Terziev A 2019 Study of the fire dynamics in a burning car and analysis of the possibilities for transfer of fire to a nearby vehicle *8th Int. Conf. on Thermal Equipment (Romania)* 01015
[4] Más F, Zhang Q and Hansen A 2010 *Mechatronics and Intelligent Systems for Off-road Vehicles* (Springer-Verlag London)
[5] Angelov I and Mitov A 2016 Test bench for experimental research and identification of electrohydraulic steering units *10th Int. Fluid Power Conf. (Dresden)* **3** pp 225–36

[6] Mitov A, Kralev J, Slavov Ts and Angelov I 2020 Model predictive control design for electro-hydraulic power steering application *9th Mediter. Conf. on Embedded Computing (Budva)* pp 658–1
[7] Ljung L 1999 *System Identification: Theory for the User* (Prentice-Hall Inc.)
[8] Zhao P, Wang S, Li X and Zhang B 2011 A novel method of parameter identification of nonlinear electrohydraulic servo systems *Proc. 2011 Int. Conf. Fluid Power and Mechatronics* pp 547–51
[9] Proca A and Keyhani A 1998 Identification of power steering system dynamic models *Mechatronics* **8** pp 255–70
[10] Kemmetmuller W, Muller S and Kugi A 2007 Mathematical modeling and nonlinear controller design for a novel electrohydraulic power-steering system *IEEE/ASME Transactions on mechatronics* **12**(1) pp 85–97
[11] Kayacan E, Kayacan E, Ramon H and Saeys W 2014 Distributed nonlinear model predictive control of an autonomous tractor–trailer system *Mechatronics* **24**(8) pp 926–33
[12] Yang Y, Zhou C and Ren J 2003 Model reference adaptive robust fuzzy control for ship steering autopilot with uncertain nonlinear systems *Applied Soft Computing* **3**(4) pp 305–16
[13] Raffo G, Gomes G, Normey-Rico J, Kelber C and Becker L 2009 A Predictive Controller for Autonomous Vehicle Path Tracking *IEEE Transactions on Intelligent Transportation Systems* **10**(1) pp 92–102

AUTHOR INDEX

Angelov, I A ... 63
Angelov, Ilcho ... 295
Barta, Dalibor ... 255
Baszczyńska, Anna ... 17
Blatnický, Miroslav ... 255
Clenci, Adrian ... 276
De La Bourdonnaye, Guillaume ... 276
Dekarchuk, O ... 132
Deligant, Michael ... 276
Dessante, Ph ... 247
Dimitrov, Evgeni ... 266
Dimitrov, E ... 181
Dimitrov, R ... 193
Dimitrova, Zlatina ... 207
Dimitrova, Z ... 247
Dižo, Ján ... 255
Dobrev, I ... 1
Fliszewska, Angela ... 17
Fomin, O ... 96, 104
Gechev, Tsvetomir ... 155
Genov, J A ... 63
Georgiev, Dimitar ... 143
Georgiev, Ivan ... 143
Gerlici, Juraj ... 85
Gieva, Elitsa ... 112
Gigov, Boyko ... 266
Gotseva, D ... 175, 200
Grozdanov, D N ... 166
Grozev, Dimitar ... 143
Gurin, D ... 132
Hassler, J ... 247
Hinov, N L ... 166
Hladchenko, Volodymyr S ... 238
Hristov, R ... 193
Ivanov, R ... 125
Ivanov, Zdravko D ... 216
Ivanov, Zdravko ... 187
J Sroka, Zbigniew ... 216
Kalinin, OA ... 73
Kamberov, Konstantin ... 227
Kovalov, M O ... 48
Kralev, Jordan ... 295
Kralov, Ivan ... 112
Kravchenko, Kateryna ... 85
Lobiak, O V ... 73
Lovska, A ... 96, 104
Madzharov, N D ... 166
Marcheva, Yordanka ... 57
Massouh, F ... 1
Matsiuk, V ... 132
Michaylov, Ph ... 181
Mihaylov, Veselin ... 187, 288
Miroshnichenko, S V ... 73
Mitov, Alexander ... 295
Mkrtychian, D ... 132
Molenda, Szymon ... 17
Nakov, O ... 175
Nedelchev, Krasimir ... 112
Nikolov, Nikolay ... 57
Nikolov, N ... 200
Nikolova, M I ... 9
Orel, Ye F ... 48
Panayotov, F ... 1
Panchenko, S ... 132
Pantchev, S ... 181
Pavliuchenkov, M ... 104
Penkov, Kamen ... 57
Petit, M ... 247
Peychev, M ... 181
Pisarski, Marcin ... 17
Plugin, A A ... 73
Plugin, DA ... 73
Podevin, Pierre ... 276
Prokhorchenko, A ... 132
Punov, Plamen ... 155, 276
Périlhon, Christelle ... 276
Rizov, V ... 35, 42
Ruskova, Ivelina ... 112
Samburenko, Ihor O ... 238
Serbezov, V S ... 9
Shevchenko, A A ... 48
Simeonov, S ... 26
Sitnik, Lech J ... 216
Sitnik, Lech ... 187
Slavov, Tsonyo ... 295
Smolyanyuk, N V ... 48
Sova, S ... 104
Sroka, Zbigniew ... 187
St'astniak, Pavol ... 85
Stefanov, S ... 193
Stoyanov, Stoyan ... 288
Stępień, Małgorzata ... 17
Timkov, Oleksii M ... 238
Todorov, Georgi ... 227
Todorov, M ... 1
Todorov, T D ... 26

Tomov, Y.......... 175
Tsonev, Veselin.......... 57
Vatulia, G L.......... 48
Vatulia, G.......... 96
Vaňko, Jozef.......... 255
Yashchenko, Dmytro M.......... 238
Zawadzki, Karol.......... 17
Zhelyazov, T.......... 125

Institute of Physics
Dirac House, Temple Back
Bristol BS1 6BE UK

ISSN: 1757-8981
ISBN 978-1-7138-3315-4

CURRAN ASSOCIATES INC.
proceedings
.com